PRINCIPLES OF AUTOMATED DRAFTING

MECHANICAL ENGINEERING

A Series of Textbooks and Reference Books

EDITORS

L. L. FAULKNER

Department of Mechanical Engineering
The Ohio State University
Columbus, Ohio

S. B. MENKES

Department of Mechanical Engineering
The City College of the
City University of New York
New York, New York

OTHER VOLUMES IN PREPARATION

PRINCIPLES OF AUTOMATED DRAFTING

Daniel L. Ryan
Clemson University
Clemson, South Carolina

CRC Press
Taylor & Francis Group
Boca Raton   London   New York

CRC Press is an imprint of the
Taylor & Francis Group, an **informa** business

Library of Congress Cataloging in Publication Data

Ryan, Daniel L., (date)
    Principles of automated drafting.

    (Mechanical engineering ; 28)
    Includes index.
    1. Mechanical drawing--Automation. I. Title.
II. Series.
T357.R93 1984   604.2'4   83-25201
ISBN 0-8247-7175-3

MARCEL DEKKER, INC.
270 Madison Avenue, New York, New York 10016

Current printing (last digit):
10 9 8 7 6 5 4 3 2 1

PREFACE

Advanced technology has recently brought
mechanical drafting into the computer age.
The purpose of Principles of Automated
Drafting is to provide a comprehensive,
practical guide that will aid both the
practicing engineer and the student in mas-
tering the essentials of producing drawings
by any automated means. The text material
is designed to introduce the reader to
each phase of the subject, step-by-step to
enable one to use the various automated
drafting devices, instruments and technique
of application. The suggested review problems at the end of each chapter will enable
the reader to develop working skill and
efficiency with digital computers, plotters,
microprocessors, video screens, magnetic
storage devices and other devices that are
common place in the industrial drafting
room of today.

The current state-of-the-art for producing engineering drawings has definite
obligations for the engineering profession.
One of the obligations is the use of the
digital computer in the production of professional maps, drawings, sketches, schedules, tables, print outs, and diagrams.
In the production of engineering drawings,
one of the engineer's professional respon-
sibilities, he or she has an opportunity
to demonstrate the ability to use the computer in a most meaningful way. It is,
therefore, hoped that the reader will ad-
here to the automated drafting procedures
and practices set forth in this book.

Much has happened in the last 15 years
to make the use of computers easier for the
draftsperson. One important event was the
formation of a group of professional users
interested in computer graphics. This was
a special interest group (SIG) for graphic
set up by the Association for Computing
Machinery (ACM) called SIGGRAPH. This
group was responsible for developing the
standards for use of computer graphics in
the United States. Automated drafting is
part of this on-going effort. Chapter 1
introduces this concept as the CORE re-
quirement. Each chapter then explains
and develops a level for the CORE. As
adopted, the CORE has three parts.

    1. INPUT

    2. OUTPUT

    3. DIMENSION

Inside the input and output parts, three
levels exist, while inside the dimension

part only the 2-D and 3-D levels exist. A
closer examination, however, reveals that
each of the three input and output levels
can be addressed in either 2-D or 3-D, or
combinations of each. Therefore, the maximum combinations are 3*3*2 or 18 levels
of computer graphics routines, procedures,
and display techniques.

Automated drafting does not employ all
18 levels, for example Chapter 5 is limited
to dimensional applications in 2-D. Chapter
11 is limited to 3-D only. In this manner
the number of levels can be cut in half
for each topic discussed. A maximum of 9
levels is rarely used in any single drafting application. Therefore, examples are
kept simple, usually with less than three
levels of applications. The figures used
in this book represent only single level
applications. More than one level may be
possible for many of the examples shown in
this book, but only one is discussed at a
time. Two procedures for doing the same
thing is very confusing for the person who
is learning a new technology. As a reader
you might wonder why more advanced treatment of certain topics was not introduced.
The rule has been, make it simple, make it
a step-at-a-time, make it easy to do, and
make it understandable and enjoyable for
readers.

Traditional topics are used throughout this book. It is important that the
reader learn how to letter, sketch, construct geometric shapes, project views of
an object, section an object, plot an auxiliary view, display a pictorial, and produce working drawings of all types. To
learn how to automate any task, the task
must be understood. The computer can not
be programmed to think, visualize, or pro-

duce designs. Humans do these things. The
computer can reproduce an inked drawing 1
million times faster than a human, however.
Computers are good at high speed execution
of small electronic tasks, however. And in
this spirit, each chapter provides those
tasks that can best be performed by both
the computer and the human.

While it is possible that the theory
of automated drafting could be learned by
just studying the examples and working the
review problems at the end of each chapter;
it is hoped that the reader will investi-
gate the common computer tools available.
For instance, most computer centers at
learning institutions have no graphic
(drafting devices) hardware items. A common CRT can be made to function like a
DVST(direct view storage tube) graphics
terminal, which is used throughout this
book. A graphics terminal is not critical
to understanding how draftspersons put an
image on a screen. A screen is critical,
but it may be of any type available.

General purpose plotters can be used
to produce drawing sheets with little modification on the part of the user. Other
equipment is available also, keyboards are
keyboards; lettering commands can be input
in a number of ways. In fact an entire
working drawing can be punched on cards and
read into large, general purpose computers
to be output on large plotters like the pen
driven CALCOMP or electrostatic VERSATEC.
The important thing is that the experience
be worthwhile and keyed to the subject being discussed in each chapter. For those
of you who are about to adventure into this
new world of computers, I invite you com-
ments and suggestions.

Daniel L. Ryan

Contents

Contents

PRINCIPLES OF
AUTOMATED DRAFTING

1

Introduction

Principles of automated drafting is a text book designed for readers who have wondered about the speed, accuracy and use of an electronic computer. The book is also for those who are presently doing any kind of graphics, drafting and artwork totally by hand. Manual drafting is very pretty and impressive if done correctly. It is also very slow and unproductive when other ways are available. This book will show you a way to produce acceptable drafting in the framework of high productivity. In order to do this, certain standards must be set.

Standards are meant to offer a common ground for both the development of drafting communication skills and the interfacing of these skills to computer hardware. One word of caution, read carefully, terms that are related to computer hardware are very colorful and often meaningless. The language of this text is simple, everyday English. In order to describe some of the more unfamiliar, computer-related terms, it may take us several paragraphs. For those readers who very familiar with computers, this may seem trite, bear with us. While

the pace of the book is slow, it is steady and it introduces a new technology to the reader who is willing to learn the standards. The skills we will be learning are the definition of a set of procedures that can be used from a computer program to generate a drafting picture, image, or detail. The hardware connection (interface) standard mentioned earlier is called the virtual device interface. This is important to a drafting user who must support several devices as part of the computer system.

1.1 AUTOMATED DRAFTING STANDARDS

Before 1976, design and development of automated drafting systems was done on a catch-as-catch can basis. There was no recognized methodology (standards). Since then, standardization efforts have set new guidlines for drafting users. Now different uses of the standard by different manufacturers will have the same capabilities and similar drafting uses.

The primary goal since 1976 has been to increase the usefulness of such draft-

Figure 1.1 Typical classroom use of automated drafting and design equipment: items from the left to right are; direct view storage terminal, magnetic tape recorder, hard copy unit, and second direct view storage terminal. (Courtesy of Tektronix, Inc., Information Display Div.)

ing systems as shown in Figure 1.1. A program to produce drafting results on equipment shown in Figure 1.1 should not be that different when placed on equipment shown in Figure 1.2. This is because of the standards developed for portability. The term portability means that a drafting program used on equipment in Figure 1.1 should be transportable with minor revisions to the equipment shown in Figure 1.2. This has come about since the standardization of a programming language for computer graphics. A program written in FORTRAN 77 for one computer should be transportable to another with minor revisions.

As with programming languages, this is the goal for automated drafting programs.

These programs must be universal in order to develop a workable drafting standard. A standard is used to produce state-of-the - art graphics. When a student graduates and enters the industrial marketplace, she or he accepts this state-of-the-art for producing modern industrial drawings by computer program. Programs have been developed for nearly all types of production work that must be done in a drafting room. This book will introduce you to each of these programs as needed. The terms automated drafting standard implies a rigidity that is not really practical in the implementation of computers. Standards should mean guidelines for proper use.

In 1976 several industrial users, pro-

Figure 1.2   Industrial use of automated drafting and design equipment: items from left to right are; computer terminal (non-graphics), magnetic disk, magnetic tape recorder and processor below, hard copy unit, direct view storage terminal (man and woman operators of each) drum plotter, keyboard with readout, second graphics processor.   (Courtesy Gerber Systems Technology, Inc.)

grammers, equipment manufacturers, and the other professional users formed the (GSPC) graphic standards planning committee under the auspices of (SIGGRAPH), the special interest group on graphics of the association for computer machinery (ACM). GSPC was charactered in 1976 to develop a methodology and set of functional capabilities for automated drafting. After three years of development the system called CORE was released in 1979.

1.2   THE CORE STANDARD

The intent of the 1979 CORE was to define an automated drafting framework for most drafting applications. The functional capabilities of the 1979 CORE included the drafting motions, line weights, lettering, geometric constructions and presentation techniques such as orthographic viewing, a choice of color and display mode. The 1979 CORE was implemented to various levels depending on whether or not the manufacturer wished to include 3D capabilities and design graphics input. In other words the 1979 planning committee recommended that all manufacturers provide a minimum level (starting package) for simple 2D drafting capability and then added several others that allowed the user to work in descriptive geometry like communication.

Table 1.1
CORE Automated Drafting Equipment Makers [*]

Manufacturer	Virtual Device Interface
Adage, Inc.	Graphics Terminals
Altek	Digitizer
Amcomp	Cathode Ray Tube
Applicon, Inc.	Graphics Systems
Auto-trol, Inc.	Plotters
Bendix	Digitizer
Calma	Graphics Systems
Calcomp	Plotters
Computervision	Graphic System
Control Data	Cathode Ray Tube
DEC, Inc.	Graphic System
Evans & Sutherland	CRT, systems component
Gerber Scientific	Graphic System
Gould	Electrostatic Plotter
Hewlett-Packard	Graphics Terminals
Houston Instr.	Plotters
IBM	Graphic Systems
IMLAC	CRT, systems components
Megatek Corp.	CRT, systems components
Numagraphics	Systems components
Summagraphics	Systems components
Talos	Digitizer
Tektronix	Graphic Systems
Varian	Plotters
Versatec (Xerox)	Plotters
Zeta	Plotters

[*] Automated drafting seminar survey respondents of November 27, 1979.

With the number of manufacturers providing CORE-like connections, shown in the Table 1.1, the 1979 CORE has become a de facto standard for automated drafting.

1.3 THE ANSI STANDARD

Overlapping with the CORE development efforts, the (GKS) or graphics kernal system was being developed in Europe. Together the CORE and GKS provided the needed stimulus for the ANSI X3/H3 graphics committee. The best features of both the GKS and CORE are now contained in the ANSI X3/ H3 standard. The GKS has become the international standard with the most recent modifications made in June 1982, in Japan.

This introduction has provided a background to the present use of ANSI standards for all automated drawings. This text is divided into information, the body of the chapter, review questions and summary. The chapters build one upon the other. You are instructed by the body of each chapter as what needs to be learned is presented. It is also part of the body of each chapter to introduce what to do, what is acceptable, . and what you should practice or study. The chapter review questions will test your understanding of the body. The summary will reinforce the presentation of material.

1.4 AUTOMATED DRAFTING DEVICES

Automated drafting is an interesting but time consuming process. Each of the new terms in Table 1.1 must now be introduced according to the ANSI standards. It may be possible that some devices are not available for you to practice review problems. In these cases it is very important that you understand the concepts being presented so that you will feel comfortable when introduced to these devices after you finish your education. Before attempting any of the review questions, however, an understanding of the basic devices used by a draftsperson is important. In order to automate the drafting functions an understanding of certain manual procedures is also required. For the most part, the procedures for the use of automated devices are identical to the use of manual devices except they are found in a computer center environment instead of the drafting room. A pen or pencil in the drafting room is used to draw a line on a piece of paper, in the computer environment a pen is used to draw a line on a video screen. For this reason a discussion of manual drafting procedures is often necessary in order to do the automated tasks. Not all manual procedures are covered, however, only those techniques are introduced that are used in

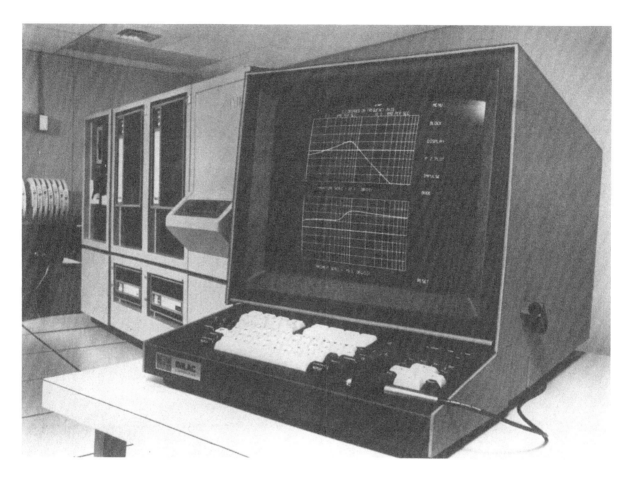

Figure 1.3 Graphics terminal with light pen.   (Courtesy of IMLAC corp.)

an automated environment.

In addition, any type digital computer can be used with the 'automated devices. A micro-,mini-, or maxi-computer may contain the computer program. The program should be thought of as the **device driver.** The device can and should be studied independent of the computer chosen to house the device driver.

1.5 GRAPHICS TERMINALS

Not all computer terminals are capable of automated drafting images. Those that can not are called alpha/numeric, A/N terminals, Figure 1.5 is an example of A/N terminals. Graphics terminals are divided into two main groups. Those that are/need constant up-dating called **refresh** and the storage devices shown in Figure 1.1 & 1.2.

Figure 1.3 is typical of the refreshed CRT graphics terminals. The screen has a quick decay coating and must be **stroked** by the writing element or the image will fade. A stroke rate of 30 times a second or faster is needed so that the fade does not occur. Refresh terminals are dynamic. By this we mean that lines or characters may change position on the screen. The writing rate is so fast that the human eye does not see the images as single pictures.

Figure 1.4 is typical of the direct view storage tube devices that are also used as graphics terminals. They are popular because they are much cheaper to purchase than the refresh terminals. They are not dynamic, they are static because once the screen is stroked, it is stored on the surface of the screen. While this makes the operation simple, it has draw - backs.

Figure 1.4 Graphics terminal with direct view storage tube. (Courtesy Tektronix Inc.)

The draw-backs are lack of movement once a line or image is painted on the screen. A technique must be learned for storage of a screen image. This will be covered in a separate chapter. For now, just remember that terminals like those pictured in Figure 1.4 have limitations.

1.ᑲ AUTOMATED DRAWING BOARDS

With every limitation there often is an advantage. Because the DVST, direct view storage tube is so simple, it will be used with a drawing board. The mainstay of the drafting activity has always been the drafting board. This surface holds a drafting sheet (paper, cloth, film, or plastic) upon which the image appears. In some applications a blank (empty) sheet is placed on the drawing board and all nec-

Figure 1.5 A/N terminal used with graphics terminal. (Courtesy Adage Corp.)

Figure 1.6 Instructor demonstrates automatic drawing board. (Photo by DLR assoc.)

Figure 1.7 Coordinatograph used as an input device. (Photo by DLR associates)

essary graphic shapes are drawn directly
added to the material. In other cases a
completed drawing is placed on the drawing
board and traced into computer memory. A
drafting board that is connected to a digital computer may take a variety of forms.
Table 1.1 lists only one type, which is a
digitizer. The drawing board shown in Figure 1.6 is called a graphics or data tablet
and is used connected to the DVST shown in
Figure 1.4. A digitizer is any device used
as an input to a computer. The coordinatograph shown in Figure 1.7 is an input device and as such is classified as a digitizer. Only manual drawn sheets can be produced from the drawing board type shown in
Figure 1.7. In order to produce automated
drawings, the image must either be contained in the device driver (computer program)
as stored information; or the image must be
converted to computer compatible information. If the image already exists it may
be traced into computer memory by either of
the devices shown in Figures 1.6 or 1.7. A
device like these measures the X-Y coordinates of each line segment. The operator
traces the drawing and stops at each line
segment -- this process is called digitizing.

The device shown in Figure 1.7 is used
to break down graphical images into X-Y locations by manual means. The operator moves
the pointing device over the start or stop
of a line segment. A button is depressed
and the X and Y data is stored for use by
another type of drawing board called a digital plotter. Both graphics tablets (Figure 1.6) and coordinatographs (Figure 1.7)
are called input drawing boards. A digital
plotter is called an output drawing board.

Before we leave input drawing boards,
you should practice digitizing with a piece
of grid paper and a pencil. Take the pen
or pencil in hand and locate the lower left
hand corner of the grid. Sketch an + there
and label it ORIGIN. Now count ten units
up and ten units over (one grid line being
equal to a unit), place a dot at this location and label it POINT 1. Now sketch a
line which is twenty units long and ends to
the right of POINT 1. Place a dot at the
end of this line, label it POINT 2. Now
sketch a line which will connect POINT 2
with POINT 3 which is located ten units to
the left of POINT 2 and five units up on
the paper. Label this new point, POINT 3.
Finish this exercise by sketching a line
back to POINT 1 to complete the polygon.

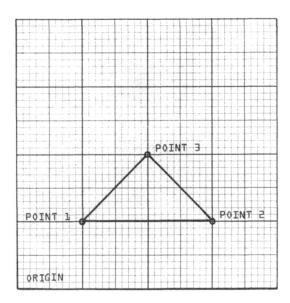

Figure 1.8   Sample digitizing exercise.

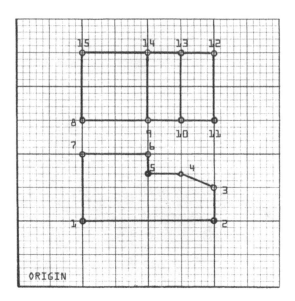

Figure 1.9   Sample digitizing exercise II.

If your digitizing practice was successful, it should look like Figure 1.8.
While this was easy to do and understand,
digitizing with computer driven devices
such as coordinatographs and graphics tablets is just as easy. In fact, sometimes
it is easier to sketch the graphics to be
input to the computer, place this sketch
on a graphics tablet and enter it as X and
Y data; than to mess with a device driver
called a **source program.** Programming or
device driver instructions should never be
written if an easier, more direct way of
entering the information is available. As
stated earlier, however, not every location
has a digitizer. Then you must search for
another method to input the graphic information.

Common input devices that can be used
in this manner are:
1.   data tablets ... Fig. 1.6
2.   coordinatographs ... Fig. 1.7
3.   interactive plotters ... Fig. 1.11
4.   joy sticks and DVSTs ... Fig. 1.12
5.   light pens and CRTs ... Fig. 1.3
Chose one of these devices and practice another example.

Begin by labeling the origin again as
we did in Figure 1.8. Move to point 1 as
before, label it, draw a line to point 2,

label it and proceed around the first polygon shown in Figure 1.9. The locations for
this first or lower polygon are 1 through 7
with a line drawn from 7 to 1 to complete
the object. Now pick up your pencil and
move to point 8 shown in Figure 1.9. Make
a dot here and label it POINT 8. Proceed
around this polygon in order of the points
shown. When you reach point 15 connect it
to point 8, now inter-connect points 14 and
9, 13 and 10. In a later chapter we will
discuss polygon construction and view placement. Both of these concepts are based upon your ability to digitize points in X and
Y. Before we leave this section, try one
more problem of your own. Use a light pencil and sketch an object on the form shown
in Figure 1.10.

Place the form on an automatic drawing
board like that shown in Figure 1.6. Use a
pen plugged into the drawing board control
box to locate the points chosen. If a data
tablet is not available, the same procedure
may be used on a digitizer like that shown
in Figure 1.7. If neither devices are used
at your location, three other options are
still available. Look for a CRT and light
pen as shown in Figure 1.3, most computer
centers have them. Still two devices remain, however. Figure 1.11 is called an

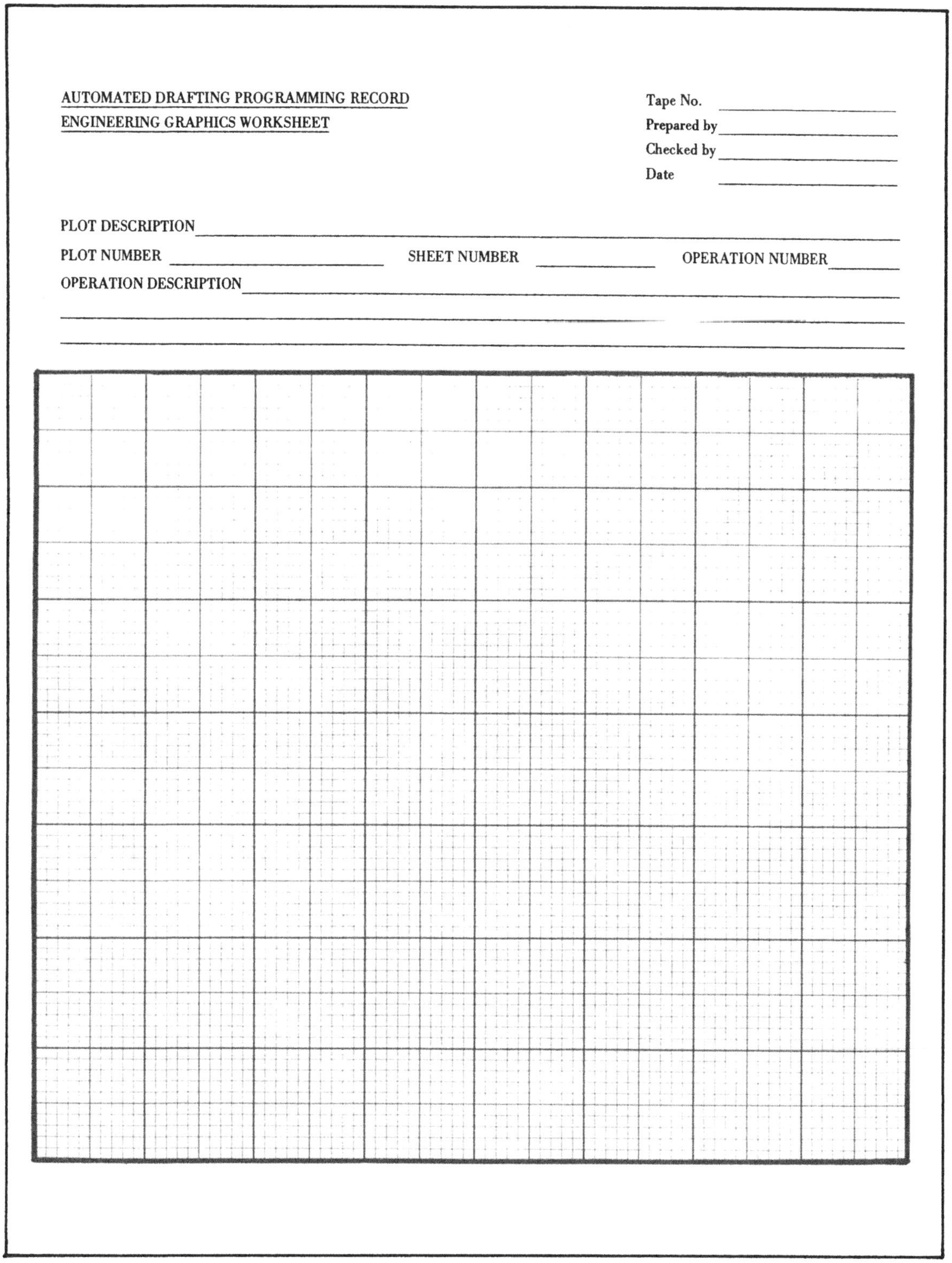

AUTOMATED DRAFTING PROGRAMMING RECORD
ENGINEERING GRAPHICS WORKSHEET

Tape No. _____

Prepared by _____

Checked by _____

Date   _____

PLOT DESCRIPTION _____

PLOT NUMBER _____   SHEET NUMBER _____   OPERATION NUMBER _____

OPERATION DESCRIPTION_____

_____

_____

Figure 1.10 Digitizing form to assist the draftsperson in preparing input data. The
size A shown here is convenient to use on a small data tablet shown in Figure 1.6.

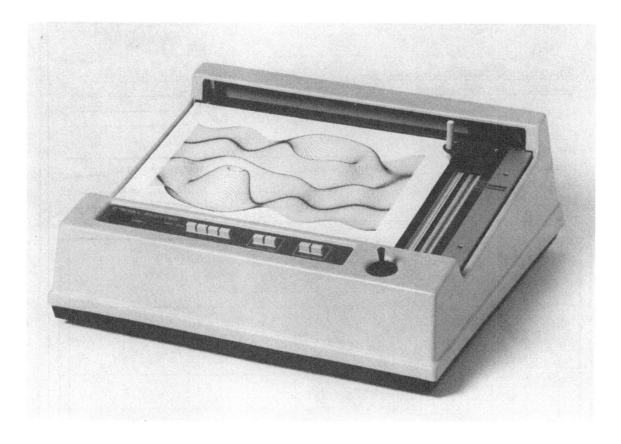

Figure 1.11 Interactive plotter used to send X-Y locations (joy stick in lower right hand) and receive data from a source program. (Courtesy Tektronix, Inc., Information Display Div.)

interactive plotter. The term interactive means that the device is capable for both sending and receiving data. To use this device, place the sketch you completed for Figure 1.10 on the plotter surface. Turn the plotter on, press sheet hold down and position the pen over the first point to be sent to the computer. The pen is position-ed by moving the small joy stick shown in Figure 1.11.

All five devices shown in Figures 1.3, 1.6, 1.7, 1.11, and 1.12 are used to create data for use in a computer program. Once the data exists it can be drawn on an out-put drawing board. These types of drawing boards are called digitial plotters. The term gets its name because it uses the dig-itized information created earlier to plot the image that was stored. Digital plot-

ters have been used for a number of years in automated drafting rooms. Figures 1.2 and 1.11 are just two types, drum (1.3), and flat bed (1.11).

The last type of digitizing that you must be familiar with is shown in Figure 1.12. This type of digitizer is called a standalone digitizer. It is composed of a DVST (Tektronix 4051) and joy stick. It is called a standalone because it does not require a host computer to store the X-Y data. The data is stored on a magnetic tape shown on the right of the screen. A terminal like this belongs to the micro-class discussed earlier. It can be pro-grammed in BASIC, an elementary level pro-gramming language. Figure 1.12 clearly shows the cursor or crosshairs that are moved by the joy stick. When a point is

Figure 1.12 Tektronix 4051 graphics terminal and positioning joy stick. (Coutesy Tektronix Inc., Information Display Division).

located by the cursors, it is sent to the local processor inside the terminal.

1.7 AUTOMATED DRAFTING MATERIALS

Now that a variety of automated drafting devices have been introduced we can begin to study how these devices produce images on paper, drafting or photographic film, and other sheet materials. In order to use the drawing boards shown in earlier figures, the draftsperson selects a paper, film or plastic material to place on the drafting surface. This was best demonstrated in Figure 1.11, where ordinary tracing paper was placed on the flat bed plotter to produce the map. The paper is held firmly in place by a hold down system, the system may be vacuum, electrostatic, or

other means. Some plotters use an electrostatic charge, others use a sproket feed mechanism to hold the paper in place.

Once the type of sheet material is chosen and the hold down is in place, the size of the sheet must be entered. This entry called the window, or page size will provide the allowable drawing area. In other words the data contained in the computer will be scaled up or down to fit the size of the paper automatically. This feature is necessary to save instruments like plotter pens from being damaged and to insure that a high quality drawing is produced. Figure 1.13 illustrates the window or paper taken from the plotter shown in Figure 1.11. Inside the window, any digitized data may be displayed, or the window may be further defined. Usually, for auto-

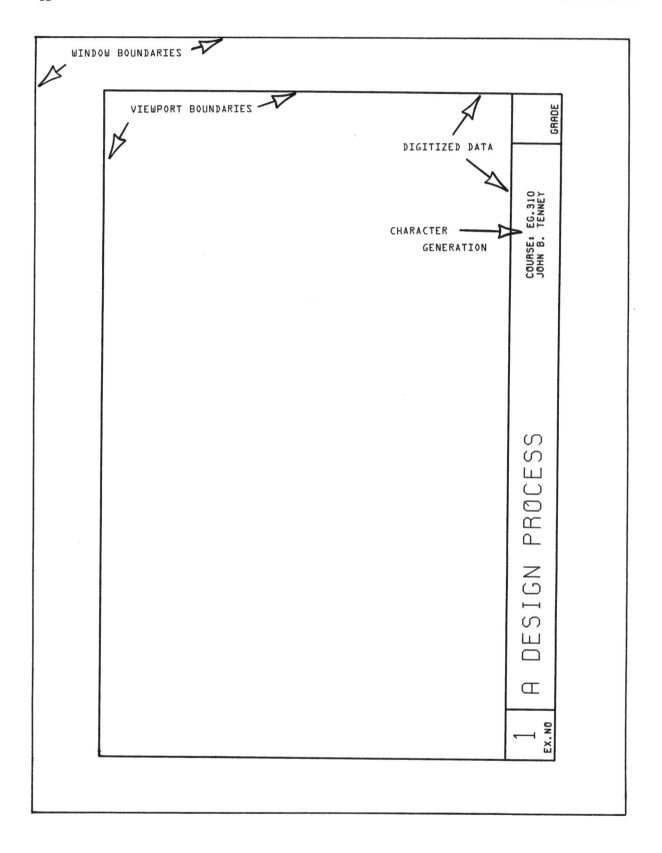

Figure 1.13 Automated drawing from flat bed plotter illustrates window and viewports plus the display of digitized data along with characters from a special purpose subprogram.

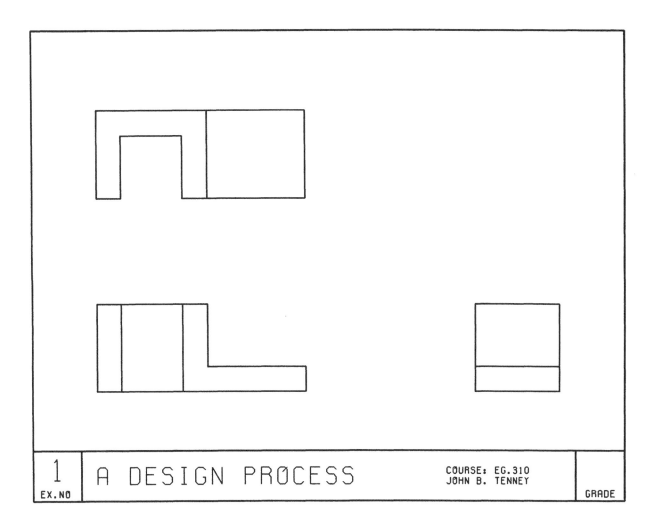

Figure 1.14 Sample drawing added to viewport. Drawing image is from digitized data.

mated drafting purposes, the viewport is a border line as shown in Figure 1.13. The window boundaries are not plotted as shown in Figure 1.13, they do not appear and are considered for reference only as shown in Figure 1.14.

Some graphical images are more difficult to digitize than others. An example is lettering. It would be very time consuming to digitize each of the small, straight line segments shown in the lettering of Figure 1.14. Therefore, the draftsperson has the option of using a pre-stored method of lettering called a character generator. While an entire chapter is devoted to lettering technique in this book, it is important at this early introduction to know that not all shapes must be digitized.

When it is convenient to digitize straight line segments, we do so. Figure 1.14 is a good example of this type of display.

Not every line on a drawing is a straight line, however. Like lettering, most lines are produced from pre-stored methods called line generators. In order to produce suitable line weights, solids, dashed, hidden, break, cutting plane, phantom and so forth; a line generator is necessary. A line generator should produce:

1. ——————————— solids
2. - - - - - - - - - - dashed
3. — — — — — — — — hidden
4. ⌒‿⌒‿⌒‿⌒ break
5. ——— — — ——— cutting plane
6. -------------------- phantom
7. ▬▬▬▬▬▬▬ newpen styles

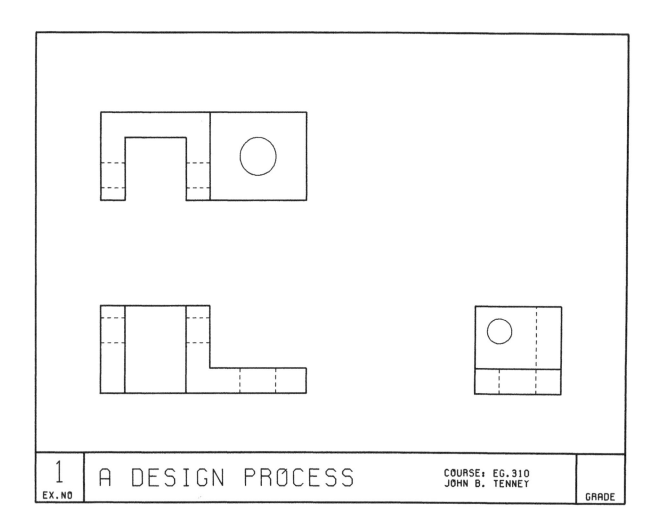

Figure 1.15 Sample drawing taken from flat bed plotter. Drawing image is from a line generator and circle generator.

In Figure 1.13, the draftsperson has the
option of digitizing the straight lines or
producing them from a line generator. The
addition of a hole in the drawing requires
that the draftsperson use a line generator
and another pre-stored method called a circle generator. The hidden lines were produced from the line generator while holes
can be represented by the circle generator.

1.8 AUTOMATED DRAFTING TECHNIQUES

Two automated drafting techniques are
line and circle generators. When these are
used they are easy to recognize. Each end
of the generator is marked with a small o.
This marking is for the draftsperson's use

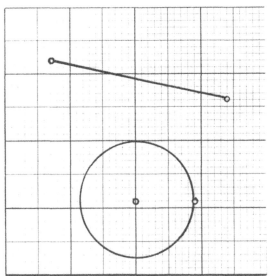

Figure 1.16 Line and circle generators.

during input. For example, suppose we are
to use a line generator to produce the line
shown in Figure 1.16? First, one of the
input methods must be chosen. These were:

    1. data tablet and stylus ... Fig. 6
    2. DVST and joy stick ... Fig. 12
    3. CRT and light pen ... Fig. 3

In each case a display area is available,
and a pointing device is used to select a
starting point. In Figure 1.16 the starting point may be either end of the line.
Once the end point is selected, the line
is displayed for you to save or delete. A
saved line is sent to the output device as
a single line image without o's at each of
the end points.

    Circles are selected in a similar man-
ner using a pointing device. First the
center is selected and then the radius is
selected. The completed image is presented as shown in Figure 1.16 for the draftsperson to save or delete. Once the circle
is saved, it can be plotted as shown in
Figure 1.15, without the o's in the center
and at the radius.

    Additional information about automated
drafting techniques is presented in chapter
2 of this book. It is important that you
understand how image generators work at
this point, not how to use them. A part
of this understanding will come from the
pointing device and how to use it.

1.9 AUTOMATED PENCILS, PENS AND POINTERS

    The basic instrument of the drafts-
person is the pencil. The pencil used depends upon the drafting board and the drawing material selected. Choices vary somewhat from model to model. For example,
Figure 1.17 and 1.18 are both used on a
data tablet as shown in Figure 1.6. In
Figure 1.7 the draftsperson may use a conventional drafting pencil because the input pointing device is contained with the
instrument and is by the operator to record the pencil marks. In Figure 1.11 a
single pen is held in the cross arm and is
moved across the face of the digital plotter by the use of the joy stick.

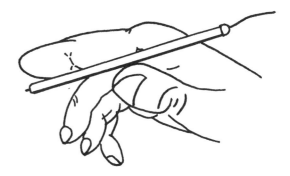

Figure 1.17   Tablet pen used in Fig. 1.6

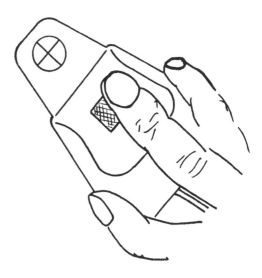

Figure 1.18   Input device used in Fig. 1.6

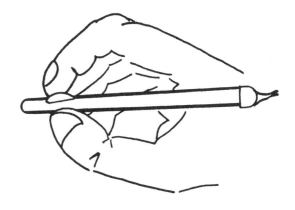

Figure 1.19   Light pen used in Fig. 1.3

Pens, especially pointing devices, are also used by draftspersons to locate graphic images on a display area as demonstrated by the line and circle generator. It should be pointed out that all graphic shapes may be located in this manner. Figure 1.20 is an example of some, but not all of the images that can be located in this manner.

Each of the various devices is demonstrated in Chapter 2. For the purpose of this chapter, any of the pointing devices can be used. A pointing aid is anything that assists the draftsperson in positioning graphic output. Pointing aids are:

1. joy sticks
2. thumb wheels
3. track balls
4. mouse

In the case of direct view storage tubes, a tracking mechanism is repositioned depending upon the adjustment of any of the devices listed above. While joy sticks have been shown in Figure 1.11 and 1.12, the remainder of the list has not been introduced.

Thumbwheels are often supplied with DVST terminals (4010 model Tektronix). A separate wheel is provided for X movement and Y movement of the display cursor. Joy sticks are preferred over the use of the thumb wheel. A track ball is a single element that replaces the joy stick, it looks like a ball sticking out of the controller and positions the cursor by rotating the ball. A mouse is a little gray device with the electrical cord representing the tail and two white buttons which look like a pair of ears. It is moved over the surface of a data or graphics tablet to input X and Y coordinate locations. Figure 1.18 is similar to this device and is often confused with a mouse. It has a single button to move the cursor on the DVST.

1.10 AUTOMATED STORAGE DEVICES

The T-square, triangle, scale, curve, and case instruments have been replaced by preprogrammed computer routines that have been stored for use by the draftsperson.

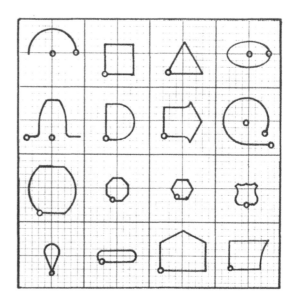

Figure 1.20 Images located by pointers

To be successful in the use of these preprogrammed routines, two things must be mastered; (1) the use of a pointer, just discussed,, and (2) the use of basic storage devices.

For a number of years (1959-1969) information was stored off the computer in a paper format, either tape or cards. No modern automated drafting system uses paper storage today. After 1969 a magnetic tape reader was used to read information into or out of a computer. While magnetic tape is becoming passe', it can still be found in second and third generation automatic drafting machines which are in use today. A magnetic tape reader of this period is illustrated in Figure 1.2. In this example the tape is reel to reel. Magnetic tape cartridges in a variety of sizes are also popular. Figure 1.1 was an example of a tape cartridge and reader. A more modern method of storing information is the magnetic disk. It was pictured in Figure 1.2 and is the fastest method of magnetic storage available.

One of the newest forms of magnetic disk is called a **floppy disk**. Floppy disk gets its name from the fact that it is thin and bendable. These types of disks come

in two popular sizes, the eight inch win-
chester and the five inch mini-floppy. A
disk is placed in a disk drive which rot-
tates the disk so that information can be
read. Care should be taken in handling a
disk. Do not touch the disk, replace it
in its protective envelope when finished.

1.11 REVIEW PROBLEMS

1. Begin a course notebook divided into
the following sections: (A) classroom notes
(B) operational instructions, (C) sample
sets of instructions for automated drafting devices, and (D) sample automated drawings.

2. After each lecture, summarize the notes
and place them in section A of your notebook. Provide a separate section called
new terms learned today.·

3. Obtain handouts for section B of your
notebook from your local computer center.
Read each pamphlet or manual carefully and
paste index tabs for easy location of each
section. Subdivide section B so each device can be described separately.

4. Prepare sample sets of instructions for
each of the devices described in problem 3.

5. Place both the instructions and drawing
produced in sections C and D of your notebook.

6. Continue the digitizing exercises on
page 8 of this text.

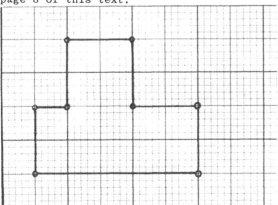

7. Continue problem 6 by digitizing the
object shown below.

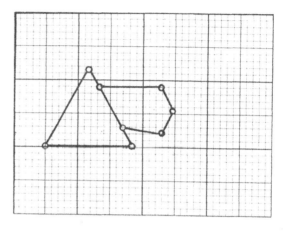

8. Continue problem 6 by digitizing the
straight line letters shown below. Save
the data on magnetic tape for use later.

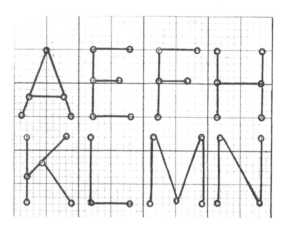

9. Continue problem 6 by digitizing the
object shown below. Save the data on tape
and route it to an output device.

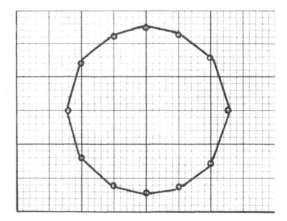

10. Prepare a simple set of instructions for each of the automated drafting devices at your location.

11. Output the set of instructions prepared for problem 10 on different materials such as paper, tracing paper, film or micro film. Place each of these in your notebook.

12. Select each of the pointing devices available at your location and practice a line or circle placement as indicated below.

1.12 CHAPTER SUMMARY

The purpose of this chapter was to introduce the reader to the automated drafting standards used in industry. The following automated drafting devices were introduced:
1. Graphics terminals
2. Drawing boards (input & output)
3. Drafting materials and techniques
4. Pointers and storage devices
A series of figures were used to illustrate the principles of automated drating.

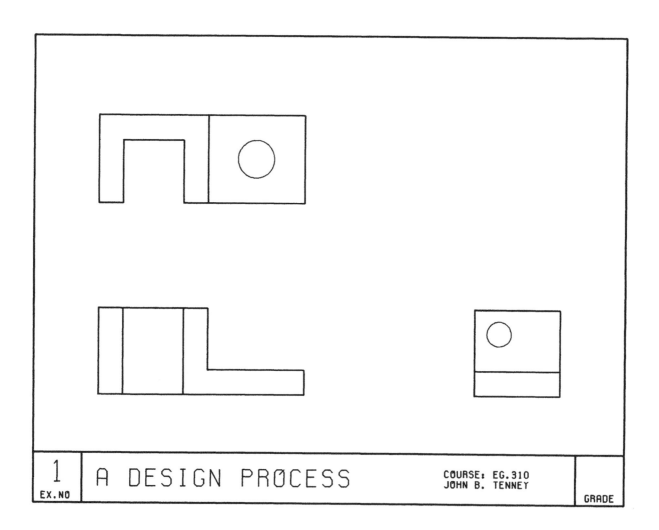

1	A DESIGN PROCESS	COURSE: EG.310 JOHN B. TENNEY	
EX.NO			GRADE

2
DRAFTING TECHNIQUES

By this point in your reading, you may be saying "Why must I study this stuff? Will I ever use it?" You may not appreciate the value of being able to communicate with the aid of a computer now but it is hoped that you will at the end of this chapter. The technique of graphical communication is a powerful one. The use of the digital computer to speed up this communication process is absolutely critical as the United States enters the last twenty years of the twentieth century.

2.1 TYPES OF GRAPHIC COMMUNICATION

Accuracy in observation is developed by sketching. There are those who learn sketching for no other reason than to see what they are considering as an idea. A laboratory set-up, for instance, will be more easily learned if it is sketched. So many people look but do not see. Also, a sketch is always an aid in the learning process. So the most important type of graphic communication is the sketch.

Can sketching be automated? Of course

it can, just as all the other types can be. The other types are:

1. Lettering (Chapter 3)
2. Sketching (Chapter 4)
3. Geometric Constructions (Chapter 5)
4. Orthographic Projection (Chapter 6)
5. Sizing and Dimensioning (Chapter 7)
6. Sectional Views (Chapter 8)
7. Auxiliary Views (Chapter 9)
8. Pictorial Representation (Chapter 11)
9. Working Drawings (Chapters 12,13 & 9)
10. Vector Analysis (Chapter 17)

2.2 LETTERING

When you have completed Chapter 3 of this book you should have developed a complete understanding of how the source program drives a character generator to produce engineering lettering. This engineering will appear on plotter outputs, on video screens, and in computer printouts. Why should we bother with lettering as the simple communication tool? Simple, how is the operator of the computer graphics systems going to understand what good engineering

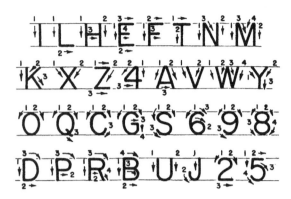

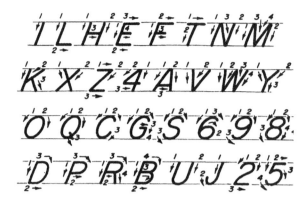

Figure 2.1 Vertical single stroke lettering for most drafting applications.

Figure 2.2 Inclined single stroke lettering for selected drafting applications.

lettering is unless you give some guidelines. Therefore, you will need to use
good engineering lettering whenever it is
required. With slight variations the lettering you should be using are single stroke commerical gothic letters, either vertical, or inclined as shown in Figures 2.1
and 2.2.

When forms, such as FORTRAN coding or
digitizing (see Chapter 1) are completed;
it is a good idea to practice these types
of letters. Correct letter form and shape
will insure that the person typing the information into the computer understandings
what you want. Draftspersons usually DO
NOT type instructions, they letter certain
pieces of information on special forms provided for them and then a person trained
in the proper typing procedures enters the
information. Figure 2.3 is an example of
one of the special forms used to enter information. If the draftsperson begins on
the first line of the special form shown
in Figure 2.3 and letters TURN KEY 1 carefully in the first row of spaces labeled
7 through 17, the computer operator will
understand that all following lines are to
be output on a digital plotter as vertical
single stroke. If the lettering is so bad
that the operator reads something else, another process might be preformed, resulting
in loss of time and draftsperson productivity.

There are three steps to learning how
to letter as shown in either Figure 2.1 or
2.2. The first step is that of obtaining
a mental picture of the shape and peculia-
rities of each letter, together with the
various strokes used in making them. Notice
that lettering in Figure 2.1 and 2.2 are
grouped by this method. In the first line
of each figure, are shown the easiest form
of single stroke letters like I L E and so
forth. These letters consist of straight
line segments connected by the small arrow
method and order of connection. The second
line is somewhat harder in that letter proportion must be considered. In the third
line curved letter sections are introduced
and in line four they are completed. This
first step in learning to letter is the
most important because a student can never
letter successfully if he does not know a
good letter shape and form when he sees one.

The second step is the one that requires the work, you must practice each
letter until you have mastered it. This
takes time, but will provide the third
step which is the ability to produce engineering lettering of high quality at will.
During the second step you will discover
several aids that will assist you in the
production of good lettering. These are
guidelines and spacing aids. Study the form
in Figure 2.3, notice that spacing aids
are provided for each letter.

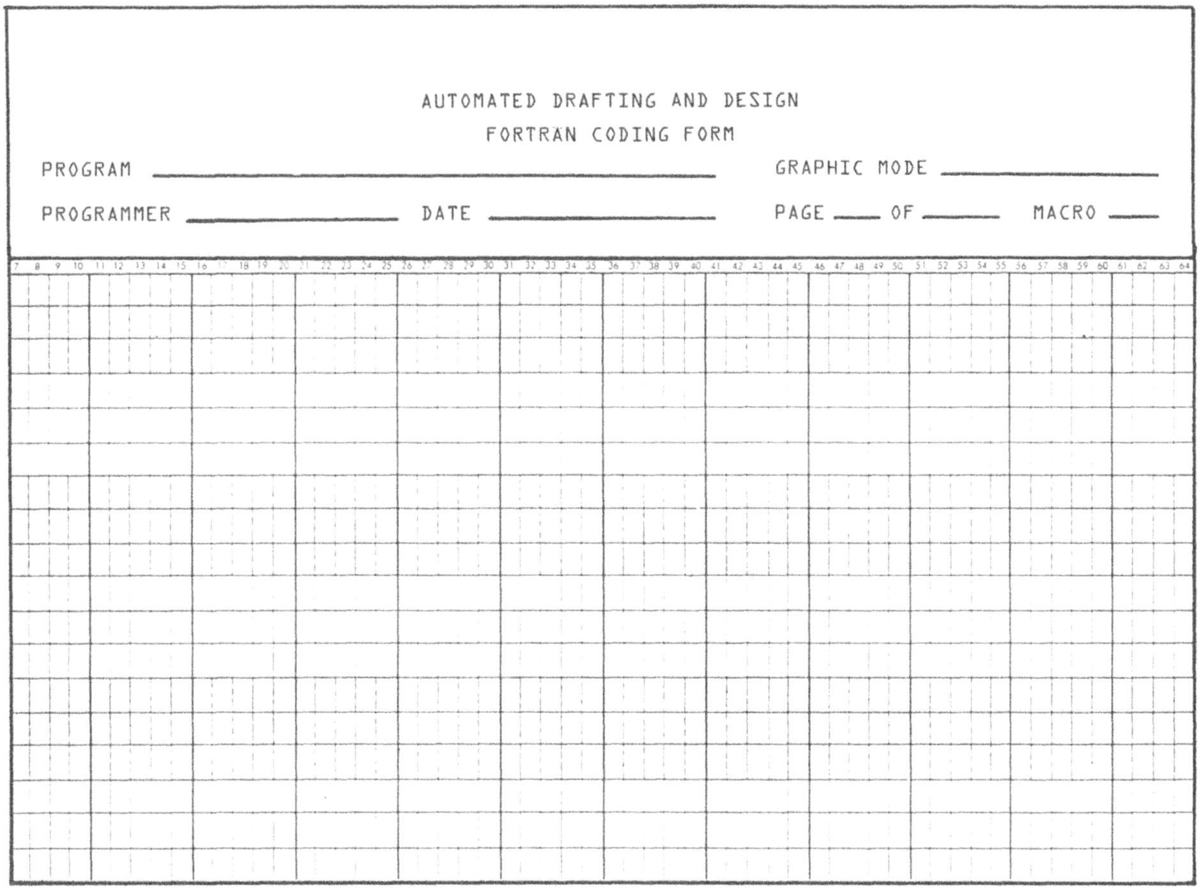

Figure 2.3 Automated drafting and design coding form. (Courtesy DLR Associates)

2.3 SKETCHING

When you have completed Chapter 4 of this book you should have developed a complete understanding of how the source program drives an output device to reproduce a freehand sketch entered from the graphics tablet. This engineering drawing will appear on plotter outputs, on video screens, and in computer output of microfilm. Why should we bother to learn this particular of automated drafting is often asked. It is the most important engineering communication tool available, that is why! As you begin to study other engineering courses, it will become more apparent to you.

Freehand sketches may be employed as an aid in visualizing and organizing problems as well as the presentation of ideas.

Engineers develop the ability to "think with a pencil", therefore, as a draftsperson you will see many sketches from engineers as part of a drafting assignment. So you will want to know how to place sketches on a graphics tablet and enter them as automated drawings, how to change them while they are still sketches and how to create new sketches of your own.

Freehand sketching is the act of executing a graphical representation of a 2- or 3-dimensional object on the surface of an input device called a **data tablet**. This is done without the aid of any mechanical means, such as a straight edge, compass, or template. Compasses and dividers with steel points will damage the data tablet surface and should be avoided.

Original sketches are never done dir-

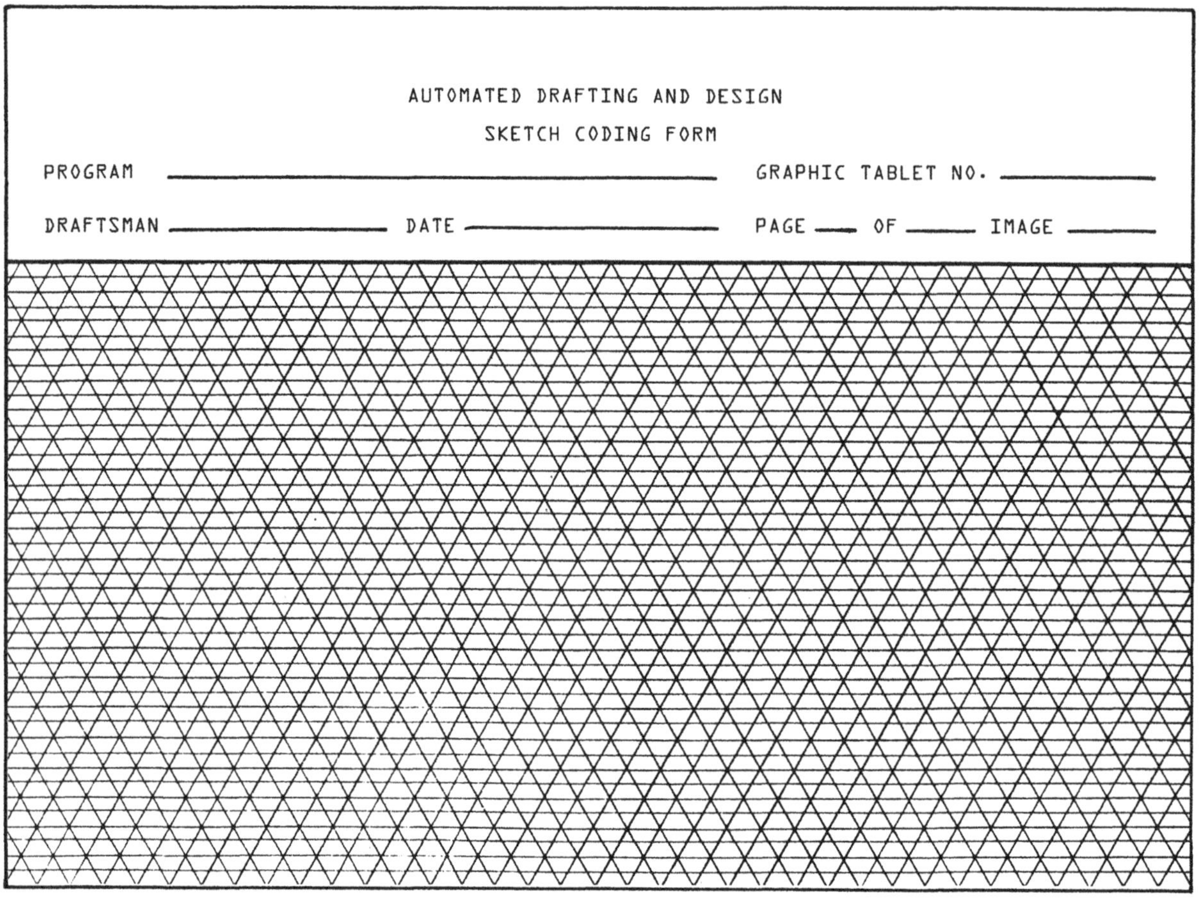

AUTOMATED DRAFTING AND DESIGN

SKETCH CODING FORM

PROGRAM ______________________ GRAPHIC TABLET NO. ________

DRAFTSMAN ____________ DATE ____________ PAGE ___ OF ____ IMAGE _____

Figure 2.4 Automated drafting and design sketch coding form. (Courtesy DLR Associates)

ectly on a graphics or data tablet. They are prepared on special forms as shown in Figure 2.4. These forms aid the draftsperson prepare a suitable sketch to enter into the computer as shown in Chapter 4.

Materials needed to sketch on the forms are simple. A pencil, eraser and supply of tissues is all that is needed. Drawing pencils are grouped into three catagories - hard, medium, soft. The harder lead is indicated by an H and varies from H to 9H. The higher the number, the harder the lead. F and HB are medium leads while the soft leads are identified by a B, and range from B to 6B. The higher the number the softer the lead in this case. Examine the leads that you have and test them by drawing lines on the form shown above in Figure 2.4. Notice that the H series does not work very well. It is too hard, it should be used to draw the background gridwork that you see. Because all layout type lines are not necessary in the use of the sketch coding form, only the two softer leads will be used.

When the type lead is chosen, sketching may begin. It is no sin to erase a line! No one is perfect and you will make mistakes, remember that the form is used so that mistakes are not entered into the computer. Use a good quality eraser and remove the lines completely so that no part of the line can be seen. It must also be remembered that no straightedges will be used, so it will not be possible to sketch straight lines like a draftsperson might produce a line. Follow the guidlines whenever possible, when not possible, do the

best that you can. Always remember that a
freehand sketch is supposed to look freehand, it not supposed to look like it was
drawn with a straightedge.

To sketch lines on the special coding
form, grasp the pencil about three inches
from the point as shown in Figure 2.5. If
the pencil is held closer to the point,
the hand will obstruct your view of the
coding form guidelines. Mark two points
approximately six inches apart anywhere on
the coding form so that a line connecting
the two points will not fall on top of a
guideline. Starting from the left point,
sketch with a series of light strokes, a
line to the right point. Let the weight
of the pencil draw the line. The hand will
only guide the pencil. The wrist is kept
in line with the forearm and the elbow is
moved away from the body as the short
strokes are sketched. See Figure 2.6.

2.4 GEOMETRIC CONSTRUCTIONS

When you have completed Chapter 5 of
this book you should have developed a complete understanding of how the source program drives a **picture generator** to produce
two-dimensional template images. This can
appear on plotter outputs, on video screen
and as parts of other automated drawings.

An example of geometric construction
is to think of a head of a bolt. It is a
circle with a hexagon inside of it. If a
draftsperson could automate this simple
construction and store it as a separate
picture to be used over and over again,
then much time would be saved during the
construction of drawings requiring bolt
heads. Suppose that we stored a circle
and a hexagon together as a single geometric construction, this is called a
subpicture. It would be used to produce
the drawing shown in Figure 2.7. Here 6
copies of a bolt head are required along
with lines and lettering. The object is
also dimensioned, but notice the use of
the geometric construction (subpicture) to
construct the drawing shown in Figure 2.7.
See Chapter 5 for other geometric construc-

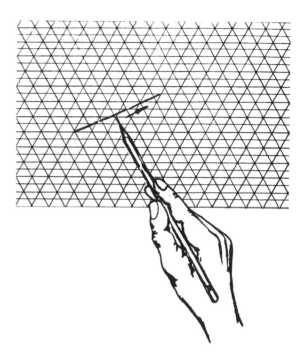

Figure 2.5 Sketching technique used on the
special coding forms for horizontal type
line construction.

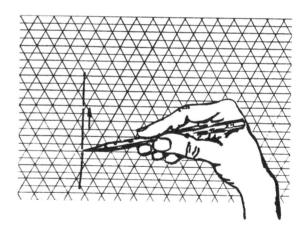

Figure 2.6 Vertical construction.

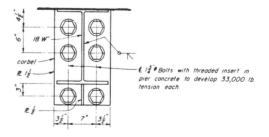

Figure 2.7 Automated drawing using the subpicture of a bolt head. (Courtesy DLR assoc)

tions that are made into subpictures. Subpictures are related to geometric constructions just as subroutines are related to
equations in computer programming.

To illustrate, lets use a subpicture
inside of another subpicture. The geometric shape circle and polygon (equations),
were used to create a simple subpicture
called a bolt head. Suppose we use that
bolt head and place it inside of two more
circles as shown in Figure 2.8? Let's call
that picture WASHER when it is stored inside the computer. Remember that WASHER
was built from a geometric construction
routine called CIRCLE and a subpicture
called BOLTHD. BOLTHD was made up from
two geometric construction routines called
CIRCLE and POLY. But wait, we aren't done
yet. We can still use WASHER to produce
another drawing or still another subpicture. Figure 2.9 is a drawing as it stands
now, but it could be part of a larger drawing by making it a subpicture.

2.5 ORTHOGRAPHIC PROJECTION

When you have completed Chapter 6 of
this book you should have developed a complete understanding of how the source program drives an **automatic view generator.**
This type of program is used to produce a
working view relationship called front,
top, side and auxiliary views. The first
three views are all 90 degrees from each
other and will be introduced in Chapter 6.
For now it is important to know that subpictures are two-dimensional in nature and
that all automatic view generators work in
three-dimensional data base. Many applications in automated drafting can be worked
in two-dimensional data, but the most useful applications require three-dimensional
data. Three-dimensional data allows the
draftsperson to generate three, four or
more views of an object by the press of a
button. Figure 2.10 is typical of the
output for this type of automated drawing.
Notice that any combination of views may be
selected, in this case top and front view.

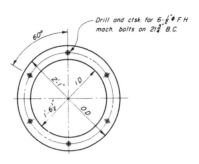

Figure 2.8 Subpicture WASHER constructed
from geometric routine CIRCLE and subpicture BOLTHD.

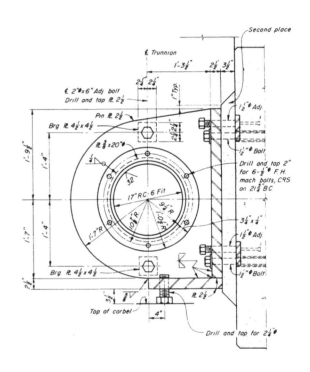

Figure 2.9 Automated drawing using subpicture WASHER.

2.6 SIZING AND DIMENSIONING

When you have completed Chapter 7 of
this book you should have developed a complete understanding of how the source program drives an automatic sizing and dimensioning package for simple objects. Figures 2.7 through 2.10 already contain examples of these types of output, study

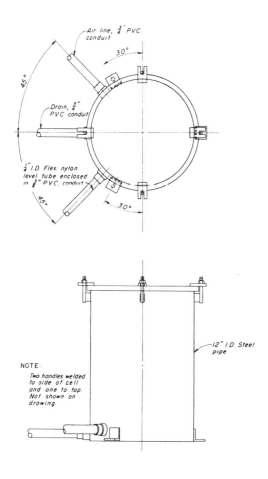

Figure 2.10 Orthographic projection of a simple object as top and front views.

the centerlines represent a bolt circle. A 60 degree angle is dimensioned and one note is included.

3. 2.9 is an advanced example of good dimensioning practice. It has all of the examples just cited plus notes for fasteners and connection techniques. In this example the art of dimensioning is clearly present. While every attempt is made to make sizing and dimensioning an automatic operation, a situation always will require the drafts-person to add, move, or delete some of the dimensions provided by the automatic package.

4. 2.10 is the minimum example of an automatic dimensioning package. Additional information will need to be added as shown in Figure 2.11 and 2.12.

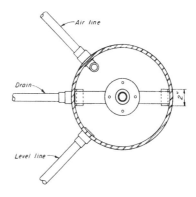

these with the following in mind:

1. 2.7 gives reference dimensions only between centerlines. Sizes are given in inches with the " mark clearly labeled, this type of dimensioning is used more for civil and structural engineering and is considered poor use in mechanical engineering drawing. Special symbols are also used such as the WF, which means wide flange, and the ^{C}L, which means centerline. These are used with ^{P}L, which means plate and a leader which contains a welding symbol. A good job of dimensioning tells the reader how the object is to be fabricated, what size it is, and the position of the various parts.

2. 2.8 gives only reference dimension between centerlines also, but in this case

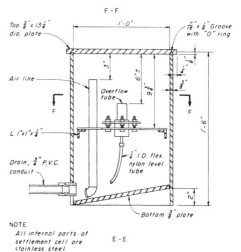

Figure 2.11 Sections of Figure 2.10.

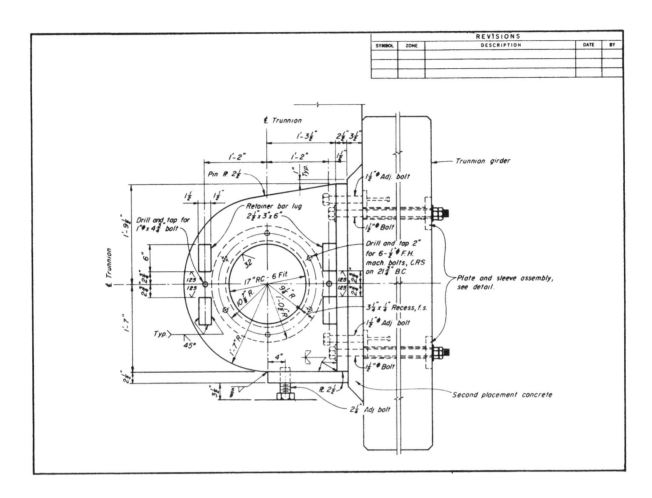

Figure 2.12 Modifications to Figure 2.9 in both sizing and dimensioning.

2.7 SECTIONAL VIEWS

When you have completed Chapter 8 of this book you should have developed a complete understanding of how the source program drives an automatic sectioning routine for simple objects. Figures 2.9 through 2.11 already contain examples of these types of output, study these with the following in mind:

1. 2.9 has a connection plate in full section. See how easy it is to read. Compare this drawing with that of Figure 2.12. Notice that without the use of a section view, many lines appear as hidden. This makes the drawing harder to understand. A good reason for sectioning is to provide clearer objects that are free from unnecessary hidden line work.

2. 2.10 and 11 are a comparison of conventional orthographic views and section views. Another use of section views is to show internal parts of a product. The use of sections F-F and E-E allow the reader to look inside the settlement cell.

3. 2.13 contains sections and auxiliary views. Notice that this drawing contains top (section view), front (orthographic view), left side (section view), bottom (orthographic view) and four auxiliaries off the bottom view.

2.8 AUXILIARY VIEWS

When you have completed Chapter 9 of this book you should have developed a procedure for constructing auxiliary views. A view that is not one of the six primary

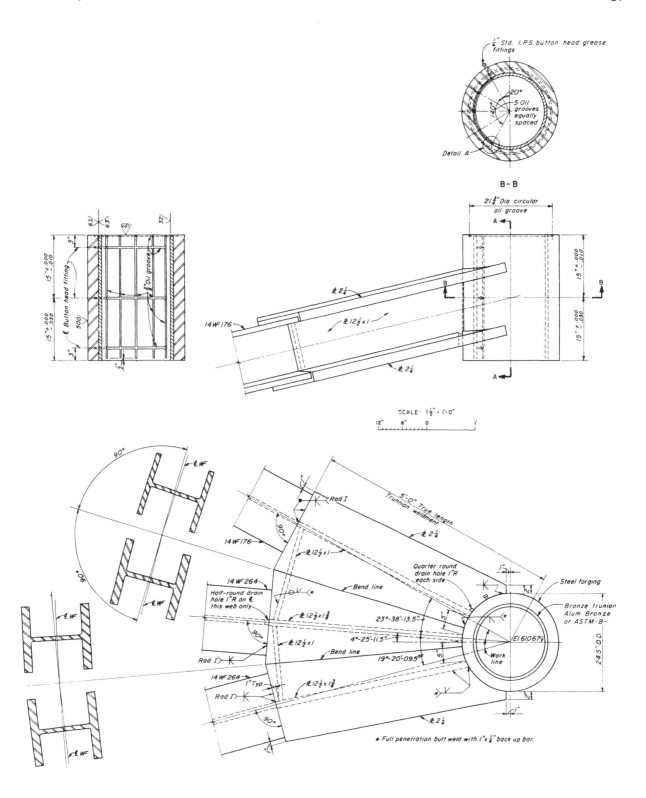

Figure 2.13   Automated drawing with sectional and auxiliary views.

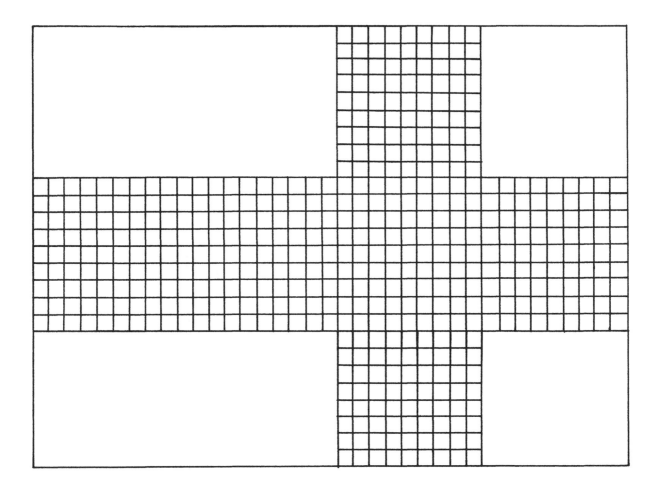

Figure 2.14 Orthographic and auxiliary view planning sheet. (Courtesy of DLR Associates)

orthographic views is considered an additional or auxiliary view. When views are
planned for computer output, it is often
convenient to locate these views on a planning sheet as shown in Figure 2.14. The
gridded section represents the locations of
the six primary views, while the open space
is used for auxiliary views. Where the
two gridded strips intersect, the front
view is located. To the top of the planning sheet is the top, to the right is the
profile or right side view. The left profile is directly left of the frontal space
and the bottom is directly below the front.
This leaves the rear which is located to
the left of the left profile. It should be
pointed out that six orthographic views are
rarely used in an automated drawing, therefore the gridded space not used, may be a

convenient space to locate a section or an
auxiliary view; and this is allowed. It is
also common practice to combine either auxiliary or orthographic views as sectional
views as shown in the many figures of this
chapter.

2.9 PICTORIAL REPRESENTATION

When you have completed Chapter 11 of
this book you should understand how a program can be used to display a pictorial object. This display may a plotter output,
video display, or microfilm image. While
Chapter 11 is a detailed treatise of pictorial representation, it is introduced in
this chapter to show how the planning form
is completed and how pictorials look if produced in an automated mode. See Figure 2.15.

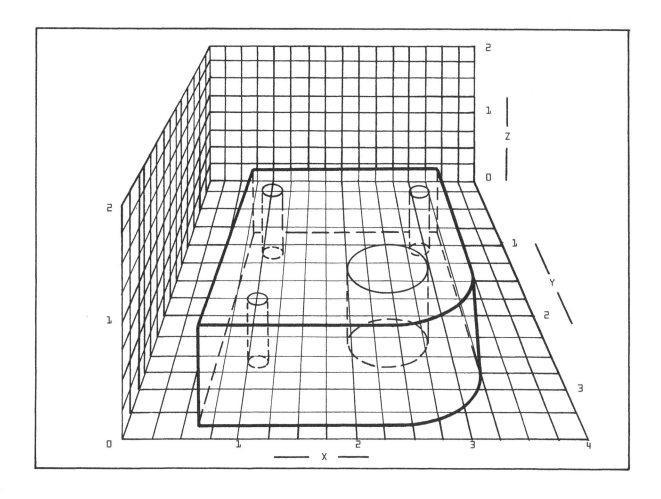

Figure 2.15 Pictorial planning form with object ready to digitize. (Courtesy DLR Assoc.)

2.10 WORKING DRAWINGS

When you have completed Chapter 13 of this book you should understand how the chapters fit together to produce a set of working drawings. The knowledge you will have gained by sucessfully working all of the chapter problems will enable you to produce finished drawings like those shown in Figures 2.16 and 2.17.

Study both 2.16 and 2.17 with the following in mind:

1. 2.16 is a top and front view of several objects. All of the drafting techniques introduced thus far have been used. Notice that sections are taken, but they do not appear of this sheet (see Figure 2. 17). A new technique is introduced on this sheet also. It is called the ballon label technique. It consists of small circles that end in a leader. Inside the circles are numbers which will appear on the second sheet in an equipment schedule. Some areas of design call these types of drawing technquies **assembly drawing** because they show how the finsihed product is assemblied or put together.

2. 2.17 contains the four sections views A-A, B-B, C-C, and D-D plus the items inside the ballons. To conserve space, not all of the equipment items are listed. A break between items 16 and 21 is noted, but in normal practice the equipment schedule is not shown in two pieces. It is listed above the title block at the right hand of the viewport as shown in Chapter 1 of this text. It is divided in this example only because of the limits of the textbook page.

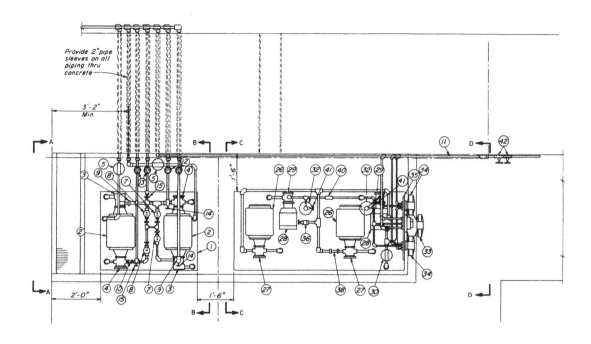

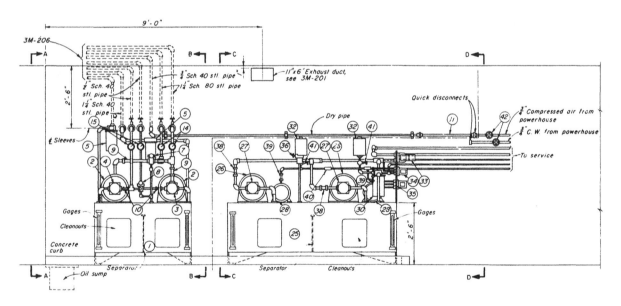

Figure 2.16 First sheet of a working drawing package showing equipment layout.

Working drawings can be developed for nearly every separate field of engineering:

1.   Architectural
2.   Civil
3.   Electrical
4.   Mechanical
5.   Structural

The most interesting examples come from the combination of these fields. The examples shown in Figures 2.16 and 2.17 are combinations of those listed. Working drawings and their production is proof that auto-drafting principles work. It is not a suggestion that you are ready to produce the working drawings shown at the end of this chapter. Just be aware that they exist.

EQUIPMENT	SCHEDULE - TAINTER	GATES			
ITEM	DESCRIPTION	QUANT.	SIZE	CAPACITY	REMARKS
1	Reservoir	1		100 Gal.	Depth as req'd
2	Electric drive motor	2			1800 RPM(Max.)
3	Pump - fixed displacement	2	2500 PSI Max	12 GPM	1800 RPM
4	Pump - fixed displacement	2	500 PSI	1.5 GPM	1800 RPM
5	Filter	2		12 GPM	Micronic
6	Strainer	2		1.5 GPM	60 Mesh
7	Pressure relief valve	2	1" I.P.S.		Set at 2500 PSI
8	Pressure relief valve	2	½" I.P.S.		Set at 500 PSI
9	Check valve	2	1" I.P.S.		
10	Check valve	2	½" I.P.S.		
11	Detachable air & water hose	1	¾"	125 PSI	Rubber
12	Plug valve	2	1¼" I.P.S.	2000 PSI	
13	Plug valve	2	¼" I.P.S.	500 PSI	
14	Pressure gage	2	4½"Dia. face	0-3000 PSI	
15	Pressure gage	2	4½"Dia. face	0-800 PSI	

EQUIPMENT	SCHEDULE - SERVICE	GATES			
ITEM	DESCRIPTION	QUANT	SIZE	CAPACITY	REMARKS
25	Reservoir	1		335 Gal.	Depth as req'd
26	Electric drive motor	2			1800 RPM(Max.)
27	Pump - fixed displacement	2	2500 PSI Max.	6 GPM	1800 RPM
28	Electric drive motor	2			1800 RPM (Max.)
29	Pump - fixed displacement	2	500 PSI	1 GPM	1800 RPM
30	Filter	2		12 GPM	Micronic
31	Strainer	2		1 GPM	60 Mesh
32	Accumulator	2		1 Gal.	500 PSI
33	4 Way directional control valve	1	1¼" I.P.S.	12 GPM	Pilot operated
34	2Way directional control valve	2	½" I.P.S.	1 GPM	Solenoid valve
35	Valve mounting plate	1			
36	Pressure relief valve	1	1" I.P.S.		Set at 2500 PSI
37	Pressure relief valve	2	½" I.P.S.		Set at 550 PSI
38	Check valve	2	1" I.P.S.		
39	Check valve	2	½" I.P.S.		
40	Pressure switch	2			
41	Stop valve	4	½" I.P.S.	1000 PSI	
42	Gate valve	2	¾" I.P.S.	125 PSI	
43	Pressure gage	2	4½"Dia. face	0-3000 PSI	
44	Pressure gage	2	4½"Dia. face	0-800 PSI	

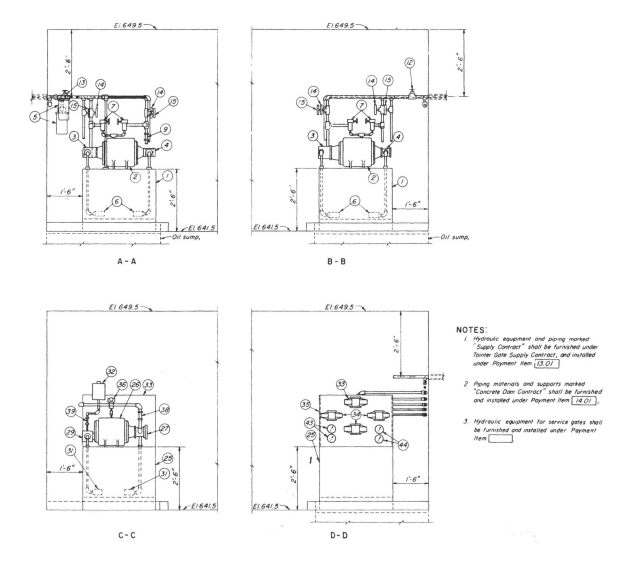

NOTES:
1. Hydraulic equipment and piping marked "Supply Contract" shall be furnished under Tainter Gate Supply Contract, and installed under Payment Item [13.01]

2. Piping materials and supports marked "Concrete Dam Contract" shall be furnished and installed under Payment Item [14.01].

3. Hydraulic equipment for service gates shall be furnished and installed under Payment Item [   ].

A - A   B - B

C - C   D - D

Figure 2.17 Second sheet of a working drawing package showing section details and schedules.

2.11 REVIEW PROBLEMS

You should complete each of the review problems at the end of Chapter 1 before attempting any of the problems which follow. In this manner you will be ready to solve, record, and store for future use each of the experiences shown.

1. Letter each of the following lines on an automated drafting and design FORTRAN coding form as shown in Figure 2.3.

```
LOGON
USERID/PASSWORD
CE .SAMPLER
INPUT
00100//JOBNAME JOB(ACCOUNT NUMBER)
00110//STEP1 EXEC FORTRAN,PLOTTER=VERSATEC
00120//C.SYSIN DD *
00130       CALL PLOTS
00140       CALL SYMBOL(1.,1.,.4,'A DESIGN
00150       + PROCESS',0.,16)
00160       CALL PLOT(0.,0.,999)
00170       STOP
00180       END
00190//G.PLOTPRAM DD *
00120//STEP2 EXEC VTECP,DEST=VER1200
00130/*
00140//
00150
SAVE
SCHEDULE
```

2. Enter the computer program lettered in problem 1 to produce:

A DESIGN PROCESS

3. Prepare freehand sketches to be entered from the data tablet for each of the following objects.

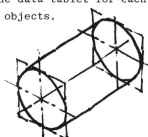

4.

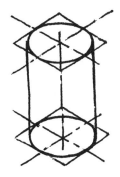

5.

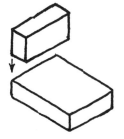

6.

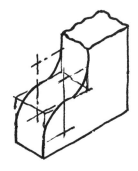

7.

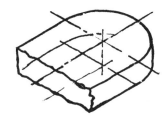

8.

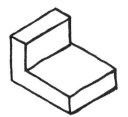

9. Prepare orthographic and auxiliary layouts for each of the following objects. A form like that shown in Figure 2.14 will be most helpful.

10.

 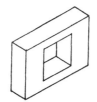

11.   12.

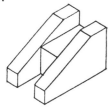 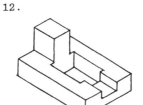

13.   14.

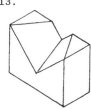 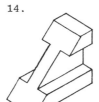

15.

16.

17.

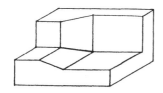

18. Prepare pictorial planning forms for each of the objects shown below.

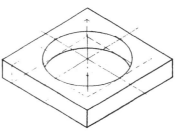

19.

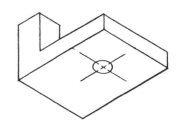

20.

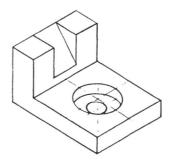

2.12 CHAPTER SUMMARY

This chapter was written so that you would understand some of the basic drafting techniques to be automated. Together with Chapter 1, you now have an idea what automated drafting equipment looks like and what type of drawings can be done. You are asked to write programs and produce computer output only after you have practiced the techniques outlined in this chapter. Each of these techniques is explained in detail in a separate chapter to follow. You are expected to practice your lettering and be able to digitize simple objects by this time in your study, however.

The skills that are learned will help you throughout the rest of this textbook and in the other computer graphics courses that you will take.

3

LETTERING

This chapter represents a new approach to computer generated lettering, stylized for the production of single stroke, Gothic as shown in Figure 2.1 and 2.2. Whereas all former types of plotted lettering were suitable for general computer graphics, none existed for engineering drawing. The CORE and ANSI standards both used general purpose character generators as described in both Chapters 1 and 2. Using the concepts discussed in this chapter the draftsperson may instruct the computer to remember his or her own style and reproduce it whenever needed. This of course requires the draftsperson to hand letter on a data tablet, each of the alphanumeric characters desired. After just one entry, however, a computers ability to generate "hand drawn" lettering is possible as shown in Figure 3.1.

The purpose of this chapter then is to develop a system which will consist of the computer requirements in both hardware (equipment) and software (programs), how to use this equipment, and finally how to code the lettering fonts for storage in a

computer. The chapter also illustrates how to use existing CORE or ANSI character generators.

3.1 USE OF AUTOMATED LETTERING

In the design of single stroke lettering, much can be learned from the use of CORE or ANSI types of automated lettering. For instance, the CALCOMP generated lettering from the CALL SYMBOL instruction produces a letter font which is inside a 21 x 21 data matrix. This square matrix contains only a 12 x 21 matrix where visible line segments are possible. The 12 x 21 sized data area is always on the left side of the matrix, while the open area is on the right as shown in Figure 3.2. The open area is provided to space the letters when placed within words. Therefore proportional spacing is not provided in CORE or ANSI lettering. As illustrated in the problems at the end of Chapter 2, this produces a rather crude looking character with equal spacing between characters. This is not acceptable for most engineering drawing work. The 9

No.	OPERATION	VALVE NUMBERS	
		OPEN	CLOSED
1	Fill dirty transformer oil tanks by gravity from a tank truck through a fill box connection (OCB tanks similar)	I-20, I-12, I-17 or I-18, I-13③	All others
2	Drain transformer by gravity to dirty transformer oil tanks (OCB similar)	I-I-2, I-II, I-17 or I-18, I-13③	"
3	Purify the oil from the transformer and transfer to clean transformer oil tanks, using purifier (OCB similar)	I-I-2, I-II, I-13②, I-2, I-3, I-5 or I-6	"
4	Purify the oil from the dirty transformer oil tanks by purifier and transfer to clean transformer oil tanks (OCB tank similar)	I-17 or I-18, I-13②, I-2, I-3, I-5 or I-6	"
5	Transfer the oil from the clean transformer oil tanks to transformer by transfer pump (OCB similar)	I-15 or I-16, I-13①, I-1, I-3, I-7, I-I-1	"
6	Circulate the oil in the clean transformer oil tanks through purifier (OCB similar)	I-15 or I-16, I-13②, I-2, I-3, I-5 or I-6	"
7	Circulate the oil of the transformer through purifier while transformer is in service	I-I-1, I-I-2, I-II, I-13②, I-2, I-3, I-7	"
8	Flush transformer lines with clean transformer oil by transfer pump (OCB lines similar)	I-15 or I-16, I-13①, I-1, I-3, I-7, I-I-3, 7-I-3, I-II, I-17 or I-18	"
9	Transfer dirty transformer oil from dirty oil storage tanks to tank truck using purifier pump or transfer pump (OCB similar)	I-17 or I-18, I-13③, I-2, (I-13①, I-1), I-3, I-9, I-20	"

✱ To initially fill the clean oil storage tanks the following steps should be taken:
   1. Fill tank (s) to overflow (by visual inspection).
   2. Stop pump.
   3. Open valves N-1 or N-2 and 3-N-1, 3-N-2, other tanks are similar.

NOTES:
1. For general notes see drawing HA5M-401.
2. All oil piping is 2½-inches unless otherwise noted.
3. Transformer oil coolers and circulating pumps are integral with transformers.
4. Valve numbers correspond to valve numbers shown on 5M 401 and 5M 402.
5. Valve operation schedule is typical for one transformer or one oil circuit breaker tank only, others are similar.
6. Valve number prefixes are designated as follows:
   I-1   Valves in the oil purification room and fill box.
   I-I-1   Valves at the transformers, the prefix indicates what number transformer
   I-I ...   Valves at the oil circuit breakers, the prefix again indicates what number oil circuit breaker.
   N-1   Valves at the nitrogen-gas seal tanks.
   7-N-1   Valves at the storage tanks, the prefix indicates what number tank
7. Gate valves at the purifier are service valves only, and are provided to prevent contamination in the system when the purifier is disconnected and being used for the governor and lubricating system.
8. Gate valves shown without specific numbers are service valves only, which are normally open or closed as shown and require no special operation.

Figure 3.1 Hand drawn engineering lettering in both upper and lower case. (Courtesy DLR Associates)

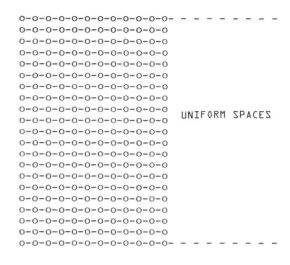

Figure 3.2 Empty data matrix for CORE and ANSI type character generators.

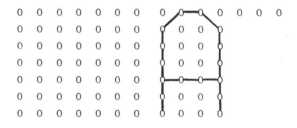

Figure 3.3 Empty point grid shown at the left which would represent a space in lettering and a visible character shown on the right.

data locations shown to the right in Figure 3.2 should be variable for proportional type spacing used in engineering drawings. Data locations are hard to visualize, often a reader is referred to a point grid so that the letter will become visible as shown in Figure 3.3. Points are measured in 3 for 1 or in other words the data matrix is a 7 x 7 point matrix. The character is vis-

ible in Figure 3.3 as the letter A. All the other letters are formed in a similar manner. Study review problems at the end of Chapter 2 for examples of these.

In order to provide for the lettering shown in Figures 2.1 and 2.2 a flexable point grid is desired. For example the letter I does not require the same space as the letter M. Therefore, not only the shape of the letters must match those shown in Figure 2.1, but also the spacing between letters must be variable. Figure 3.4 compares the CORE or ANSI generated letters directly with the single stroke letters designed for use in this chapter. Notice the expanded use of the grid points for certain

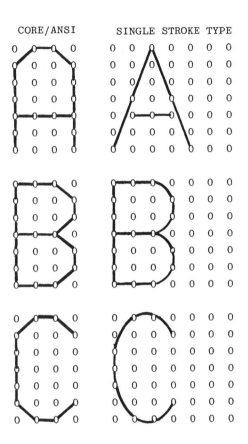

CORE/ANSI   SINGLE STROKE TYPE

Figure 3.4 The ABC's of computer generated
lettering CORE/ANSI vs single stroke.

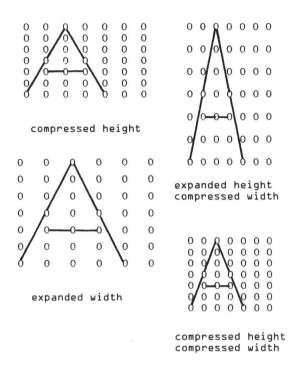

compressed height

expanded height
compressed width

expanded width

compressed height
compressed width

Figure 3.5 Variations in character plots.

letters like A.

A full set of characters for automated
drafting have been designed as listed in
Table 3.1. This set of characters differs
from the CORE or ANSI in a number of important ways. The area between points may be
enlarged or reduced independently as shown
in Figure 3.5. These variable zones between grid intersections allow considerable
freedom in the display of the items shown
in Table 3.1. Another difference is the
use of upper and lower case letters. The
last important item that should be included
in every character generator is a rotation
procedure. Figure 3.6 indicates that the
rotation is counterclocksise, the same as
CORE and ANSI. The height of the character
is set by HT, while the rotation is set by
DEG. In this example, shown in Figure 3.6,
the height (HT) was 2.8 and the rotation
(DEG) was 20.

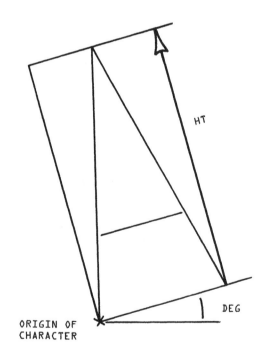

HT

DEG

ORIGIN OF
CHARACTER

Figure 3.6 Rotation of characters.

Table 3.1

Special Single Stroke Engineering Lettering Available

Special Characters	Numerics	Communication Code	Alpha	Notation Code
+	0	:	A	a
Ψ	1	;	B	b
♪	2	⌐	C	c
⊣	3	=	D	d
$	4	{	E	e
%	5	}	F	f
!	6	?	G	g
&	7	.	H	h
*	8	'	I	i
–	9	"	J	j
▮			K	k
▮			L	l
—			M	m
			N	n
			O	o
			P	p
			Q	q
			R	r
			S	s
			T	t
			U	u
			V	v
			W	w
			X	x
			Y	y
			Z	z

3.2 COMPUTER REQUIREMENTS

Two types of computer generated lettering have been introduced, CORE/ANSI used by CALCOMP, Versatec, and Tektronix; and a single stroke engineering lettering. Both are considered excellent forms of output, however, several improvements were used to develop the second method, they were:

1. The need to provide for proportional lettering.

2. Elimination of the bunching where lines join at an acute angle such as V, K, M or W.

3. Linear elements of CORE severely distort some letter shapes, so a provision for circular elements was maintained.

4. 21 x 21 data matrix characters are formed from straight lines, dots and quarter circles. However, when factoring is applied to produce a grid point image, the arcs become elliptical yielding a pleasing, more engineering oriented form.

5. Because any combination of straight line, dot or arc may be used, the dra-

ftsperson must indicate the connection desired. Figure 3.1 was only one possible combination.

There are at least two different methods to indicate the connections desired. Figure 3.7 is called a source description because it is written in computer language and describes the motion of a plotter pen or CRT beam movement. Figure 3.8 is called data base construction because it was done from a data or graphics tablet as described in section 1 of this chapter. Both Figure 3.7 and 3.8 are for the character A so that you might compare this method with the outputs shown in Figures 3.5 and 3.6. If you are not trained in computer languages then the data base construction method is easy to learn and understand.

One thing to keep in mind is that only one of the two methods described needs to be used and that the examples shown in Figures 3.7 and 3.8 are ends in themselves. They must be used inside a larger piece of software called the character generator. Section 4 of this chapter has more details.

```
SUBROUTINE A(X,Y,HT,DEG)
WD=HT*.7
THETA=.017453*DEG
X6=X-SIN(THETA)*HT
Y6=Y+COS(THETA)*HT
X1=X+COS(THETA)*WD
Y1=Y+SIN(THETA)*WD
X3=X-SIN(THETA)*HT/8.*3.
Y3=Y+COS(THETA)*HT/8.*3.
X4=X3+COS(THETA)*HT/8.
Y4=Y3+SIN(THETA)*HT/8.
X5=X4+COS(THETA)*HT/4.
Y5=Y4+SIN(THETA)*HT/4.
X2=X6+COS(THETA)*HT/4.
Y2=Y6+SIN(THETA)*HT/4.
C DATA IS NOW READY TO BE PRINTED FOR A
C MEMORY ARRAY FOR POINT CONNECTION.
PRINT 1, X,Y
PRINT 2, X,Y
1 FORMAT F5.3,1X,F5.3,1HH
2 FORMAT F5.3,1X,F5.3
PRINT 2,X2,Y2
PRINT 2,X3,Y3
PRINT 1,X4,Y4
PRINT 2,X5,Y5
RETURN
END
```

Figure 3.7 Source description of letter A.

```
C  CHARACTER CODE 65(A)
   DATA ID(65)/471/
   DATA IC(471),IC(472),IC(473),IC(474)
  +/11512,21215,20815,20512/
   DATA IC(475),IC(476),IC(477),IC(478)
  +/20508,20805,21205,21508/
   DATA IC(479),IC(480),IC(481),IC(482)
  +/21518,21221,20321,20018/
   DATA IC(483),IC(484),IC(485),IC(486)
  +/20003,20300,21200,21503/
   DATA IC(487)/32100/
```

Figure 3.8 Data base construction for A.

3.3 DESCRIPTION OF EQUIPMENT USED

An IBM 370 model 3081 with a workstation by Tektronix, model 4010 shown in Figure 3.9 was used. Input and output were obtained for both CORE/ANSI and the newly designed lettering fonts. The lettering from the CORE subroutine SYMBOL was modified and placed in the PLOT-10 software provided by Tektronix Corporation. In this manner the plot-10 preview routines could be avoided. This provided direct output to the DVST as shown in Figure 3.10.

To convert CORE type software routines for direct display on a DVST, model 4010 follow the simple steps listed below.

1. List the routine currently stored as a CORE subroutine on your terminal.

2. Look at each line of source code and identify those lines containing calls to subroutines not presently contained in PLOT-10 software. The most common of these is CALL PLOT.

3. If call PLOTS are found, list the following lines at the bottom of the routine after the RETURN, END statements:

```
SUBROUTINE PLOT(XVAL,YVAL,IPEN)
IX=XVAL*130.
IY=YVAL*130.
IF(IPEN.EQ.2)CALL DRWABS(IX,IY)
IF(IPEN.EQ.-2)CALL DRWREL(IX,IY)
IF(IPEN.EQ.3)CALL MOVABS(IX,IY)
IF(IPEN.EQ.-3)CALL MOVREL(IX,IY)
RETURN
END
```

This is a simple line generator to convert CORE plotter statements to CORE screen graphics.

4. If CALL WHERE is found, change it to CALL SEELOC. This will convert pen movement to beam position.

In the case of the CORE CALL SYMBOL, this is all that is necessary to display characters on the DVST screen. To test a character shape as shown in Figure 3.10, a draftsperson enters:

```
CALL SYMBOL(X,Y,CHT,'F',0.,1)
```

where the CORE instruction CALL SYMBOL has arguments of X to represent the distance across the screen and Y represents the vertical screen distance. CHT is the character height of F, 0. is the amount of rotation and 1 is the number of characters to be displayed. The screen shown in Figure 3.10 contains several lines of characters. More than one character may be displayed per line by merely expanding inside the ''s. The first line of Figure 3.10 was generated by 'SYMBOL FORM PRELIMINARY VERSION',0.,31 along with the proper X and Y locations.

Figure 3.9 DVST (Courtesy Tektronix Inc.)

3.4 CODIFICATIONS OF LETTERING FONTS

Because the IBM 370 has a great deal of storage, the fonts defining the lettering style currently being used are held in a device called a subroutine. The subroutine is read into the computer memory from its library storage location. In order to save more than one set of coded lettering fonts, more than one stored subroutine is required. In general purpose computer graphics systems there is only one lettering subroutine and it is called SYMBOL. Because special purpose fonts are required in automated drafting, multiple subroutines are used such as SYMBOL, LETTER, and NOTE. In this chapter, SYMBOL contains the CORE subroutine, LETTER contains single stroke, and NOTE contains lower case.

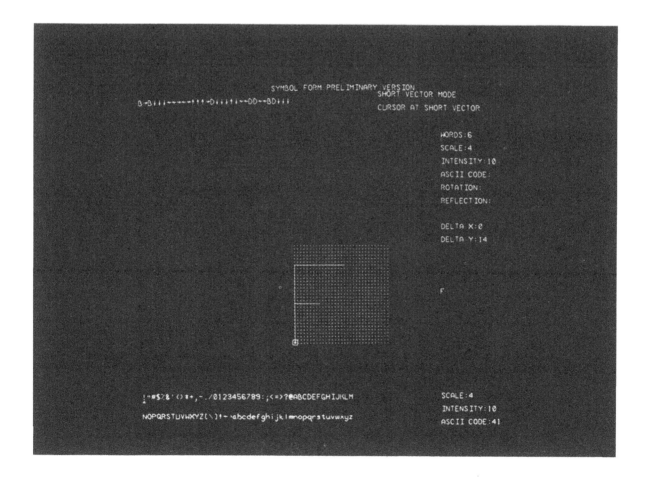

Figure 3.10 Modification of SYMBOL routines displayed on DVST (Courtesy DLR Associates)

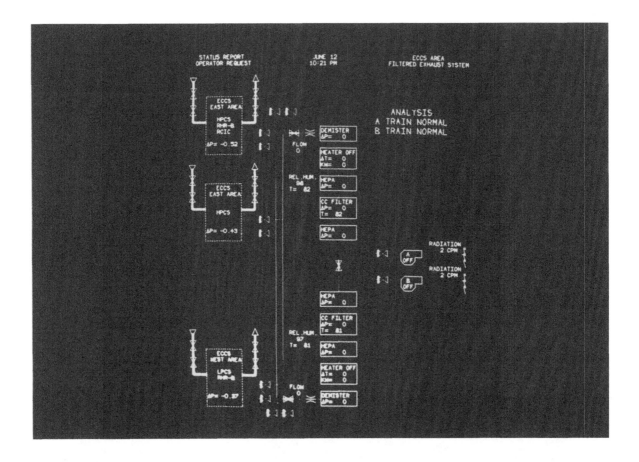

Figure 3.11   Display of LETTER routines and line generator from page 38. (Courtesy DLR Assoc)

It should be pointed out SYMBOL is an easy subroutine to use. LETTER requires a knowledge of subroutine construction and NOTE requires a knowledge of data base construction. As a present or future draftsperson you should be able to recognize each of the graphic outputs. Figure 3.10 is an example of SYMBOL, 3.11 is LETTER, and 3.12 is note. The use of each depends upon the computer system used for automated drafting and the type of drawing produced. In most cases the output from SYMBOL is considered adequate by drafting offices.

The use of all three is somewhat related in that the draftsperson uses the subroutine name (SYMBOL, LETTER, or NOTE) inside the call to the character generator as CALL _____(X,Y,CHT,'CHARACTERS',DEG, NCHRS). One of the three character generators is inserted in the blanks and the items inside the parentheses describe the line of lettering to be displayed.

3.5   METHOD OF STORING CHARACTERS

How will each of the subroutines find the order of letters in the line 'CHARACTERS'? NCHRS must be set equal to 10. The line 'CHARACTERS' is 10 characters not 12. The 's do not count in the NCHRS. Each character is located by the subroutine, in the case of SYMBOL a temporary location is used as IC(1)=C, IC(2)=H, IC(3)=A .... IC(10)=S until all ten storage locations have been filled. The subroutine checks each IC(I) location by the order of Table 3.1. In other words the routine checks IC under the special characters column and does not find it, then the numerics column, then the communication column, then the alpha

Figure 3.12 Display of NOTE routines.

column where it locates how to form a "C". Nine more trips through the columns of the Table 3.1 and the entire line of lettering is complete. This happens at the speed of electronics so the draftsperson does not have to wait, the display appears to be instantaneous.

Three methods for storing characters presently exist, the ones shown in Figure 3.7 and 3.8 and the following:

```
C       aaaa
341  CONTINUE
     CALL MOVER(.2,0.)
     CALL DRAWR(.1,0.)
     CALL DRAWR(.1,.033333)
     CALL DRAWR(.066666,.066666)
     CALL DRAWR(.033333,.1)
     CALL DRAWR(0.,.2)
     CALL DRAWR(-.033333,.1)
     CALL DRAWR(-.066666,.066666)
     CALL DRAWR(-.1,.033333)
     CALL DRAWR(-.1,0.)
     CALL DRAWR(-.1,-.033333)
     CALL DRAWR(-.066666,-.066666)
```

```
     CALL DRAWR(-.033333,-.1)
     CALL DRAWR(0.,-.2)
     CALL DRAWR(.033333,-.1)
     CALL DRAWR(.066666,-.066666)
     CALL DRAWR(.1,-.033333)
     CALL MOVER(.8,0.)
     GOTO 99
C       bbbb
```

The grid methods of storing characters is done by data locations because the grid points never change they are fixed in space. The character data can therefore be compressed or packed in DATA statements to take up less space. The packing scheme is used because the shape of the letter is not changing (Figure 3.8 only) and represents a compromise between the amount of subroutine space saved and the simplicity of the character (Figure 3.7). This last method of storage requires more lines of intersections inside the subroutine. In fact, another subroutine DRAWR is used to give the

appearance of the letters shown in Figure
3.12. This method requires seven times
longer to display a simple character. The
time involved is usually only a few thou-
sandths of a second but computer times are
charges against running times (display periods), therefore, the NOTE subroutine is
seven times more expensive than the SYMBOL
subroutine.

3.6 CHARACTER PLOTTING SUBROUTINES

Do draftspersons develop lettering subroutines? Of course not, but they use one
of the three types just described on a constant basis. In some automated drafting
offices, lettering in one form or another
consists of 40% of the output from plotters.
In fact, line generators and character generators can account for nearly 90% of all
automated drafting output. It is, therefore, important that you understand how to
use character generators, not how to invent
new ones.

keyboard shown in Figure 3.13 is typical,
but not the only type of keyboard available
so check how your keyboard is arranged. A
keyboard that meets CORE/ANSI standards can
be used with only a few minutes of instructions. Some simple steps to assist you are:

1. Let your hands hang loosely at
your sides. Your fingers will relax in a
curved position as shown.

2. With fingers curved in this relax-
ed position, lightly place the fingertips
of your left hand on keys F D S A. This is
called the home postion.

3. Similarly, lightly place the fingertips of your right hand on keys J K L ;.

4. Repeat steps 1,2 and 3 several
times until you can place them without looking at the keyboard. Learn to keep fingers
curved and upright as shown. Only the fingertips should touch the keys.

5. Now study the Figure 3.13 to see
which fingers are assigned which key. Only
a very few instructions are typed by draftspersons, heavy typing is always done by a

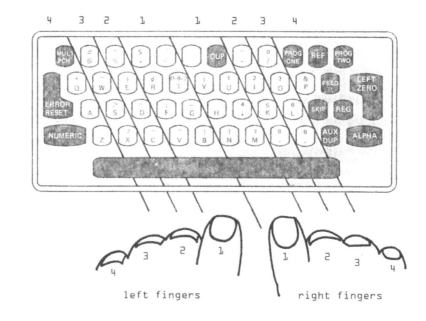

Figure 3.13 Use of a keyboard to enter simple instructions for character generators.

The first thing you will need to do,
in order to use a character generator, is
familiarize yourself with a keyboard. The

person called a keypunch operator.

6. Practice the drafting instructions
that you will be using. They are:

LOGON	LOGOFF	STOP	END
LIST	LISTM	LISTA	LISTC
CALL	SYMBOL	LETTER	NOTE
ACS.	USERID	PASSWD	CE .

Let's take the first instruction, LOGON.
It is entered by selecting the L key with
third finger of the right hand. The key
is struck with a quick, sharp finger stroke
as shown in Figure 3.13. You snap the finger toward the palm of the hand as the key-
stroke is made with each finger to complete
the instruction LOGON. L and O are made
with the same finger, while G is made with
the index (1) finger of the left hand. A
letter N is made with the index (1) of the
right hand.

In this manner, a minimum number of
instructions may be learned so that the
draftsperson can enter them quickly. How
can the draftsperson select the character
subroutine from others stored in the computer? Simple, he asks the computer which
are stored and where. This is done from
the keyboard by entering:

        LOGON
        USERID/PASSWD
        LISTA

The computer responds by listing all of the
subroutine libraries available that are
available for you to use. You notice that
all three subroutines may be contained in
the plotter library, but you will need to
check this by listing each of them.

3.7   SOFTWARE LISTINGS

You can begin the search by asking for
the subroutine SYMBOL to be listed for you.
This is done by:

        LIST 'ACS.CC.WATLIB(SYMBOL)'

The computer responds by listing the subroutine as shown in Figure 3.14. In this
example, the entire subroutine is not shown
only the first page or screen size is listed for you. The fact that the computer

```
SUBROUTINE SYMBOL(X,Y,CHT,IC,DEG,NCHRS)
COMMON/CLOMP/XFSET,YFSET,XACUM,YACUM,
+SKIP,NSKIP,IOPT,XLEN,YLEN,XFAC,YFAC,NHA
+RD,XPOSIT,YPOSIT,FILE,PLOTR,CFILE
DIMENSION KADE(100),ISTRNG(100),IDSPL(6
+4),ID(128),IC(920)
LOGICAL SKIP,FILE,PLOTR,CFILE
DATA IDSPL(1)/32/
DATA IDSPL(2)/65/
DATA IDSPL(3)/66/
DATA IDSPL(4)/67/
DATA IDSPL(5)/68/
DATA IDSPL(6)/69/
DATA IDSPL(7)/70/
DATA ISSPL(8)/71/
DATA IDSPL(9)/72/
DATA IDSPL(10)/73/
DATA IDSPL(11)/74/
DATA IDSPL(12)/75/
DATA IDSPL(13)/76/
DATA IDSPL(14)/77/
```

Figure 3.14   Subroutine SYMBOL.

listed the routine is fact that it can be
used by a draftsperson to output lettering
to a plotter or DVST screen.

Suppose the draftsperson wanted to output a rather long line of lettering like:

1. FOR GENERAL NOTES SEE DRAWING HA5M-401

See Figure 3.1 for this line of lettering.
Notice that it is located below the shorter
line:

NOTES:

The X and Y plotter locations must be used
inside a call to the symbol subroutine as:

CALL SYMBOL(X,Y,CHT,'1. FOR GENERAL NOTES
SEE DRAWING HA5M-401',0.,42)

Figure 3.15 is the listing for the subroutine LETTER when the command:

        LIST 'ACS.CC.WATLIB(LETTER)'

is entered from the keyboard. In another
example, Figure 3.16 is the listing for the
subroutine NOTE when a similar command is
entered.

```
  SUBROUTINE LETTER(X,Y,CHT,IC,DEG,NCRS)
  DIMENSION X(1),Y(1),NBUF(6),ITG(2),IBUF(
 +13),ITEXT(1), IT1(2)
  COMMON/PLTCOM/SAME,PREF(2),LDC(2),C(2),
 +RORG(2),CMAT(10,3),LMT(2),XYDOTS(2),SPX
 +,SPY,MIX,NSKP,NBAD,NPLOT,MINX,MAXX,NDX,
 +NDY,LTYPE,LWDTH,DEGRAD,NBITS,NBTM1,NBYT
 +E,NBYM1,NCRS,NCHM1,MSK(7),IBT(16)
  EQUIVALENCE(C(1),ITG(1))
  EQUIVALENCE(T1,IT1(1))
  DATA IBUF(1)/:40/, IOV2 /'# '/
  DATA IDASH/:55/,IOV1/'##'/
  DATA IDOTT /:56/,NTXT /:60/
  XO=X(1)
  YO=Y(1)
  IF(XO.EQ.SAME) XO=PREF(1)
  IF(YO.EQ.SAME) YO=PREF(2)
  IF(NCRS)200,300,5
5 ND=NCRS-1000
  IF(ND.GE.0)GOTO 400
```

Figure 3.15   Subroutine LETTER.

Figure 3.17   Typical output from LETTER

```
  SUBROUTINE NOTE(X,Y,CHT,IC,DEG,NCRS)
  DIMENSION IC (1),IBUF(120)
  N=NCRS
  IF(N.EQ.0)N=1
  IF(N.EQ.-2)CALL PLOT(X,Y,2)
  CALL PLOT(X,Y,3)
  CALL SEEREL(COS,SIN,SCALE)
  ANG=ATAN2(SIN,COS)*180./3.14159
  CALL ROTATE(ANG1+DEG)
  CALL RSCALE(CHT*SCALE)
  IF(N.GT.0)GOTO 5
  GOTO(100,101,102,103,104,105,106,107,10
 +8,109,110,111,112,113,114),IC(1)
5 NN=(N+1)/2
  J=0
  DO 9 I=1,NN
  J=J+1
  IBUF(J)=RS(LS(IC(I), 8),8)
  J=J+1
  IBUF(J)=RT(IC(I),8)
```

Figure 3.16   Subroutine NOTE.

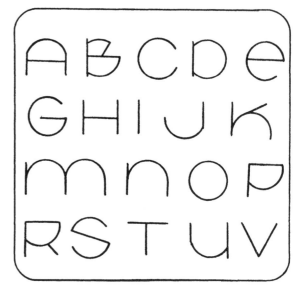

Figure 3.18   Typical output from NOTE.

All three subroutines used for lettering have common elements. Each use tables
of values (two index tables and one stroke
table). Subroutine SYMBOL shown in Figure
3.14 contains fixed grid locations to generate the characters shown in Table 3.1. A
variation of SYMBOL is used to develop the
logic of subroutine LETTER shown in Figure

3.15 and 3.17. Here the table 3.1 characters are driven by the NCRS value and sorted
before plotting. Subroutine Note, shown in
Figure 3.16 and 3.18, contains the point
locations for each of the characters in IC.
This method contains actual stroke vectors
for each single character used as:

    1. vector type (move,draw,rotate,etc)

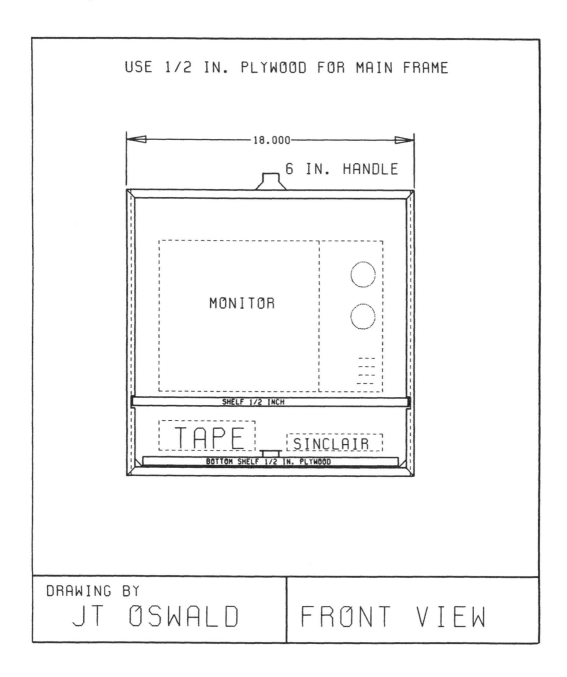

Figure 3.19 SYMBOL generated lettering output on a digital plotter and used as part of the information for an automated drawing. (Courtesy DLR Associates)

2. X location on a display area, and

3. Y location on a display area.

Generation of the display data requires any method (SYMBOL, LETTER or NOTE) take place in the following stages:

1. The data for the character is located by the two outer index loops from the Table 3.1.

2. The grid formation is connected by predetermined individual point locations on the rows and columns of the grid pattern.

3. The width-height ratio is computed or read from CHT.

4. Transformation DEG is calculated for rotation of the lettered line and spacing is done by checking the preceding char-

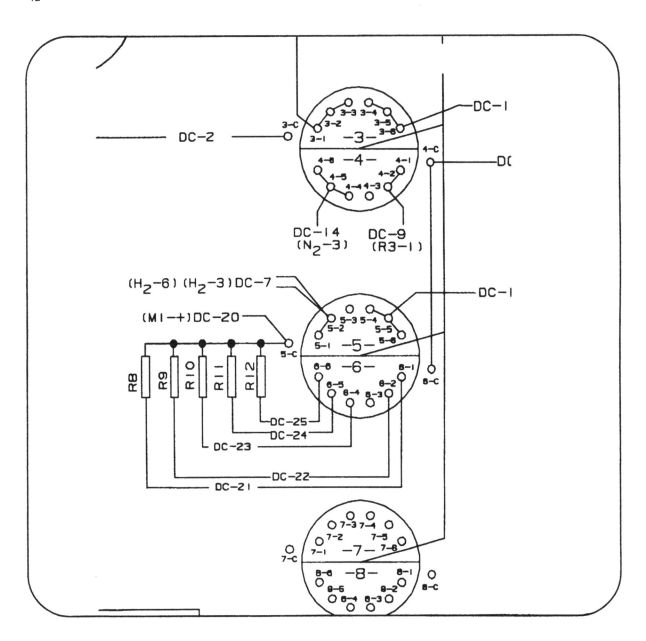

Figure 3.20 LETTER generated lettering output to a DVST graphics terminal and used as part of the information for an automated drawing. (Courtesy DLR Associates)

acter in the line. If DEG = 0., this stage is omitted, as shown in Figure 3.19. If a DEG value is greater than 0., then stage 4 is completed as shown in Figure 3.20. See R8 through R12 for examples of this in Figure 3.20.

5. The required number of points in the stroke table may also be transformed, this produced inclined – lettering as shown in Figure 3.21. All letters are indexed by

NCRS, and the last index completes the required line defined by IC(NCRS).

3.8 CHARACTER MANIPULATION

In addition to the obvious character manipulation such as grid height, rotation, grid transformation (inclined), carriage return, line feed and space commands which all three contain; special features also

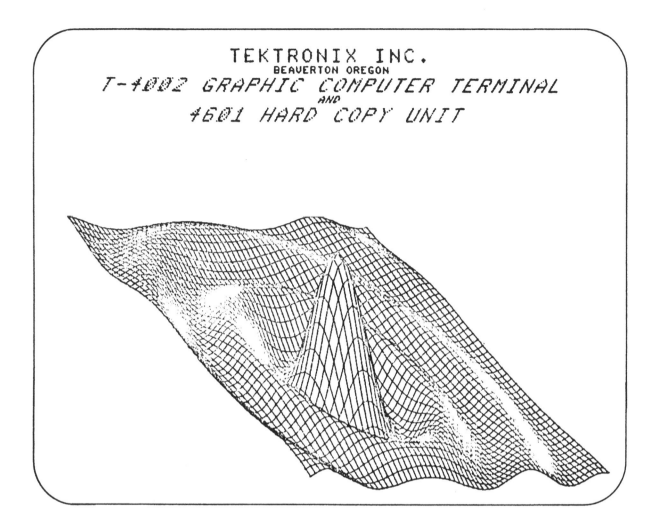

Figure 3.21 Inclined lettering output to a DVST. (Courtesy Tektronix Inc.)

exist. They are:

1. Factored heights. It is possible to produce lettering where the first letter of a word is larger as shown in Figure 3.1.

2. Extension or compression. The use of factoring inside of a stroke table will cause pleasing results as illustrated in Figures 3.12, 3.18 and 3.21.

3. Slope. The favored inclined lettering is used in Figure 3.21 shown above.

4. Weight. The thickness of the line that forms the character is not controlled by the shape subroutine but is a function of the character generator. The line weigths may be changed by CALL NEWPEN(KIND). KIND is a number 1 through 5 and compares directly to the line weights shown in Chap-

ter 1.

3.9 REVIEW PROBLEMS

Automated, computer-generated engineering lettering is easy if the proper equipment is made available to the draftsperson. In each of the problems outlined in this section of Chapter 3 a special piece of equipment is required. If no equipment is available, then the special forms shown in Chapter 2 should be used so that a keyboard operator may enter the instructions for you. Figure 3.22 is typical of the keyboards an operator will use to produce engineering lettering on an automated drawing. Select one of the types shown and enter the CALL's

Figure 3.22 Typical keyboards used to produce engineering lettering. (Courtesy Tektronix Inc.)

to SYMBOL, LETTER or NOTE as required.

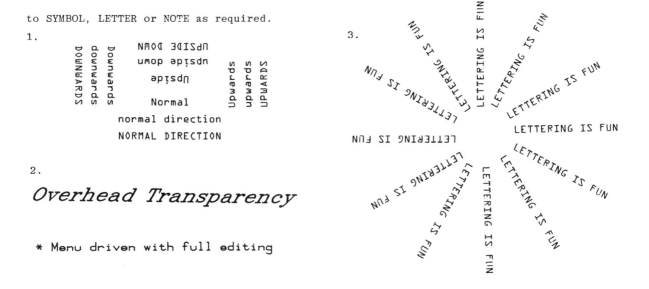

4.

MARKET DIVERSIFICATON AND GROWTH

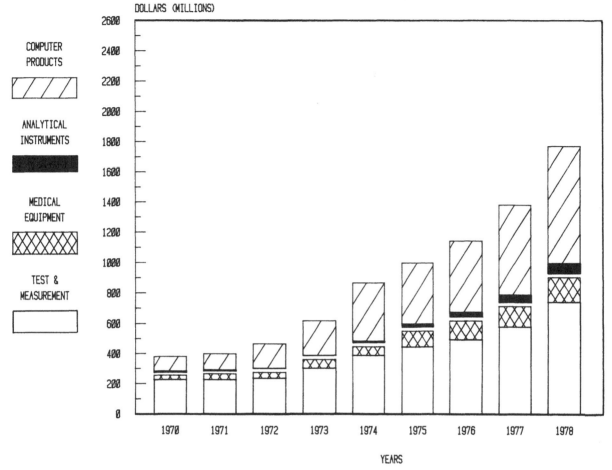

DOLLARS (MILLIONS)

COMPUTER
PRODUCTS

ANALYTICAL
INSTRUMENTS

MEDICAL
EQUIPMENT

TEST &
MEASUREMENT

YEARS

5.

OPERATING SCHEDULE

No.	Operation	System	Closed-Valves-Open	
1	Normal		A4·A5	A1·A2·A3
2	By-pass aftercooler		A1·A4	A2·A3·A5
3	By-pass air receiver	100 psi. station service system	A2·A3·A5	A1·A4
4	By-pass aftercooler and air receiver		A1·A2·A3	A4·A5
1	Normal		B2·B3·B4	B1
2	By-pass aftercooler	100 psi draft tube depression system	B1·B3·B4	B2
3	Draft tube depressing		B2	B1·B3·B4
1	Normal		C3	C1·C2
2	By-pass air receiver	300 psi system	C1·C2	C3

Note: During draft tube depression, valves B3 and B4 (diaphragm valves) are opened by means of solenoid valves in conjunction with pressure switches in the gallery at El. 466.5 and a float switch located in the draft tube access at the draft tube liner.

Note: All unnumbered gate and globe valves are service valves and are normally open. Drain and relief valves are normally closed. Gate valves at the governor pressure tanks are normally closed, except for charging of the tanks. Cooling water gate and control valves are normally as shown.

6.

VALVE AND SWITCH OPERATION SCHEDULE

NO.	OPERATION	VALVE AND SWITCH POSITIONS
1.	Fill draft tube, Unit 1 (Other units similar)	a) Close valves U-5, U-6, U-7, 2-DT, 3-DT, and 4-DT. b) Set selector switch No.1 to "OFF" and selector switch No. 2 to "Complete". c) Open valves 1-DT, U-1, U-2, (or U-3, U-4) and U-8. d) Raise draft tube gates.
2.	Fill spiral case and penstock Unit 1 (Other units similar)	a) Close valves U-1 thru U-4, 1-DT thru 4-DT, and U-7. b) Set selector switch No.1 to "OFF" and selector switch No. 2 to "Complete". c) Open valves U-5, U-6, and U-8. d) Open penstock by-pass valve. e) Open intake gate.
3.	Normal setting of all valves (4 units operating with seepage from future Unit 5)	a) Valves closed are 1-DT thru 4-DT, U-1 thru U-4, U-7, and penstock by-pass valve. b) Set selector switch No.1 to "OFF" and selector switch No. 2 to "Complete". c) Valves open are U-5, U-6, and U-8.
4.	Unwater Unit 1 (Other units similar) A. Closed system B. Wet sump system "Complete" C. Wet sump system "Partial"	a) Close valves U-2 thru U-6, U-8, 2-DT thru 4-DT, and penstock by-pass valve. b) Open valves 1-DT, U-1, and U-7. c) Set selector switch No.1 to "Unit 1" and selector switch No. 2 to "OFF". a) Set selector switch No.1 to "OFF" and selector switch No. 2 to "Complete". b) Open valves U-8, U-6, and 1-DT. c) Close valves U-7, U-2 thru U-5, 2-DT thru 4-DT, and penstock by-pass valve. a) Set selector switch No.1 to "OFF" and selector switch No. 2 to "Partial". Valves as "Complete."
	Special Note	Turn both selector switches to "OFF" while changing valve positions.

7

```
**------------------------------------------------------------**
**                 INTERACTIVE VECTOR DISPLAY                 **
**.............................................................**
**        BY RICHARD E. COTTON - C.E. - APRIL 1981           **
**                                                            **
**        INPUTS FROM KEY BOARD:                              **
**                                                            **
**                 CHARACTER          COMMAND                 **
**                 =========          =======                 **
**                     M        MOVE(NO LINE DRAWN)           **
**                     D        DRAW FROM LAST POINT          **
**                     E        ERASE SCREEN                  **
**                     H        COPY DRAWING                  **
**                     Q        QUIT PROGRAM                  **
**                     C        DRAW ARC                      **
**                     P        PIN JOINT                     **
**                     F        FIXED AXIS                    **
**                     L        TITLE OF DRAWING( 0 THRU 40)  **
**                     A        ARROW HEAD                    **
**                     S        SAVE INPUTS NOT WRITE TO FILE **
**                     $        RE-DIRECT TO TEK. 4662 PLOTTER**
**                     R        RE-INPUT LABELS               **
**                     W        WRITE TO EXTERNAL FILE        **
**                     B        SHADE RECTANGULAR AREA        **
**                     1        GIVES DASHED LINE             **
**                     Z        FITS 3 POINTS TO A CURVE      **
**                     X        SMOOTHES UP TO 99 POINTS TO CURVE **
**                     V        SHADES ANY PLOYGON            **
**                     I        DRAWS CENTERLINE              **
**                     G        DRAWS A LINEAR GRID           **
**                     K        DRAWS A RECTANGLE             **
**                     O        DRAWS A MULTI-SIDED POLYGON   **
**                     !        HELP COMMAND                  **
HIT RETURN -----------------------------------------------------**
IHO001A PAUSE      0
```

8.

CABLE TRAY SUPPORTS - LIST OF MATERIAL			
Item No.	Description	Manufacturer	
		Unistrut	Power Strut
1	Straight Length	P-1000	PS-200
2	Straight Length	P-1001	PS-201
3	Straight Length	P-1001A	PS-202
4	"Z" Fitting	P-1045	PS-611
5	Flat Plate Fitting	P-1065	PS-601
6	Flat Plate Fitting	P-1066	PS-602
7	90°Angle Fitting	P-1331	PS-2546
8	90°Angle Fitting	P-1332	PS-2547
9	90°Angle Fitting	P-1355	PS-2539
10	90°Angle Fitting	P-1069	PS-605
11	90°Angle Fitting	P-1934	PS-603
12	22½°Open Angle Fitting	P-2102	PS-641
13	22½°Closed Angle Fitting	P-2107	PS-627
14	45° Angular Fitting	P-2452	PS-926
15	Wing Shape Fitting	P-2472	PS-983A
16	12"Bracket	P-2494R	PS-641R
17	6"Bracket	P-2491R	PS-838R
18	18"Bracket	P-2497R	PS-844R
19	22"Bracket	P-2499R	PS-846R
20	28"Bracket	P-2502R	PS-849R
21	Clamping Nut	P-1010	PS-10
22	½"x 1¼" H.H.Cap Screw		
23	90°Angle Fitting	P-1026	PS-603
24	"Z" Fitting	P-1453	PS-756
25	"Z" Fitting	P-1347	PS-647
26	Pipe Clamp	P-1114 to P-1119	PS-1129 to PS-1133
27	Maple Cable Clamp	U-371 to U-375	PS-1610A to PS-1614A
28	½"x 1⅛" H.H.Cap Screw		
29	Flat Plate Fitting	P-1380	PS-750

9.

SYMBOLS:

◯ Indicating Instruments: A=Amps, V=Volts, VAR=Vars W=Watts, F=Frequency, S=Synchroscope, WH=Watthour SI = Speed Indicator

◯ Recording Instruments: VAR=Vars, W=Watts, T=Temperature F=Frequency, V=Voltage

(TC) Thermal Converter

☐ Instrument Switch: AS=Ammeter Sw. VS=Voltmeter Sw. SS=Synchronizing Sw.

◯ Relay: Number=ASA device function number.

(PMG) Permanent Magnet Generator

(FBM) Fly Ball Motor

(VR) Voltage Regulator

(LDC) Line Drop Compensator

(CC) Cross-Current Compensator

(RR) Rectifier-Resistor Pack

C◯ Air Circuit Breaker, S= Solenoid Operated, A= Pneumatic

§ Current Transformer

§ Bushing Type Current Transformer

10

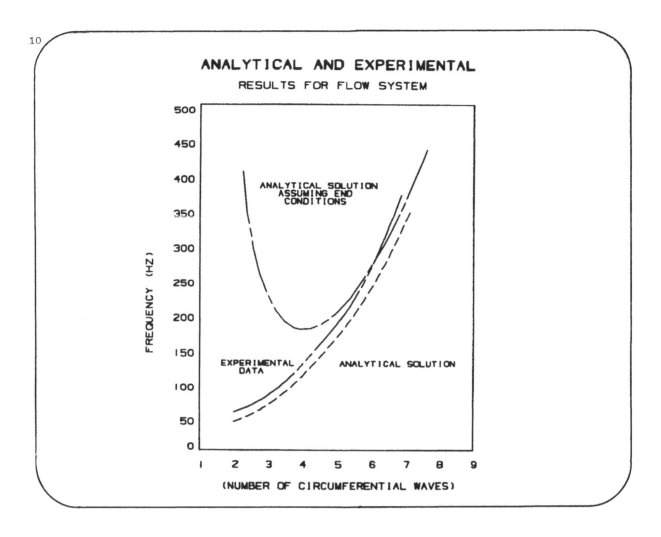

ANALYTICAL AND EXPERIMENTAL
RESULTS FOR FLOW SYSTEM

11.

SYMBOLS:

◯	Indicating Instruments : A - Amps. W-Watts, V-Volts,-Watthour
[CS]	Control Switch
(SI)	Overcurrent Relay
(R)	Protective Relay
[C]	Motor Controller
[▯]	Pushbutton Station
(×)	Power Receptacle - 3∅
◯	Instrument Switch VS=Voltmeter AS=Ammeter
[ALT]	Alternator
[T]	Thermostat
[SS]	Hand-Off-Auto selector Switch.
⎍F	Air Circuit Breaker: Electrically Operated ST- Shunt trip only
⎍+⎍	Mechanically interlocked Circuit Bkrs.
⎍F	Float Switch
(5)	Motor (Numeral indicates Horse Power)
⎍P	Press: Switch

12.

LEGEND:

▽◻	Carrier Current Telephone - Desk
▽◻	Telephone - Public (Local Commercial Co.)
▽W◻	Wall Phone - Indoor or Outdoor (Private)
▽◻	Desk Phone
▽◻E	Desk Telephone - Executive Right of Way
◻S	Desk Telephone (3 lines) 1. Executive Right of Way 2. Fire Alarm, 3. PAX Telephone
⌂	Code Call Bell
⌂	Code Call Chime
▷◻◁	Code Call Horn
◎	Sound Power Telephone jack
[R]	Code Call Relay
◻◁	Carrier Current Speaker
[K]	Switching Key Box
[B]	Buzzer
[▯]	Push Button
[T]	Turret on Control Room Operator's Desk for patching Carrier Current and PAX Telephone lines
[JB]	Junction Box _ FT- Floor Outlet Box, W-Wall Outlet Box PB -Pull Box
[TT]	Telephone Terminal Box
[m]	Electrical Door Release
CCU	Code Call Unit

3.10   CHAPTER SUMMARY

This chapter has represented an approach to computer generated lettering. It
was based upon the draftsperson's ability
to capture hand lettering and store it in
a computer for use later. Figure 3.23 re-
views some of the pieces of equipment that
can be used for this purpose. They are:

1. Data or graphics tablet.
2. Flat bed plotter.
3. Hard copy unit.
4. Joy stick.
5. Matrix printer
6. File manager

All of these items were used with a keyboard attached to a direct view storage
tube.

Once the lettering style is captured it
is stored in a mass storage device as shown
in Figure 3.24. Both data and instructions
are stored for each of the character gener-
tors described in this chapter. Whereas
all former CORE and ANSI lettering styles
existed in most automated drafting systems,
none existed for the single stroke, Gothic.

Two additional character generators
were discussed called LETTER and NOTE. A
complete set of software listings were provided for use on automated drawings. The
purpose of the software was to provide information on the automation of a heretofore
difficult area of computer graphics. It is
now possible for a draftsperson who is not
a computer programmer, to make use of these
styles in the production of professional
hand-drawn display images.

Figure 3.23 Peripherals for DVST (Courtesy
Tektronix Inc.)

Figure 3.24 Mass storage module. (Courtesy
Tektronix Inc.)

4
SKETCHING

Much of the design information is passed to draftsmen or women in the form of a free-hand sketch. As introduced in Chapter 2, this sketch may or may not be a computer-assisted sketch. If it was computer-aided, then a storage file for that sketch exists and you can display the sketch in a variety of ways. If the sketch is hand drawn, then it must be entered into computer storage. Once entered, it may be modified, scaled, rotated, or saved for use later in a set of working drawings. Computer-aided sketching is introduced in this chapter directly after the automated lettering so that the quality of the lettering, sketching and working drawings will be equal.

Throughout the material in this chapter the concept of computer assistance is stressed. Beginning with the tools (hardware) necessary and instructions for successful operation -- to the applications and purpose of design sketch construction. The chapter ends with review problems that you may place in computer storage and recall to get first hand experience in computer-aided sketching.

4.1 FREEHAND DATA BASE CONSTRUCTION

Many technical reports have been written about the computer and its role in modern drafting and design office practice, a major portion of these studies have either tried to justify or ridicule the computer as a replacement for a draftsperson. This chapter will not agree with either extreme. Draftspersons will not be replaced by a computer, but they may be assisted by one! In an age of increasing automation it is only natural that uses for the computer will probably occur in all fields of engineering. The assumption made by some engineers that the computer is not capable of freehand sketching or that the software is too difficult to write for the end results will be cheered by this chapter. This chapter was written soley for the purpose of filling the gap in engineering graphics and drafting known as freehand data base construction.

What has been lacking is a technique which can achieve a degree of coupling for the human input (sketching) and machine out-

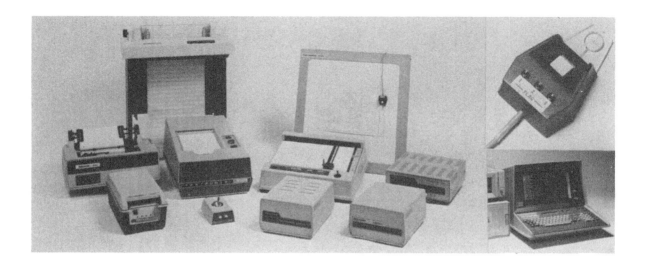

Figure 4.1 Continuation of equipment from Chapter 3 with special emphasis on the data tablet. (Courtesy of DLR Associates and Tektronix Inc.)

put (computer display and enhancement of a freehand sketch). This coupling must not force the draftsperson to change thinking habits, language habits, or graphic response habits. In the course of the examples shown in this chapter, the reader will understand how this coupling is possible. The equipment manufacturers listed in Table 2.1 have been striving to make this coupling by graphic input devices shown in Figure 4.1. When used correctly, these devices will give a draftsperson an electronic medium close to the natural or professional frame of reference. How to use this equipment is shown throughout this text.

4.2 USE OF THE GRAPHICS TABLET

If one studies our short but dynamic history in the computer graphics industry, it is possible to note a few manufacturers who have understood the problem of freehand data base construction. This situation prompted some unnamed researcher to invent the data tablet, shown in Figure 4. 2. This drawing board like device allowed a draftsperson to simply point at a spot on its surface and record this spot as an X and Y location in computer memory. The

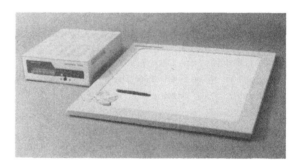

Figure 4.2 Data tablet, stylus (electronic pencil), and control box used in freehand sketching. (Courtesy Tektronix Inc.)

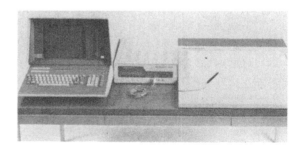

Figure 4.3 Data tablet is connected to the direct view storage tube so that the freehand sketch is seen. (Courtesy Tektronix Inc.

memory location could then be used to display a dot on a video screen or locate a pen on a plotter, hence the name data tablet became graphics tablet out of this use.

The graphics tablet will input graphic data to a digital mass storage device shown in Chapter 3, through a video display terminal shown in Figure 4.3. Graphic tablets are identical except for size as shown in Figure 4.4. The basic function of the tablet is to convert the position of the stylus on the surface to a coooresponding digital position that is usable on the DVST. The digital position may be simultaneously transmitted to the mass storage device, a plotter and DVST (direct view storage tube) for visual checking. See Figure 4.4 which shows this relationship. Starting at the left of Figure 4.4, the items are:

1. Large graphics tablet
2. DVST and keyboard
3. Mass storage device
4. Small graphics tablet
5. Hard copy device
6. Digital plotter
7. Matrix printer

A graphics tablet extends the draftspersons ability to input freehand sketches in one of 4 different methods. The input is done by typing a three character code from the keyboard (see Chapter 3 on how-to use a keyboard). The three keys are pressed at the same time. The first two keys are always the same, they are the ESC key and the ! key. The third key is the number to select one of the 4 different methods mentioned.

A graphics tablet has four main components that you should become familiar with, they are: (See Figure 4.5)

1. The writing surface
2. The writing pen or stylus
3. The control box and
4. The tablet control card that is placed inside the DVST. Each will now be outlined for you.

1. **Tablet Surface.** The tablet surface is slightly recessed, magnetically prebiased, on which gridded layout forms (illustrated in Chapter 2) must be placed

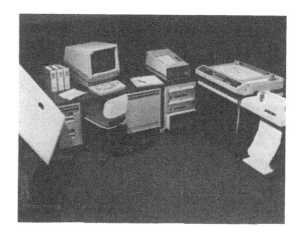

Figure 4.4 Graphics tablets used as input devices. (Courtesy Tektronix Inc.)

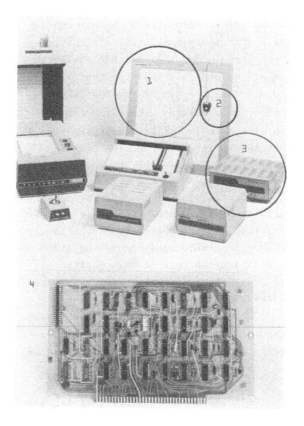

Figure 4.5 Main components used with a data tablet. (Courtesy DLR Associates and Tektronix Inc.)

before points can be input. Under the surface of the tablet are two grids of magneto-

strictive wires, one set for the X axis
and one set for the Y axis. An acoustic
wave is sent along these wires and detect-
ed by the writing pen or stylus; the time
between when the wave is sent and when it
is received allows data points in the grid
of 1024 x 1024. This means that identifiable points are 0.01 inches apart.

2. Writing pen or stylus. The pen
used on a graphics tablet is shown in Figures 4.2 and 4.3, if differs from the stylus shown in Figure 4.1, upper right-hand
corner. Both function in the same manner,
however, each contains a sensitive pick-up
coil which, when moved onto the tablet surface, detects the change in magnetic field
caused by the acoustic wave. The signal is
then converted into digital information.
The pen can be demonstrated by placing it
anywhere on the tablet surface, lightly,
do not press down. Notice that the location of the pen is present on the DVST as
shown in Figure 4.6. If the pen is moved
on the tablet surface, this symbol moves
with it. This is called a tracking symbol.
Once the pen is depressed, the tracking
symbol goes away and a bright dot is present as shown in Figure 4.7.

3. Control box. The control box contains the power supply, the tablet pulser
and preamp, and the connections which tie
the graphics tablet components together.
In addition, the control box contains several front panel indicators such as; power
on, ready, pen, data, and off.

4. Tablet control card. The tablet
control card contains those circuits necessary to convert the pen signal to its digital equivalent.

4.3 SKETCHING PROCEDURES FOR DATA TABLETS

Four example sessions for freehand input will now be given; in each case the control box and DVST must be turned on 'ready'
to function.

Single point/pen method

1. Press the ESC, !, l keys together

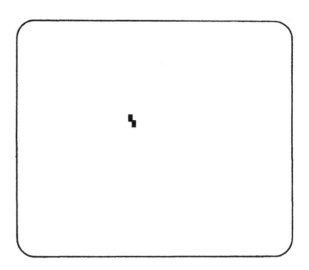

Figure 4.6 Tracking symbol to indicate a
pen presence, shown on DVST screen. (Courtesy DLR Associates)

Figure 4.7 Pen is depressed on tablet surface and point appears as part of sketch.
(Courtesy DLR Associates)

on the keyboard of the DVST shown in Figure
4.3. The ready light on the control box
should light-up.

2. Bring the pen over the lower left
corner of the tablet surface and press the
pen once. The data and pen lights should
blink, and the ready light should go out.

3. Without lifting the pen, move it

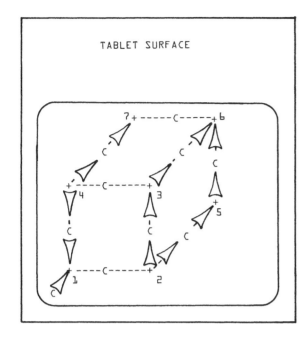
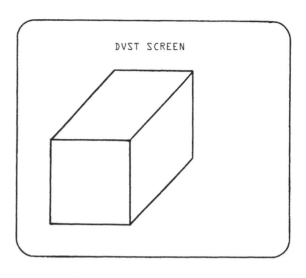

Figure 4.8 Single point/pen method of sketching an object and saving the data. (Courtesy DLR Associates and Tektronix Corp.)

to the right proceeding from point 1 to 2 as shown in Figure 4.8. Press ESC,!,1 again and the 'ready' light should come on.

4. Press the pen at the + mark shown for point 2. The 'data' and 'pen' lights should blink once, and the 'ready' light should go out. On the DVST screen, a line should be displayed as shown to the right in Figure 4.8. This technique is called the rubber band, because the line appears to snap into place.

5. Without lifting the pen continue the process through points 3,4 and returning to 1. (See Figure 4.8)

6. Lift the pen off the tablet surface, press ESC,!,1 and place the pen over point 2, now move to point 5 without lifting the pen. Keep the pen down and proceed to point 7 through point 6, be sure to stop and 'snap the rubber band' between the points.

7. Lift the pen again and move to 3, stop at 6, repeat the process by moving to 4 and stopping at 7. Check your sketch at the right of Figure 4.8. Does it look like the 3-D box shown on the DVST screen?

Single point/on-the-fly method

1. Press the ESC,!,5 keys together as shown in Figure 4.8. As long as the pen is held depressed the C function (ESC,!,5) will not be repeated. In other words, the pen was depressed and then released in Figure 4.8 for each single point. If the pen was removed from the tablet presence a space is produced, otherwise a rubberband line is produced. The on-the-fly method is more like freehand sketching because you do not have to break away from the surface of the tablet.

2. Repeat the steps in the last example, but keep the pen depressed. The C function is now ESC,!,5 so that fewer commands are sent.

3. Build the figure as shown in Figure 4.8, add to this figure if you wish by adding your initials on the front face.

Multiple point/ pen method

1. Press the ESC, !, 3 keys together and follow the steps shown in Figure 4.9.

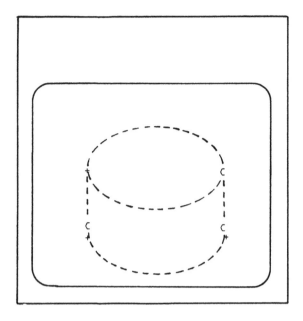

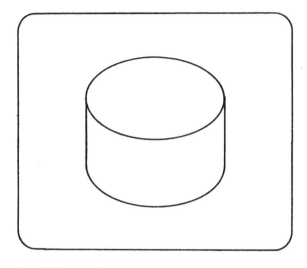

C = ESC, !, 3
+ = pen depressed
----- = pen path during depression

Figure 4.9 Multiple point/pen method of sketching an object and saving the data. (Courtesy DLR Associates and Tektronix Corp.)

2. Repeat all of the steps in the other sketching modes, but this time do not send a C between pen moves. Follow the procedure shown in Figure 4.9. Notice that the examples are getting more like natural drawing and sketching motions.

Multiple point/on-the-fly method

1. Press the ESC,!, 7 keys together and follow the instructions shown in Figure 4.9. In this example, the pen may be left down throughout the sketching action. The C function is given only once. The DVST output is completely natural as shown in Figure 4.10. In Figure 4.9, the curved lines were really .01 inches apart and made up of tiny straight lines on a grid pattern which formed a closed sided polygon. In Figure 4.10 the lines are completely freeform.

4.4 TYPES OF AUTOMATED SKETCHES

Almost any type of sketch can be completed on a graphics tablet. The examples given thus far have been pictorial, but an

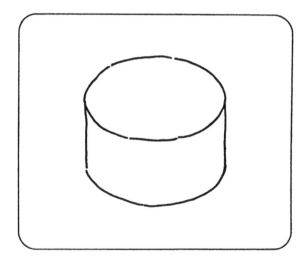

Figure 4.10 Multiple point/on-the-fly output display. (Courtesy DLR Associates)

orthographic sketch is often necessary in communciation between the engineer and person responsible for the finished drawing. Figure 4.11 is a typical design sketch. It contains multiple point/on-the-fly data and automated lettering from Chapter 3.

The characteristics of a design sketch

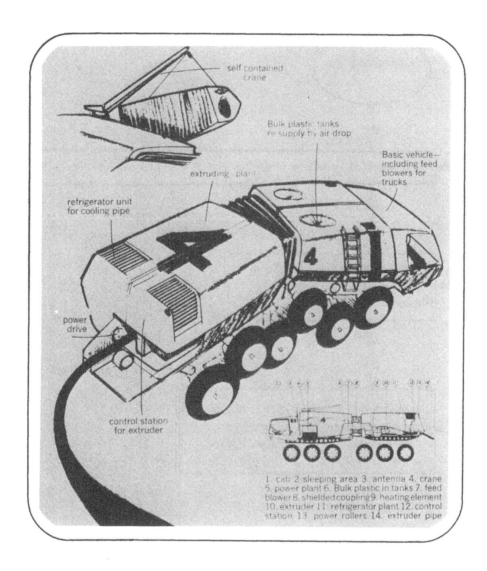

Figure 4.11 Automatic plastic pipe installer, design for desert irrigation vehicle. Design by Donald Deskey Associates. (Courtesy Bruning)

depend upon which stage of the design process is communicated. The design process consists of several stages that mark the development of an idea into a product. A listing might be:

1. Demand which redults in a market product sketch. See Figure 4.12.

2. Concept sketch. See Figure 4.11.

3. Engineering analysis sketches. See Figure 4.13.

4. Engineering description sketches. See Figure 4.14.

5. Production sketches. See Figure 4.15.

6. Sketches for publication. See Figure 4.16 for marketing, sales and distribution sketches.

7. Modification sketches are used at all stages of design.

This modification of the design process best illustrates the areas where sketching plays an important role in communicating design information. Each stage will now be examined to determine the correct use of a graphics tablet - freehand sketching system. The four tablet techniques outlined in section 4.3 are ideally suited to each of the design stages.

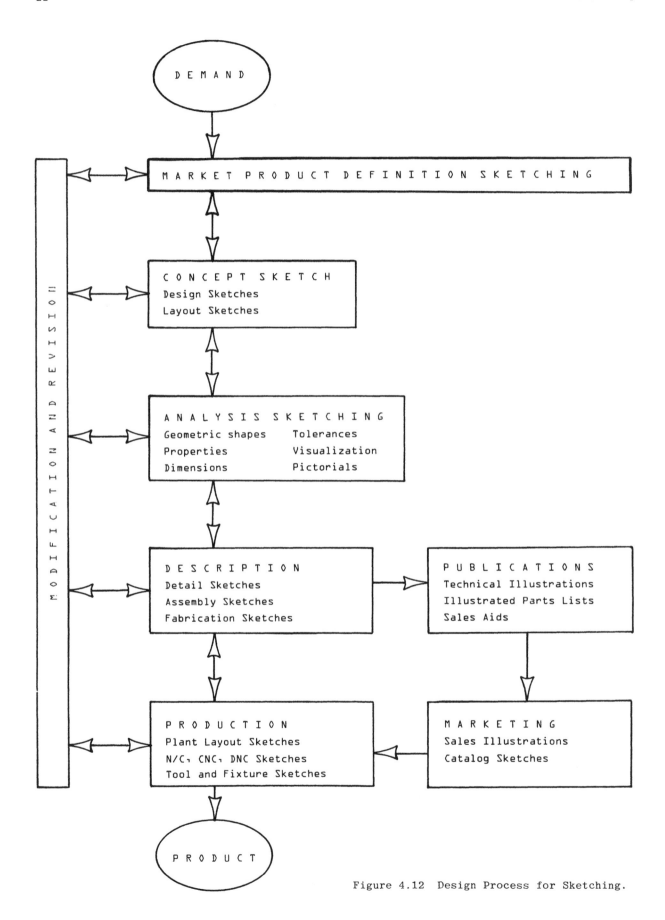

Figure 4.12 Design Process for Sketching.

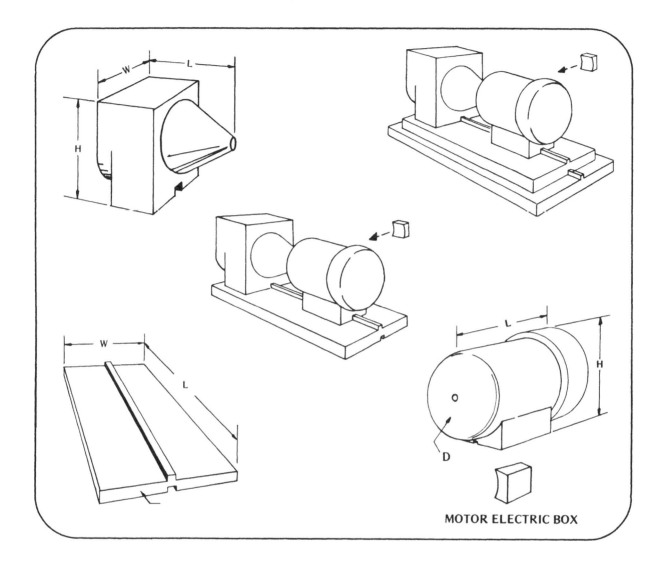

Figure 4.13 Engineering analysis sketch for cooling unit onboard desert irrigation vehicle.

4.5 THE CONCEPT SKETCH

At this stage of the sketching process, the engineer begins to collect first impressions of the solution to a design problem. Most often this solution is based on past experiences of the engineer in working with similar problems. Therefore the solution, in the form of a concept sketch, can be produced by a modification of past sketching elements already stored in computer memory (mass storage from Figure 4.4). This situation of creating a 'new' concept by modification is clearly diagrammed in Figure 4.12. The prior sketches are retrieved on the DVST shown in Figure 4.3, and the new features are added by the use of the data tablet. A general concept sketch can be constructed in this manner and returned to mass storage under a 'new' label (storage location). This is the most desireable procedure because it will allow all the successive sketches to be constructed by draftspersons specially trained in the use of the data tablet. Each person assigned the task of design process sketching can recall the concept from mass storage and begin their design based upon the overall concept. The work-hours required to create a concept in this manner is not much different from those

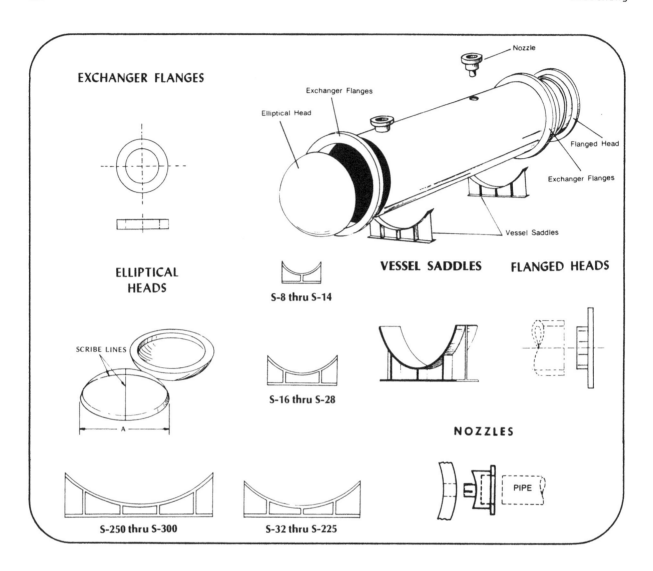

Figure 4.14   Engineering description of storage vessel for desert irrigation vehicle.

for manual methods. The advantages of the data tablet and mass storage consist of greater flexiblity through computer applications, ease of manipulation of parts, and assembly sketching. If the design requires many standard components or redundant geometries such as mirrored outlines or repeated patterns, the total design sketching can be greatly reduced by the techniques presented in this chapter.

4.6   ENGINEERING ANALYSIS SKETCH

After the creation of the concept sketch by the engineer, the draftsperson will experiment with the best method of developing the description of each element within the concept. Once the description is known the engineer will begin the analysis of each element. One of these analyses is the geometric verification of how the design will fit together. This involves the determination of allowance and interference; on a DVST and data tablet system, this can be greatly improved through the use of visual inspection. The inspection of the design parameters can be accomplished quickly, indicating what dimensions or distances are critical. This concept is much faster than manual methods, for here the computer may

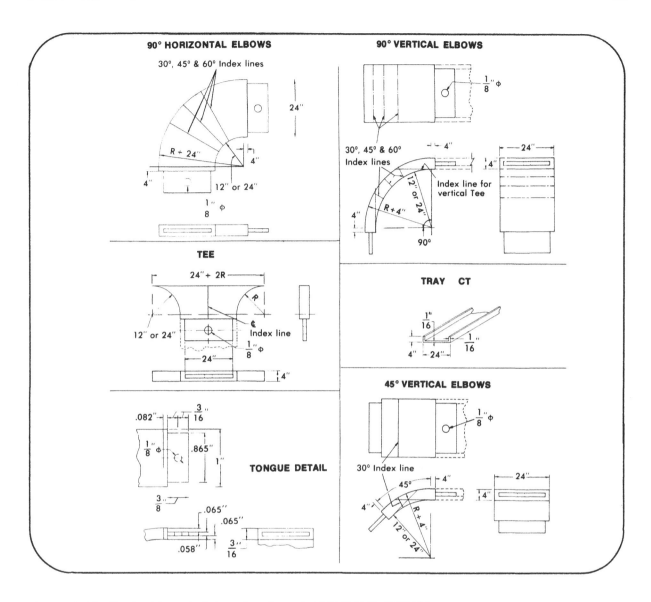

Figure 4.15   Production sketches for desert irrigation vehicle.

use programs to assist the designer in the analysis.

4.7   ENGINEERING DESCRIPTION

The sketches for this stage of the design are completed by draftspersons at the DVST and data tablet. They consist mainly of assembly, shown in upper right hand screen of Figure 4.14; and the necessary fabrication techniques to complete the assembly process. The ability of the draftsperson to create axonometric, perspective, or orthographic sketches greatly enhances the ability to visualize the object before produc-

tion sketches are made. All questions regarding assembly must be answered here.

4.8   PRODUCTION SKETCHES

The production stage of the design process is devoted to the creation of working drawings for manufacture. This work is entirely done by a drafter since the design information is communicated to manufacturing through the use of detailed drawings. The information for a detailed drawing is contained in the production sketches like those shown in Figure 4.15. Each detail is stored under its own label such as TEE.

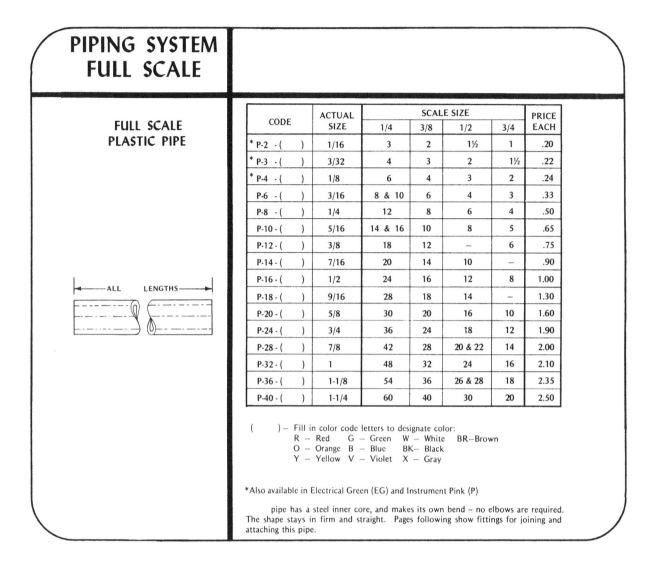

Figure 4.16   Typical sketch used on marketing catalog layout.

4.9   SKETCHES FOR PUBLICATION

Sketches from each of the other stages are used as needed to produce catalog pages as shown in Figure 4.16. Here a simple and easy to understand sketch is placed below the full scale plastic pipe header to represent all lengths of pipe. You will notice that pages following this catalog page will show fittings for joining and attaching this pipe. Further practice in producing sketches can now be introduced. Suppose that an engineer wanted you to produce the sketches necessary to create the pages following Figure 4.16?

4.10   REVIEW PROBLEMS

Work each of the following problems so that it can be stored as a sketch to be used in support of the concept sketch shown in Figure 4.11.

1.   **RECTANGLES**

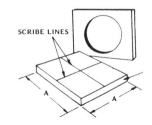

2.   **RECTANGLES**

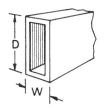

3.   **SQUARES**

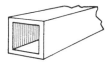

4.   **DISCS**

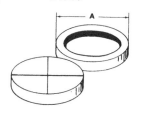

5.   **LARGE HEMISPHERES**

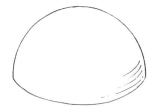

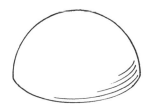

6.   **ELLIPTICAL HEADS**

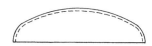

7.   **THREE DIAMETER BENDS**

RADIUS = 3 x DIA.

8.   **CONCENTRIC REDUCERS**

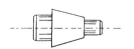

9.   **ECCENTRIC REDUCERS**

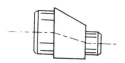

10.   **WELD CAPS**

11.   **CONTROL VALVES**

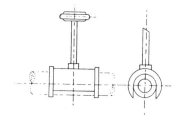

12.   **150 LB. GATE VALVES**

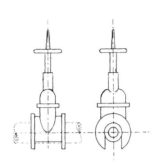

13.   **PLUG VALVES**

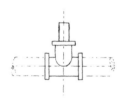

14.   **TEES**

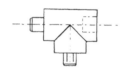

15.   **STUB-IN TEES**

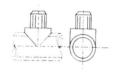

16.   **SNAP-ON TEES**

PIPE SIZES 1/16", 3/32" & 1/8"

17.   **INSULATION SLEEVES**

18.   **VALVE STEM COMPONENTS**

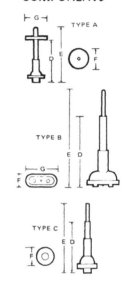

19.   **PLUG-CHECK VALVE COMPONENTS**

20.   **CONTROL VALVE DIAPHRAMS**

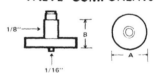

21.   **COUPLING SLEEVES**

22.   **L. R. 90° ELBOWS**

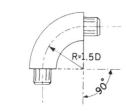

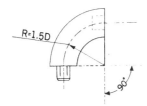

23.   **URETHANE**
       FOAM
       VESSELS

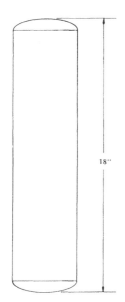

24.   **CHANNELS**

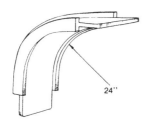

25.   **CABLE   TRAY**

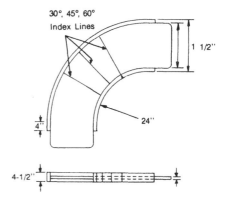

26.   **90° VERTICAL BEND**

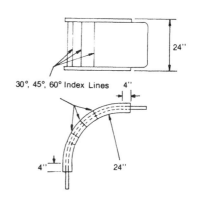

27.   **TEE**

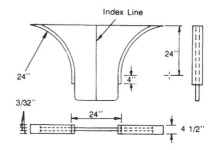

28.   **45° VERTICAL ELBOWS**

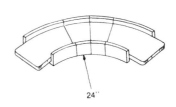

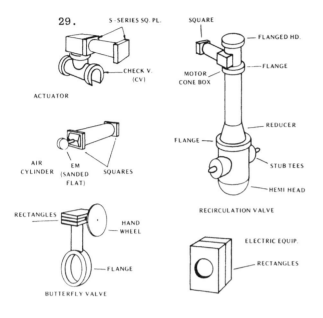

29.

ACTUATOR

S-SERIES SQ. PL.

CHECK V.
(CV)

AIR
CYLINDER

EM
(SANDED
FLAT)

SQUARES

RECTANGLES

HAND
WHEEL

FLANGE

BUTTERFLY VALVE

SQUARE

FLANGED HD.

MOTOR
CONE BOX

FLANGE

REDUCER

FLANGE

STUB TEES

HEMI HEAD

RECIRCULATION VALVE

ELECTRIC EQUIP.

RECTANGLES

tice you complete at the graphics tablet. Like lettering, freehand sketching requires the use of a keyboard, stylus and tablet surface. Once you have logged the necessary hours practicing the useful sketches that are required in the various stages of design, you will master the ability to produce freehand sketches. Skill building of any kind requires time to develop your technique and knowledge of the computer operation.

4.11 CHAPTER SUMMARY

After working the review problems you understand how much of the design information is passed from engineer to drafter. It is stored in a computer file as data so it may be modified, scaled, rotated, or saved for use later. Computer-aided sketching is a modern tool of todays designers and includes equipment, materials and knowledge. One is of little use without the other two. You should be familiar with the equipment shown in Figures 4.1 through 4.5. You should be familiar with the materials necessary for freehand sketching from chapter 2 and 3, and you have a great deal of knowledge about how freehand sketches are used from this chapter.

The ability to produce excellent freehand sketches comes from the amount of prac-

30.

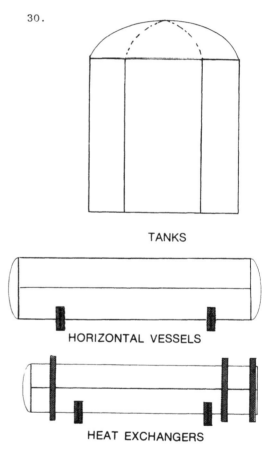

TANKS

HORIZONTAL VESSELS

HEAT EXCHANGERS

5

IMAGE PROCESSING

Geometric and template image processing for automated drafting applications is generally thought of as being either passive or active in nature. When a draftsperson selects a computer-assisted mode rather than manual documentation, a decision is made if the images can be produced with computer memory modules. The use of these memory modules is called computer-assisted rather than computer-aided. The two terms aided and assisted do not carry the same meaning. Computer-aided is an active process, while computer-assisted is considered passive.

An example of this might be the use of a Tektronix model 4051 with memory module plugged into the back as shown in Figure 5.1 (passive - computer-assisted); or contrast to that, the use of a model 4012 on-line to a large digital computer. The use of a 4012 with large computer is considered active - computer-aided. Computer-aided is considered an active process because it appears that the graphics terminal has the ability to display graphic images without prior input by the draftsperson.

Both types of terminals are used to display graphic images. Passive terminals such as the 4051 have a microprocessor built into them for quick display from a memory module. And they may be operated off-line from a central computer, this makes these types of devices useful in small drafting rooms. There is a limit to the amount of images that can be processed, however. And for this reason, large industrial drafting rooms have adopted the on-line terminals to a large central computer. For the purposes of this chapter either method may be used.

5.1 IMAGE CONSTRUCTION

Images are constructed directly on the screen of the DVST as shown in Figure 5.2. The drafter follows a set procedure whereby the image is presented on the DVST screen so the drafter may approve, or modifiy before sending it to a hard copy unit, or pen plotter as shown in Figure 5.3. Modifications are done with the joy stick shown in Figure 5.4. To begin the image construction, the drafter places the memory module in place as shown in Figure 5.1 or types a

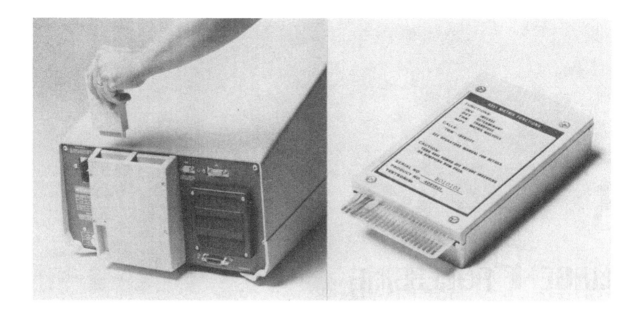

Figure 5.1 Memory module used with Tektronix 4051 terminal. (Courtesy Tektronix Inc.)

series of commands from the keyboard if an
on-line terminal is used. These are:

    LOGON
    USERID/PASSWD
    CALL 'ACS.USERID.DEMO(PROG)'

These commands locate the memory modules
that are stored in the large digital computer and place them on the screen for you
to review. Figure 5.5 is the unit for the
DVST screen contents. A hard copy can be
made from the terminal keyboard by pressing
the 'make copy' key.

Once the listing is complete you know
which graphic images are available for your
use. These images may be used to create
other shapes as desired.

Figure 5.2 Model 4051 DVST (Courtesy Tektronix Inc.)

5.2 LINE IMAGES

The first image construction element
listed on the screen was the MOVE function.
A move is made possible by the display of
crosshairs on the DVST. The crosshairs are
moved by the proper use of a joystick as
shown in Figure 5.6. The draftsperson will
operate the joytick to position the cross-

hairs on the screen. The crosshairs appear
when a command, CALL SCURSER(IX,IY) is typed
from the keyboard. Any position between 0
and 1024 may be entered for IX and any position between 0 and 761 may be entered for
IY. If IX=0 and IY=0 then the crosshairs
will appear in the lower left hand corner

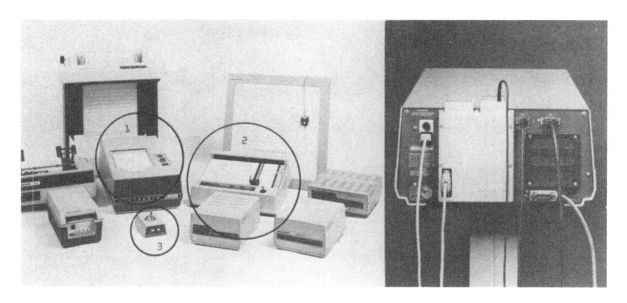

Figure 5.3 Continuation of equipment from Chapter 4 with special emphasis on the hard copy
unit (1), pen plotter (2), and joy stick (3). (Courtesy of DLR Associates and Tektronix Inc.)

of the DVST screen. To move the crosshairs
connect the joy stick shown in Figure 5.6
and move the stick. The crosshairs follow
the motion of the stick. In this manner a
screen position may be sent by pressing a
key from the keyboard.

For example, suppose we wanted to indicate that a move to the upper part of the
screen is desired? Simply adjust the joy
stick and watch the placement of the crosshairs on the screen, when they are placed
where you want them, press the M key from
the keyboard of the terminal. The screen
will flash, the crosshairs will reappear,
ready for the next image to be processed.
Position the crosshairs at a new location
by adjustment of the joy stick, press the
D key from the keyboard and the crosshairs
blink. A bright line appears from the last
position of the crosshairs to the present
position of the crosshairs as they reappear
on the screen. Continuous line images may
be constructed in this manner (closed polygon) or intermixed with spaces.

Three keys are used for line construction, the M and D just explained and the E
key which is used to erase lines from the
storage screen. A flash indicates an erase

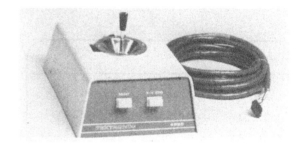

Figure 5.4 Joy stick.
(Courtesy Tektronix Inc.)

Figure 5.5 Hard copy unit. (Tektronix Inc.)

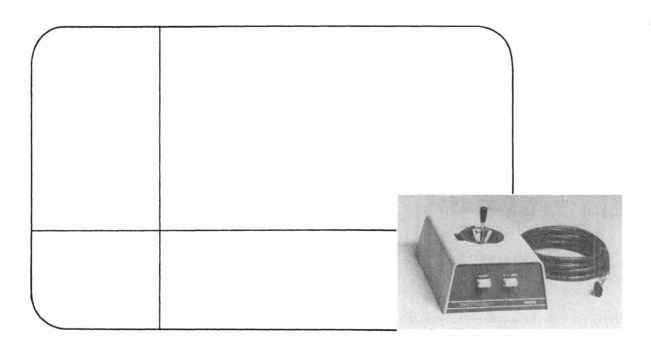

Figure 5.6 Joy stick controls position of crosshairs on DVST screen. (Courtesy DLR Associates and Tektronix Inc.)

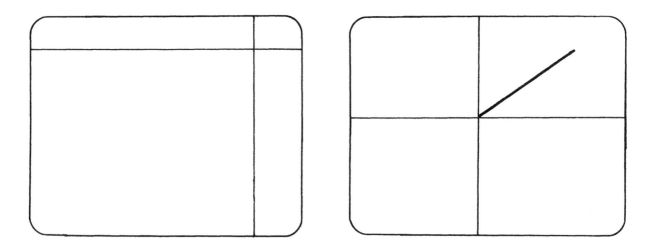

Figure 5.7 Crosshairs are moved by M key. (Courtesy DLR Associates)

Figure 5.8 Line is displayed by D key and movement of crosshairs. (Courtesy DLR Assoc)

has taken place. If the drafter presses a E key than the entire screen contents are removed, not just the last line segment. A M key will position the crosshairs anywhere on the DVST screen as shown in Figure 5.7. The joy stick must be used to position the placement of the crosshairs before the M key is depressed. Any line segment may be

placed by positioning the crosshairs as in Figure 5.8. You will notice that the joy stick has been used to reposition the crosshairs from Figure 5.7. A lower and left position is chosen, the key D is depressed and a line segment is displayed from computer memory. A predefined program to display the line is used in this case.

5.3 ARCS AND CIRCLES

Circles and segments of circles are a C key function. Figures 5.9 through 5.11 demonstrate how this image is processed. In Figure 5.9 the center of the circle is located by the joy stick and the C key is depressed. The crosshairs blink and a small dot appears at this point. In Figure 5.10 this dot is clearly visible. The dot is not part of the display image. It is there so that the drafter may use it to locate centerlines or other objects. If a hard copy or plotter drawing is made of the display image the dots will not appear.

A second dot is also placed at the end of the radius of the arc, as shown in Figure 5.10. The drafter moves the crosshairs in any direction from the center to establish the size of the radius. Once this is done another C function (depress C key) is sent to the computer from the terminal keyboard. The crosshairs blinks again and a second dot appears at the radius location.

A third piece of information is needed for the circle generator before the arc or circle is displayed. The arc's ending position is required, so the drafter moves the joy stick again and depressed the C key. A circle or arc now appears on the screen.

Figure 5.11 represents a full 360° arc or circle because the joy stick was not moved from Figure 5.10's placement. Figure 5.12 represents several arcs that are not full circles, and in each case the joy stick was moved to the ending point for the arcs.

5.4 RECTANGLES

In a similar manner, rectangles are images that are displayed in three key depressions. The K key has been assigned the rectangle function. Figure 5.13 represents the three key function necessary to display a rectangle on the DVST screen. First the lower left-hand corner of the image is located with the joy stick. The K key is depressed and a dot is placed on the screen. This dot represents the starting position of the rectangle. Like the center of a circle,

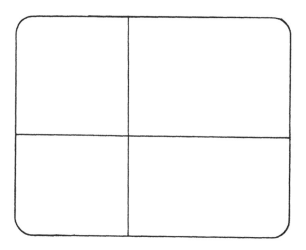

Figure 5.9   Center location for a circle or arc display image.   (Courtesy DLR Associates)

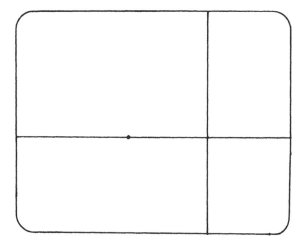

Figure 5.10   Radius location for circle.

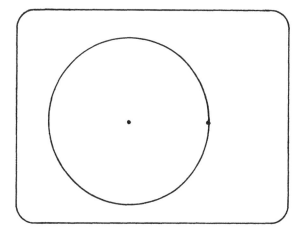

Figure 5.11   Circle image display.

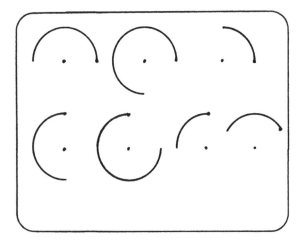

Figure 5.12 Arc image displays.

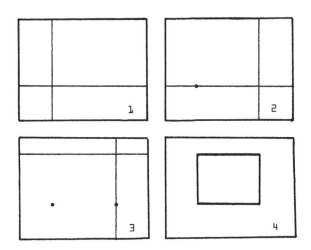

Figure 5.13 Rectangle image display.

the dots are not reproduced when saved in
computer memory or when plotted. Figure
5.14 represents how the set-up points on
the rectangle can be used to rotate the
display image. Notice that the second dot
is not in alinement with the starting position, this causes the rectangle program
to display the image in a rotated position.

5.5 THE SAVE FUNCTION

With just these few display images, a
certain amount of drafting can be constructed by the draftsperson. Figure 5.15 is an
example of MOVE, DRAW, CIRCLE, and RECT
memory modules used to create an automated
drawing. The process is time consuming if
the ERASE (E key) and SAVE (S key) are not
used. You will remember that the E key removes everything from the screen. What if
you are near the end of an entire drawing
and a mistake is made? Do you remove all
the screen contents by pressing the E key?
If you do, all is lost, nothing is saved.
To avoid this, press the S key whenever an
image is correct. This places the image
in a saved position so that it may be recalled whenever you need it. More than
one image may be placed in the SAVE file,
so at the end of a completed drawing - all
images are in the SAVE location in computer
memory. By pressing the S key, an entire

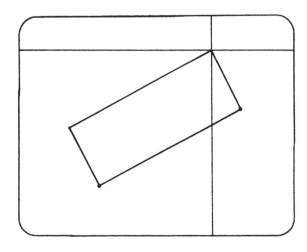

Figure 5.14 Rectangle rotated.

drawing can be saved. The drawing is represented by a list of data that appears as:

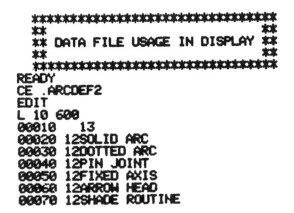

```
****************************************
**                                    **
** DATA FILE USAGE IN DISPLAY **
**                                    **
****************************************
READY
CE .ARCDEF2
EDIT
L 10 600
00010    13
00020 12SOLID ARC
00030 12DOTTED ARC
00040 12PIN JOINT
00050 12FIXED AXIS
00060 12ARROW HEAD
00070 12SHADE ROUTINE
```

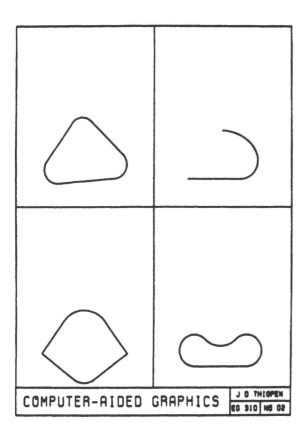

00370	86	0.20	2.59
00380	86	1.33	1.78
00390	86	2.09	2.17
00400	69	3.38	2.68
00410	83	0.0	0.0
00420	71	4.46	0.24
00430	7	0.86	3.63
00440	10	0.86	3.63
00450	71	0.35	0.23
00460	71	4.82	0.24
00470	83	0.0	0.0
00480	79	6.55	3.38
00490	2	0.86	3.63
00500	-5	0.86	3.63
00510	79	6.85	3.35
00520	83	0.0	0.0
00530	90	4.60	3.63
00540	90	5.36	3.37
00550	90	4.64	2.89
00560	83	0.0	0.0
00570	83	0.0	0.0
00580	83	0.0	0.0
00590	77	4.56	4.78
00600	68	4.56	4.31

Figure 5.15 Automated drawing by image
construction. (Courtesy DLR Associates)

00080	12DASHED LINE		
00090	12FIT 3 POINTS		
00100	12SMOOTH SEVERAL POINTS		
00110	12SHADE ANY AREA		
00120	12CENTERLINE		
00130	12LINEAR GRID		
00140	9DRAW A STAR THRU "O"		
00150	135		
00160	67	1.65	4.18
00170	67	0.42	5.43
00180	67	3.02	5.43
00190	67	1.65	4.18
00200	67	0.38	4.75
00210	68	2.95	4.75
00220	70	1.78	4.17
00230	83	0.0	0.0
00240	77	1.52	4.17
00250	68	0.86	3.63
00260	80	0.86	3.63
00270	86	0.34	0.73
00280	8	0.38	4.75
00290	75	2.95	4.75
00300	86	0.66	1.51
00310	86	1.19	0.18
00320	86	1.32	0.93
00330	86	1.94	0.93
00340	86	2.22	0.37
00350	86	3.11	2.08
00360	69	3.42	0.28

5.6 THE PLOT FUNCTION

Once the data is saved it may be sent
to a pen plotter by depressing the $ key.
The drawing shown in Figure 5.15 may be a
plotter drawing, because the $ function is
a set of commands for the Tektronix 4662.
This is shown in Figure 5.3, item 2. It is
not always necessary to route image constructions to pen plotters, in some cases
a screen copy like that shown in Figure 5.
15 is all that is required. In these cases
the H key can be depressed and a hard copy
is made from the unit shown in Figure 5.3,
item 1.

5.7 IMAGE MODIFICATIONS

While a great deal of drawing information can be constructed from just the items
shown (MOVE, DRAW, CIRCLE, RECT, SAVE, PLOT,
ERASE, and COPY); there is more to image
process than simple construction. For example DRAW produces a solid line. It would
be nice to have a dashed line between points
in some cases, therefore, a modification to
DRAW is the 1 key. If the M key is used to
position the crosshairs at one end of a line
then the 1 key will produce a dashed line to
the next point. The 1 modification works

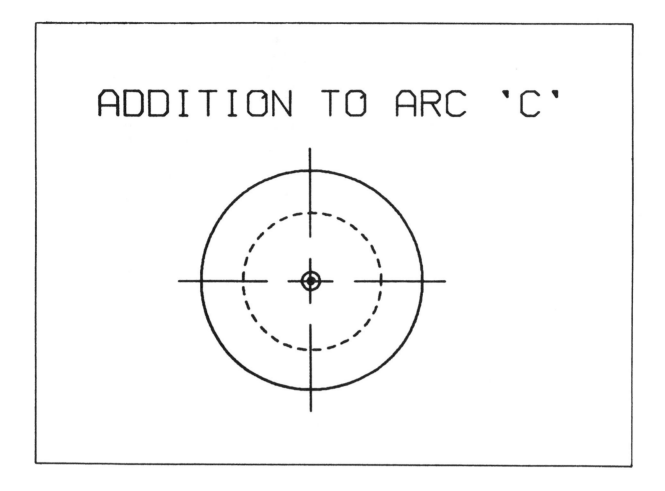

Figure 5.16 L type modifications to CIRCLE. (Courtesy DLR Associates)

for all graphic images. Figure 5.16 shows these types of modifications for CIRCLE. A C key is used to locate the center of the first outside circle shown in solid line. The procedure you will remember was:

    1. Center crosshairs at center of arc and enter C from the keyboard,

    2. Position at the starting point of the arc and enter C again.

    3. Locate the ending point of the arc and enter C again.

    Now let's construct the inside (dash) circle. Repeat the first two steps above and enter a D for the third step. Let's now add the centerlines. Repeat the first step above and enter D for both step 2 and 3. The last step will be to indicate that this is the end of a shaft that is to be used as an axis. Locate the center of the

two circles by using the joy stick to position the crosshairs over the center dot. Now press the F key and the center symbol is displayed for you.

    The only items shown in Figure 5.16 not produced from C, its modifications and F, are the characters displayed across the top of the screen. You will remember from Chapter 3 that character generators produce most types of lettering. The image display used for this chapter has a character generator also. It is the L key. The joy stick may be used to position the crosshairs anywhere on the screen and the L key is depressed to produce the lettering prestored in the memory module. Therefore, a drafter must load all the lettering to be used in a drawing prior to using the L function. If you forget to load the module, nothing is

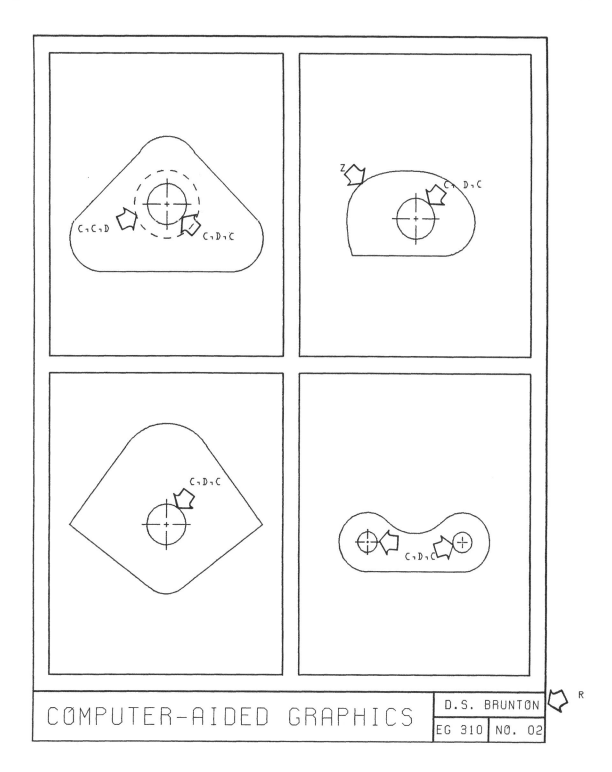

Figure 5.17   Modifications added to automated drawing displayed in Figure 5.15.

displayed from the L function.   To avoid
this problem, most image processes programs
ask the operator to type all lines of lettering to be used on the drawing before an-

other function can be used.   Figure 5.17 is
an example of modifications to CIRCLE and
lettering from the R function.   You will notice that even the K function can be modi-

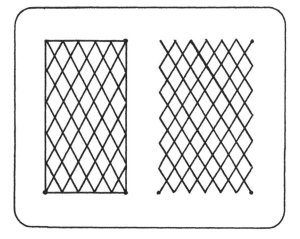

Figure 5.18 Modification of RECT.

Figure 5.19 Shading from the V key.

fied by use of the B key. This produces
a crosshatching for any rectangle. It may
be used separately or inside a K function
as shown in Figure 5.18. Often a rectangle
needs to be crosshatched to show sectional
relationships. If cross hatching is not
suitable then a more general section shade
can be used as shown in Figure 5.19. Not
only rectangles, but any shaped area can be
shaded with the V function.

5.8 IMAGE ENHANCEMENTS

Both key functions B (shown in Figure
5.18) and V (shown in Figure 5.19) are considered enhancements. Several enhancements
exist for lines. They are:

1. Arrowheads (Figure 5.20)
2. Fits (Figure 5.21)
3. Flairs (Figure 5.22)
4. Centerlines (Figure 5.23)
5. Points on lines (Figure 5.24)
6. Polygons (Figure 5.25) and
7. Linear Grids (Figure 5.26).

5.9 ARROWHEADS

An arrowhead may be placed at the end
of a line image by use of the A key on the
terminal keyboard. The drafter moves the
joy stick to the base of the arrowhead, end

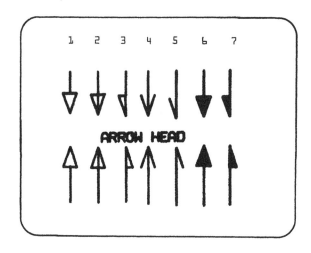

Figure 5.20 A function key for arrowheads.

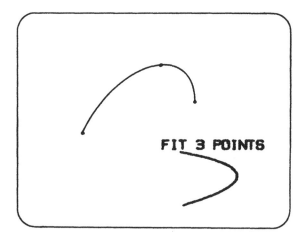

Figure 5.21 Z function key for fits.

of the line, depresses the A key and then
moves to the tip of the arrowhead. Next a
type of arrowhead is selected from those
listed in Figure 5.20, 1 through 7. This
number is then depressed and the selected
type is displayed.

5.10 FITS THROUGH THREE POINTS

Many times a flexible curve through 3
points will enhance the line quality of a
displayed object. Figure 5.17 has a Z key
function as a modification, and is a good
example of a fit through three points. Another is shown in Figure 5.21. You will
notice that the crosshairs are positioned
by the joy stick over the first point and
the Z key is depressed. The crosshairs
blink and a dot appears at this first location. The second point is input by moving
the crosshairs again over the second point
and depressing the Z key again. The second
dot now appears and you may enter the last
point as shown in Figure 5.21. After the
last point has been entered, a fit is displayed through the three points. If the
fit is saved by pressing the S key, the
screen flashes and the line is replotted
without the location dots, as shown below
'FIT 3 POINTS' in Figure 5.21.

5.11 SMOOTHED LINES THROUGH MANY POINTS

Topo lines for maps and many other
types of lines run through several points.
To avoid the straight line effect, these
points may be smoothed with the X key function shown in Figure 5.22. This process
is like the Z key function, except more
than 3 points are input by moving the crosshairs and pressing the X key. At the last
point to be smoothed, the drafter presses
the E key to end the sequence. The smooth
line may be stored, dropping the location
dots which appear in Figure 5.22.

5.12 CENTERLINES

Centerlines have been shown as a modification in Figure 5.17, but they may also

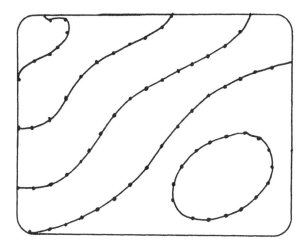

Figure 5.22   Smoothed lines through X points.

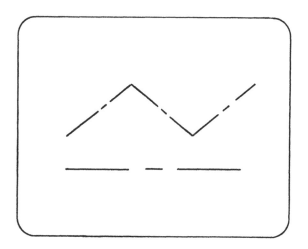

Figure 5.23   Centerlines from I key.

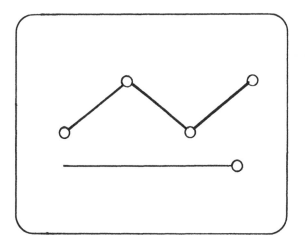

Figure 5.24   Points on a line from P key.

be displayed by themselves. Move to the
end of the centerline with the M function,
move to the opposite end of the centerline
and depress the I key. A centerline proportional to the distance is displayed as
shown in Figure 5.23. Centerlines may be
displayed end to end by positioning the
joy stick at each point and entering the I
key function.

5.13 POINTS ON A LINE

Small, open dots many be displayed at
any location on the DVST screen by placement of the crosshairs and depressing the
P key. To duplicate Figure 5.24, we would
move to the lower line segment with M and
then display the line with D and then move
slightly to the right of the end point. By
pressing P, a small pin joint is displayed.
To complete the figure, we move up to the
left hand portion of the screen, press the
P key, move slightly to the right, press M,
move to the end of the first line, press D,
and continue the same process again until
the figure is completed.

5.14 POLYGON CONSTRUCTION

Any equal sided polygon may be constructed by the use of the O key. As shown
in Figure 5.25, the drafter moves to the
starting position with the M key, presses
the O key and enters the number of sides
for the polygon to be constructed. A dot
is placed at the starting point and will be
erased if the object is saved. Any number
of sides may be selected, if a minus number
is used, then the memory module is directed
to construct the object inside-out. In an
example, +5 with display a pentagon, while
a -5 will display a star, as shown in Figure 5.25.

5.15 GRID CONSTRUCTION

Grid backgrounds may be constructed,
sized, and displayed from the G key. The
drafter moves to the lower left hand corner
of the grid and depresses the G key. The

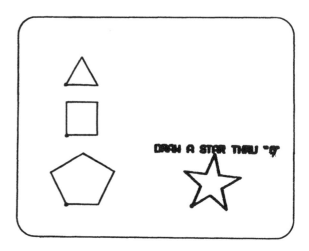

Figure 5.25 Polygon construction.

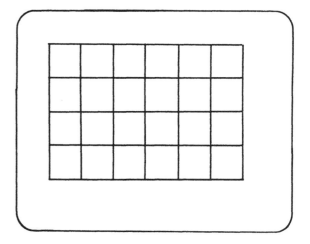

Figure 5.26 Grid construction.

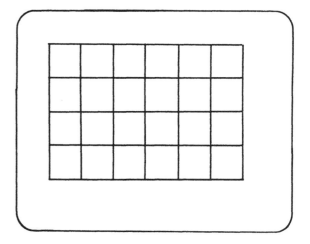

Figure 5.27 ! command - HELP.

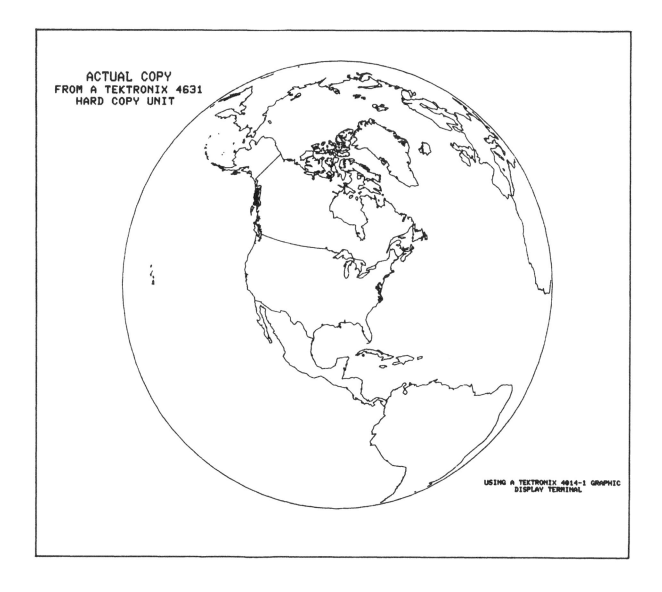

ACTUAL COPY
FROM A TEKTRONIX 4631
HARD COPY UNIT

USING A TEKTRONIX 4014-1 GRAPHIC
DISPLAY TERMINAL

Figure 5.28   Typical image processing using C and X functions.   (Courtesy DLR Associates and Tektronix Inc.)

crosshairs blink and place a small dot at this location. A prompt now appears in the upper left hand corner of the screen and asks for the number of grid elements in the X direction (5 in Figure 5.28) and the number of grid elements in the Y direction (4 in this case). Next the size (X and Y) of a single grid box is typed. Once this information has been entered, the grid pattern appears as shown in Figure 5.26.

The grid (G) function, along with the shade (V), label (L), relabel (R), polygon (O), and W (write to an external file), all require that information be given through screen prompts. For this reason, special help procedures are provided in the memory modules. These are available through the ! key and are shown in Figure 5.27. In fact Figure 5.27 appears on the DVST if the ! key is entered during the image processing program execution. If the ! and E key are both sent, then a help message for the ERASE function is returned to the screen. This is very helpful to drafters who are just begining, and those who want to produce fairly detailed images as shown in Figure 5.28.

```
E KEY = ERASE DISPLAY
   PROCEDURE—
      IF AN UNDESIRED DISPLAY IMAGE
   IS KEYED IN, THIS COMMAND WILL
   ERASE ONLY THAT DISPLAY IMAGE
   WITHOUT ALTERING ANY OF YOUR
   SAVED DATA.


   HIT RETURN
IH0001A PAUSE
```

Figure 5.29 E help message.

```
C KEY = ARC AND CIRCLE DISPLAY
   PROCEDURE—
      1. CENTER CURSOR, ENTER C
      2. POSITION END OF ARC, ENTER C
      3. LOCATE END OF ARC, ENTER C
DASHED ARC OR CIRCLE: C,C,D
SOLID LINE ARC OR CIRCLE: C,C,C
SOLID ARC OR CIRCLE WITH C_L: C,D,C
DASHED ARC, CIRCLE WITH C_L: C,D,D

   HIT RETURN
IH0001A PAUSE
```

Figure 5.30 C help message.

```
A KEY = ARROW HEAD DISPLAY
   PROCEDURE—
      1. PLACE CURSOR AT BASE, SEND A
      2. PLACE CURSOR AT TIP, SEND 1
         TO 7 FOR HEAD TYPE — SEE
         FIGURE 5.20 IN YOUR TEXT.



   HIT RETURN
IH0001A PAUSE
```

Figure 5.31 A help message.

```
S KEY = SAVE DISPLAY
   PROCEDURE—
      THIS COMMAND PERMITS THE USER
   TO SAVE THE DISPLAY.  ONCE A DIS-
   PLAY IS SAVED NO OTHER COMMANDS
   WILL ALTER IT.  THE CURSOR WILL
   BE SET AT THE LAST POINT OF THE
   SAVED DATA.

   HIT RETURN
IH0001A PAUSE
```

Figure 5.32 S help message.

```
$ KEY = ROUTE TO PLOTTER
   PROCEDURE—
      THIS COMMAND PERMITS THE USER
   TO PLOT THE DVST DISPLAY.  ONCE
   A DISPLAY IS PLOTTED NO OTHER
   COMMANDS WILL ALTER IT.  THE
   CURSOR WILL BE SET AT THE LAST
   POINT OF PLOTTED DATA.


   HIT RETURN
IH0001A PAUSE
```

Figure 5.33 $ help message.

```
W KEY = WRITE X,Y POINTS TO A FILE
   PROCEDURE—
      FROM THE READY MODE, ENTER A
   DATA SET BY: 'CE VECTORS' AND
   INSERT SOME BLANK SPACES IN ONE
   LINE, THEN SAVE BY: 'S'.  NOW
   LEAVE THE DATA SET AND ALLOCATE
   DATA SET TO FILE 10 BY: 'ALLOC
   DA(USERID.VECTORS) FI(FT10F001)'
   HIT RETURN
IH0001A PAUSE
```

Figure 5.34 W help message.

```
B KEY = CROSSHATCHING FOR RECTANGLE
   PROCEDURE—
      THIS COMMAND PERMITS THE USER TO
SHADE A RECTANGLE IN 3 STEPS:
1. B IS ENTERED FOR LOWER/LEFT
2. L,R, OR B IS ENTERED FOR LOWER
RIGHT, WHERE:  L= LEFT TO RIGHT
               R= RIGHT TO LEFT
               B= BOTH WAYS
3. 1 THRU 9 IS ENTERED FOR UPPER
RIGHT, WHERE NUMBER INDICATES LINES
PER INCH.
```

Figure 5.35 B help message.

Figure 5.29 is the screen image provided from the !E keyboard input. Likewise all the other keyboard inputs are shown in:

Figure	Function
5.30	C circle
5.31	A arrowhead
5.32	S save
5.33	$ plotter
5.34	W external file
5.35	B crosshatching
5.36	Z 3 point fit
5.37	X smooth
5.38	V shading
5.39	I centerline
5.40	G grids
5.41	K rectangle
5.42	O polygon

5.16 CORE PROCEDURES

The CORE system for image display has three main parts. They are:
1. Output
    1.A temporary display files.
    1.B buffered-saved files with a visibility attribute as shown in this chapter or Chapter 4.
    1.C dynamic-full transformation attributes (not used in this text).
2. Input
    2.A no input at all (passive)
    2.B synchronous input

```
Z KEY = HYPERBOLIC FIT OF 3 POINTS
   PROCEDURE—
      THE USER MERELY INPUTS 3 POINTS
USING THE Z KEY FOR ALL 3.  A REF.
WILL BE DISPLAYED BUT NOT KEPT IN
S KEY FUNCTION.



   HIT RETURN
IH0001A PAUSE
```

Figure 5.36 Z help message.

```
X KEY = HYPERBOLIC FIT FOR MULTIPTS
   PROCEDURE—
      THE USER MAY INPUT AS MANY POINT
AS DESIRED, UP TO 99.  EACH POINT
IS CONNECTED UNTIL THE LAST INPUT
WHICH IS ANY KEY OTHER THAN X.



   HIT RETURN
IH0001A PAUSE
```

Figure 5.37 X help message.

```
V KEY = B FUNCTION FOR POLYGONS
   PROCEDURE—
      A SHADING FUNCTION FOR SHAPES
OTHER THAN RECTANGLES.  THE POLY-
GON MUST BE INPUT BEFORE CROSS-
HATCHING CAN OCCUR.  THIS IS DONE
WITH A SCREEN PROMPT _ _  _ _ _.
INSERT NUMBER OF LINES PER INCH
AND THE ANGLE DESIRED.

   HIT RETURN
IH0001A PAUSE
```

Figure 5.38 V help message.

```
I KEY = CENTERLINE
   PROCEDURE—
      POSITION THE CURSOR AT THE 1ST
   POINT AND ENTER I, MOVE TO THE
   NEXT POINT AND SO ON.  AT THE LAST
   POINT ENTER ANY OTHER KEY TO END
   THE CENTERLINE FUNCTION.


   HIT RETURN
IH0001A PAUSE
```

Figure 5.39   I help message.

```
G KEY = GRID PATTERN
   PROCEDURE—
      POSITION THE CURSOR AT THE LOWER
   LEFT CORNER OF THE PATTERN.  ENTER
   G, SCREEN PROMPT WILL APPEAR AT THE
   UPPER RIGHT. INPUT NX,NY AND SIZE
   OF EACH GRID WITH THE CURSOR AND
   G KEY.


   HIT RETURN
IH0001A PAUSE
```

Figure 5.40   G help message.

2.C asynchronous input for inter-
action between drafter and display processor.

   3. Dimension
      3.A 2-D graphics only as shown
throughout this chapter.
      3.B 3-D graphics (not part of
this chapter).

   There is, therefore, 18 different application levels for the full CORE system
of computer graphics. This chapter does
not employ all of these. It is only a 2-D
graphic image presentation. Refer to sections in Chapters 6, 8, 9, 10 and 11 for
3-D applications. Only level 1.B is shown
for output capability for this chapter, however 1.A and 1.C are illustrated in detail
in other chapters. The input level is high-
ly interactive as shown throughout this
chapter.

5.17 USING AN IMAGE PROCESSOR

   Draftspersons can do a great deal of
their work by understanding how an image
processor works. A draftsperson is not a
computer programmer and as such does not
write or develop programs - they just use
them. The first 16 sections of this chapter explained a rather simple image processor for the graphics equipment shown in
Figure 5.3. Before a drafter can use a
program it must be written and stored for

```
K KEY = RECTANGLE
   PROCEDURE—
      3 KEY INPUTS ARE REQUIRED.  POS-
   ITION THE CURSOR AT THE LOWER LEFT
   CORNER OF THE RECTANGLE AND ENTER
   K.  MOVE TO LOWER RIGHT, K.  MOVE
   TO UPPER RIGHT, K.  LOCATION DOTS
   WILL NOT BE SAVED.


   HIT RETURN
IH0001A PAUSE
```

Figure 5.41   K help message.

```
O KEY = MULTISIDED POLYGON
   PROCEDURE—
      2 KEY INPUTS ARE REQUIRED.  1ST
   INPUT SETS LOWER LEFT HAND CORNER
   AND SCREEN PROMPT FOR SN AND N AT
   SN=1 OR 2 AND N = NUMBER OF SIDES
   DESIRED.  MOVE CURSOR TO INDICATE
   LENGTH OF SIDE AND INPUT WITH O.


   HIT RETURN
IH0001A PAUSE
```

Figure 5.42   O help message.

```
**----------------------------------------------------------------**
**                    INTERACTIVE IMAGE PROCESSOR                  **
** ................................................................**
**      WRITTEN TO CORE SPECIFICATIONS        APRIL 1981           **
**      INPUTS FROM KEYBOARD:                                      **
**           CHARACTER            COMMAND                          **
**              M            MOVE — NO LINE DISPLAY                **
**              D            DRAW — FROM LAST POINT                **
**              E            ERASE SCREEN                          **
**              H            COPY DRAWING                          **
**              Q            QUIT PROGRAM                          **
**              C            DRAW ARC                              **
**              P            PIN JOINT                             **
**              F            FIXED AXIS                            **
**              L            LETTERING FOR DRAWINGS                **
**              A            ARROWHEADS                            **
**              S            SAVE INPUTS IN BUFFERED FILE          **
**              $            DIRECT TO 4662 PLOTTER                **
**              R            RELABEL LETTERING                     **
**              W            WRITE TO EXTERNAL FILE SPACE          **
**              1            DASHED LINE IMAGES                    **
**              Z            FITS 3 POINTS TO A CURVE              **
**              X            SMOOTHES UP TO 99 POINTS TO A CURVE   **
**              V            SECTION LINES                         **
HIT RETURN FOR NEXT PAGE OF COMMANDS -----------------------------**
IH0001A PAUSE     0
```

Figure 5.43 Listing of the image processor used in this chapter.   (Courtesy DLR Associates)

recall. In section 5.1, the instructions to do this were given and later each command shown in Figure 5.43 was explained in detail. To see if the program is loaded, the draftsperson may use a variation of the instructions shown in section 5.1, they are:

```
LOGON
USERID/PASSWD
LIST 'ACS.USERID.DEMO(PROG)'
```

When this is done, a page listing as shown in Figure 5.43 is produced. This means the image processor is in place and ready to use as explained in sections 5.2 through 5.15. You are now ready to begin to use an image processor.

5.18 REVIEW PROBLEMS

Each of the following problems have been selected to give experience in the use of image processing. Some problems use just a single key function, while others require nearly every key function to be used. Your instructor will assign those problems that are most suited to your application needs. Within the range of the problems selected, almost every field of engineering and technology is represented. Simple problems can be worked in a matter of minutes, while a few require hours to complete. To test your own skill, simple problems should be worked before the more advanced examples are tried. Once success is attained with an image processor, any 2-D type problem may be worked.

1. L & P KEY FUNCTIONS

SAMPLE FAMILY OF PARTS
  E=ERASE
  R=RETURN
  1=BALL BEARINGS
  2=LARGE DOT
  3=SHAFT AND KEY
  4=SHAFT FOR HUB

SAMPLE FAMILY OF PARTS
  E=ERASE
  R=RETURN
  1=BALL BEARINGS
  2=LARGE DOT
  3=SHAFT AND KEY
  4=SHAFT FOR HUB

2. L & C KEY FUNCTIONS

SAMPLE FAMILY OF PARTS
  E=ERASE
  R=RETURN
  1=BALL BEARINGS
  2=LARGE DOT
  3=SHAFT AND KEY
  4=SHAFT FOR HUB

SAMPLE FAMILY OF PARTS
  E=ERASE
  R=RETURN
  1=BALL BEARINGS
  2=LARGE DOT
  3=SHAFT AND KEY
  4=SHAFT FOR HUB

3.   L, M, D, & K KEY FUNCTIONS

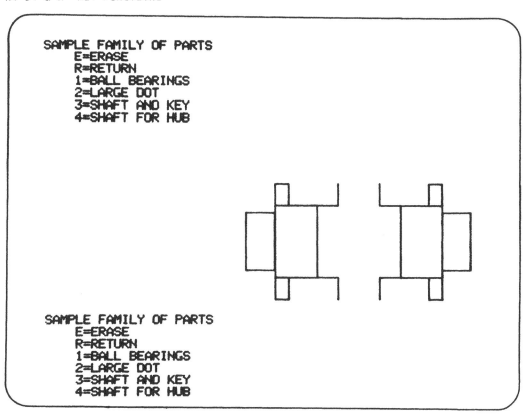

4.   L, M, D, C, & K KEY FUNCTIONS

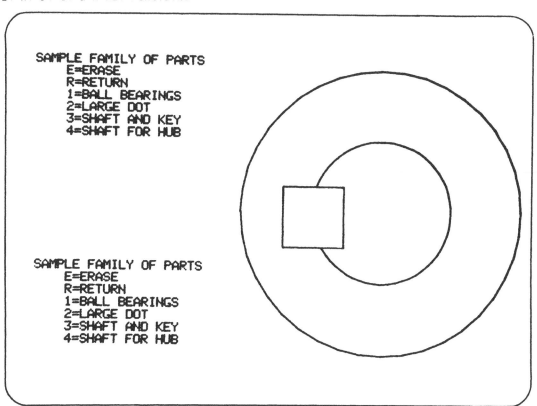

5. M, D, C KEY FUNCTIONS

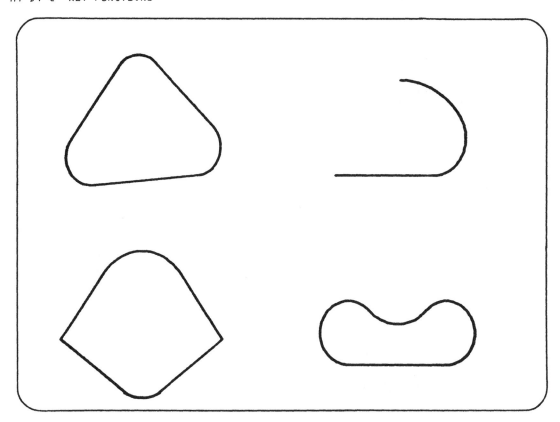

6. M, D, C, Z KEY FUNCTIONS

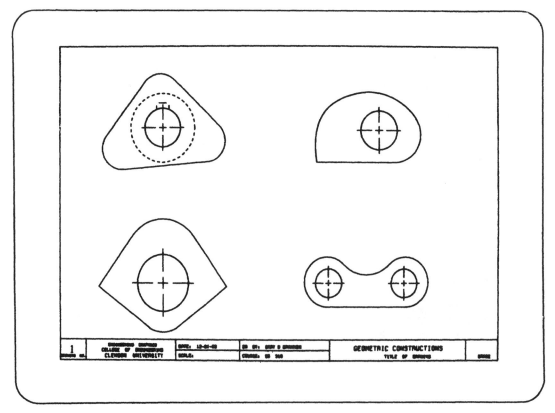

7.   M, D, K, 1, C KEY FUNCTIONS

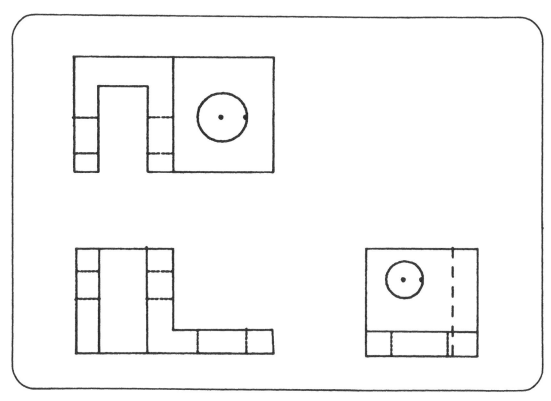

8.   A, D, K, 1, C & L KEY FUNCTIONS

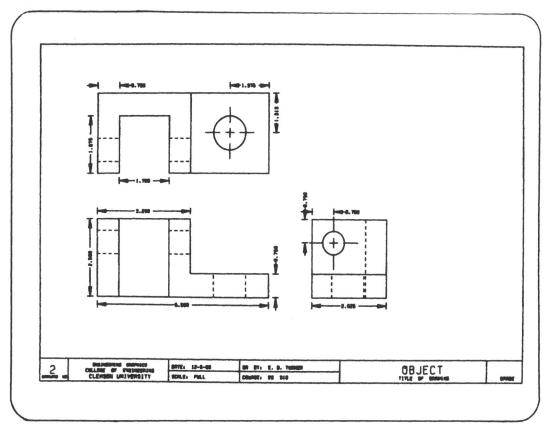

9.   M, D, L, K, A, C, O   KEY FUNCTIONS

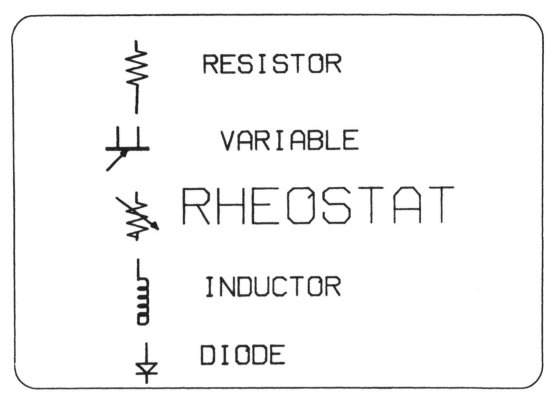

10.   M, D, L, K, A, C, O   KEY FUNCTIONS

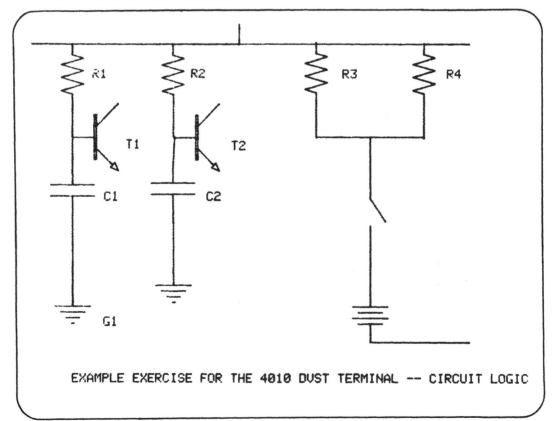

11.   G, M, D, 1, L   KEY FUNCTIONS

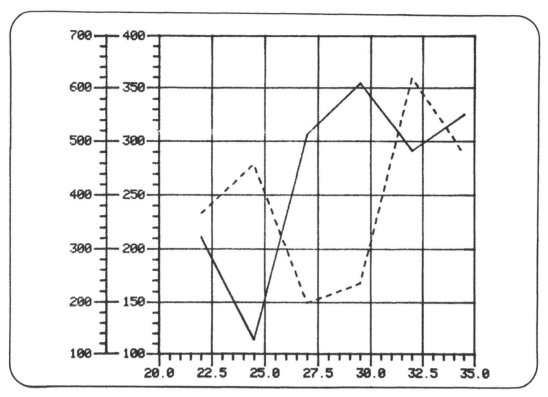

12.   G, M, D, 1, L, X   KEY FUNCTIONS

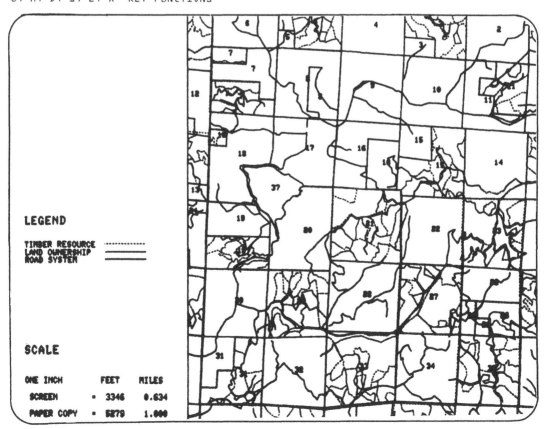

13.   Simulated X functions with M and D.

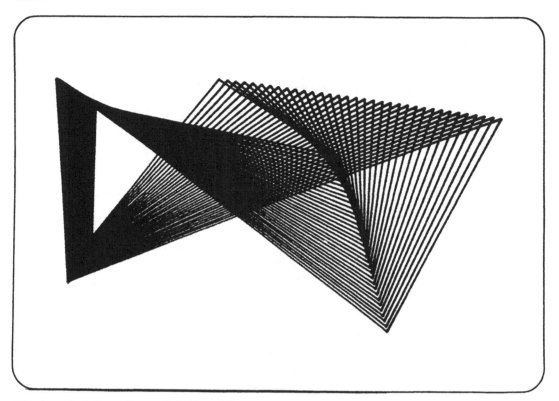

14.   Simulated X functions with M and D.

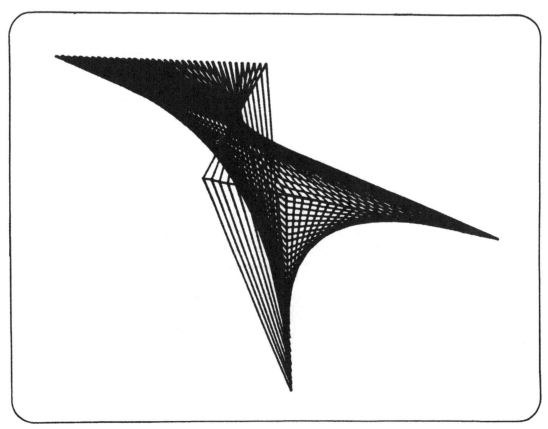

15. ALL FUNCTION PROBLEM

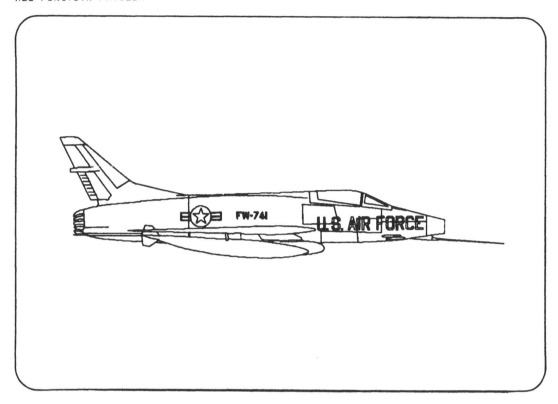

5.19 CHAPTER SUMMARY

Image processing is basic to automated drafting and design. This chapter has presented most of the CORE concepts for 2-D image input and output. This chapter, like all chapters, presented the concept, the equipment necessary to present this concept and the steps involved with learning this new concept.

Images are constructed from memory modules which are CALLED, as shown in section 5.1. Each element listed was then demonstrated as:

    lines -- section 5.2
    arcs and circles -- section 5.3
    rectangles -- section 5.4
    saving images -- section 5.5
    plotting images -- section 5.6
    modification of images -- section 5.7
    enhancing images -- section 5.8
    arrowheads -- section 5.9
    fits -- section 5.10

    smoothed lines -- section 5.11
    centerlines -- section 5.12
    points on a line -- section 5.13
    polygons -- section 5.14
    grids -- section 5.15

The use of these key functions was then presented through a series of separate figures as shown throughout the chapter. The combined used of image construction is shown in Figure 5.28. To assist the beginning draftsperson, a series of screen helps was introduced for each of the key functions.

A discussion of the CORE requirements followed this with an introduction to the use of an image processor type program. A student who has read, studied the examples and reviewed the help messages was then presented with a series of review problems to be completed. In this manner, a person may review the concept of image processors if needed, by finishing the chapter summary.

6

Multiview Projection

Multiview projection is a technique used by drafters to represent three dimensional objects. This representation is sometimes called orthographic projection, and the projections used are usually the top, front, and right side views. You will recall from Chapter 2.5, orthographic projection that a total of six views may be selected. All can be placed on the special worksheet form shown in Chapter 2.

Engineers, designers, and drafters all use the multiview approach to describe the objects that will be part of an automated, computer generated working drawing. These drawings are produced on the DVST screen and routed to a plotter as shown in Figure 6.1. In this case a larger model DVST is chosen so that more than a single view may be displayed at a time and with reasonable size. Changes are made on the screen after the 3-D image processor has displayed the chosen views. Any combination of views may be displayed as shown in Figure 6.1. Because a multiview drawing is produced from a 3-D image processor, it is not limited to 2-D representations. Notice that in the upper

right hand corner of Figure 6.1, a database may be represented as a simple 3-D object. The concept of a simple child's block is a good example to start with, to understand all the possiblities of a 3-D image process. You will notice that the child's block has letters on the faces called X, Y, and Z. A simple explaination might be to think of the block as a drawing of any object. The X face would be the top or horizontal (plan) view. The Y face would be the front and Z would be the right side or profile view.

6.1 PROJECTION THEORY

The child's block is displayed from a 3-D database as shown in Figure 6.2. This is an enlargement of the upper right hand corner of Figure 6.1 with the data reference axis X, Y, and Z. Every object has a list of data. This object's list would be:

X	Y	Z	IPEN
0.0	0.0	0.0	3
1.0	0.0	0.0	2
1.0	1.0	0.0	2

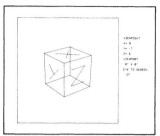

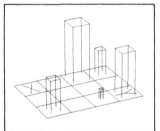

Figure 6.1 Multiview projection from 3-D database. (Courtesy Tektronix Inc.)

0.0	1.0	0.0	2
0.0	0.0	1.0	2
1.0	0.0	1.0	2
1.0	1.0	1.0	2
0.0	1.0	1.0	2

This means that the child's block is one
unit by one unit by one unit. In Figure
6.1 middle right hand example, each of the
faces must be mapped or sent to the plotter
as flat projections. This is done by the
order of the data list. For example, suppose we wanted to plot the entire data list?
The result would be as shown in Figure 6.2
or a simple pictorial. Suppose that only
the X and Y part of the data list was sent
to the plotter? This would result in a
front view as shown is Figure 6.3. Suppose
that the X and Z data list was plotted? A
top view would then appear as shown in Figure 6.4. A right side view would appear if

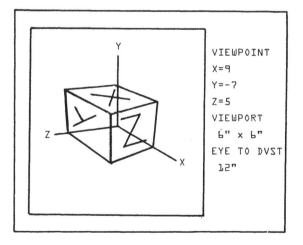

Figure 6.2 Entire data list sent to plotter
would result in a pictorial of the object.

the Z and Y lists were sent as shown in Fig-

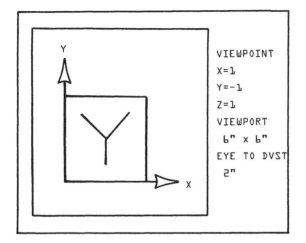

Figure 6.3 Frontal projection.

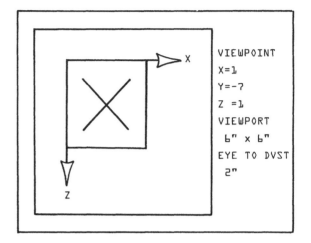

Figure 6.4 Horizontal projection.

ure 6.5. You will notice that three terms
are used with 3-D image processors, they
are:

1. Viewpoint. This is the relationship of the projected plane of the object
to the output area of the DVST or the plotter surface. In our example of a front projection, the projection will be located at
plotter units 1,1 inside the viewport set
aside for the front view.

2. Viewport. This is the display
area set aside for each projection. The
entire plotter surface is called a window,
but separate viewports may be used inside
a single window. Figure 6.1 middle right
is a good example of this.

3. Eye to DVST. This sets the scale
of the projected view. If the eye to DVST
is zero, the object is projected full size.
The further away from the object, the size
decreases. (There is no provision to size
projected views larger than life size in
this chapter). If the eye to DVST were a
minus number, than you are inside the object and it would appear much larger than
normal. However, orthographic projection
does not provide for this type of viewing.

The lower right corner of Figure 6.1
represents the total pen movements required
to plot each viewport contents. While this
seems rather silly to mention at this point,
it will become very apparent why we need
this information in the next sections.

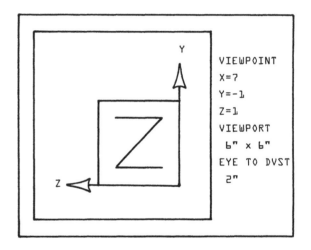

Figure 6.5 Profile projection.

6.2 VIEW CONSTRUCTION

Views are constructed directly on the
screen of the DVST as shown in Figure 6.6.
The drafter follows a set procedure whereby
an entire view is presented on the DVST so
the drafter may approve, or modify before
sending it to the plotter as shown in Figure 6.7. Modifications are done with an
image processor in 2-D as shown throughout
Chapter 5. To begin the view construction
process, the drafter places the memory modules in place as shown in Figure 6.8 or
uses the synchronous communications interface to a large host computer(item 4, Figure 6.6).

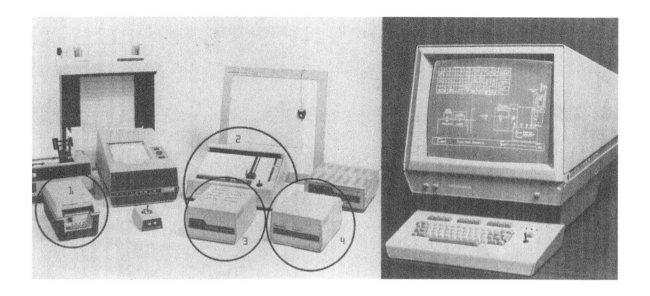

Figure 6.6 Continuation of equipment from Chapter 5 with special emphasis on the magnetic tape unit (1), pen plotter (2), ROM expander (3), and synchronous communication interface (4). (Courtesy DLR Associates and Tektronix Inc.)

You will remember that a series of keyboard commands were used to recall the image processor from host computer storage. Another, similar set is used as:

```
LOGON
USERID/PASSWD
CALL 'ACS.USERID.DEMO(CUPID)'
```

These commands locate the memory modules that are stored in the large computer and place them on the large screen for you to review. Models 4014,4054 and 4080 series are used in this step depending upon the items selected from Figure 6.6. For instance, if you used a 4051 (small screen-stand alone) pictured in Figure 5.2 as an image generator, then you select a 4054. If a 4010 (small screen - host dependent) with thumbwheels was used, then a 4014 is the correct choice. If a 4012 with joy stick was used in chapter 5, then a 4080 or a 618 model will be your choice. It is not absolutely required that a larger size screen be used, however. A small screen will serve nicely for single viewport displays. Three or more viewports may be previewed and sent to the plotter shown in

Figure 6.7 Plotter. (Courtesy Tektronix Inc.)

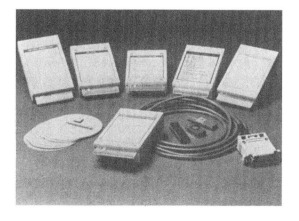

Figure 6.8 Memory modules. (Courtesy Tektronix Inc.)

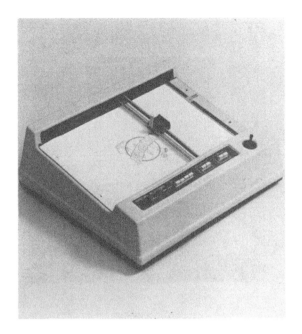

Figure 6.9 Size B plotter. (Courtesy Tektronix Inc.)

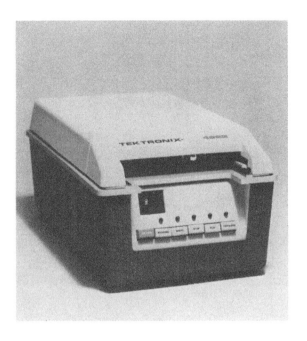

Figure 6.10 Digital cartridge tape recorder
(Courtesy Tektronix Inc.)

Figure 6.7 or Figure 6.9. The size of the
finished drawing is limited to 12 x 18 inches if a plotter like that pictured in
Figures 6.6 and 6.9 is selected.

The selection of equipment depends upon the final use of the drawing. Is it an
industrial drafting room that requires a
larger size or is it a training session -
where a ROM extender is not needed to complete a library of drawing functions. The
final storage of a drawing might be fairly
simple. A magnetic tape copy may be all
that is needed, if so, Figure 6.10 represents the type of equipment needed. If a
faster, larger system is needed than Figures 6.11 and 12 might be needed depending
upon the model of DVST selected. Figure
6.13 represents the DVST used with the ROM
extender and Figure 6.14 represents the
DVST used with Figure 6.12.

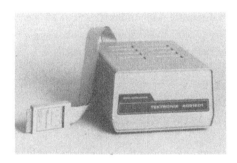

Figure 6.11 ROM extender. (Courtesy Tektronix Inc.)

6.3 VIEW MANIPULATION

Depending upon the equipment selected,
several manipulations can now be introduced
for you to try. Single viewport construction and execution is the most important,

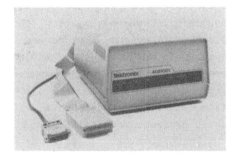

Figure 6.12 Communication interface to IBM
protocol. (Courtesy Tektronix Inc.)

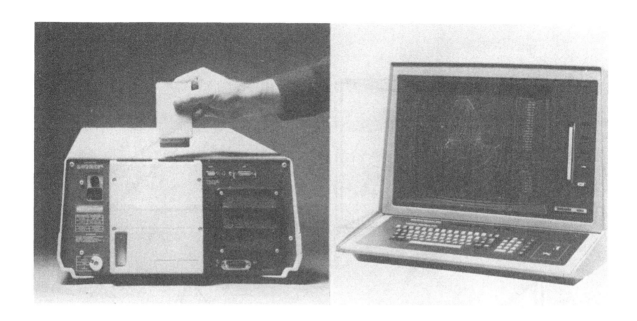

Figure 6.13   ROM extender used with DVST.   (Courtesy Tektronix Inc.)

because the lowest order DVST and pen plotter can be used to demonstrate this concept
as shown in Figure 6.15.

In this example, four viewports have
been displayed. The first, in the upper
right hand corner of the screen is the viewport containing X, Y, Z, and IPEN data listing. This is not usually displayed with
other multiviews, but is done here for explaination purposes. Any position could
be used to display this data list, but if
other views such as the front, top and side
are to be displayed before plotter output,
the upper right hand corner is the only open
space available. Next the second viewport
is plotter in the lower left hand portion
of the screen. Once this viewport is selected, it automatically locates the baselines for the right side and the top views.
The X baseline for the side and the Y baseline for the top.

The two remaining viewports are displayed from these baselines by simple trans-
lation. In other words, the horizontal data
is moved to the space directly above the
frontal viewport (see Figure 6.4) and the
profile data is moved to the space directly
to the right of the frontal viewport (see

Figure 6.14   Model 4014 DVST. (Courtesy Tektronix Inc.)

Figure 6.5). Using this technique, any 3-D
object can be translated into the normal
multiview patterns. You will notice that
a screen prompt appears at the bottom of

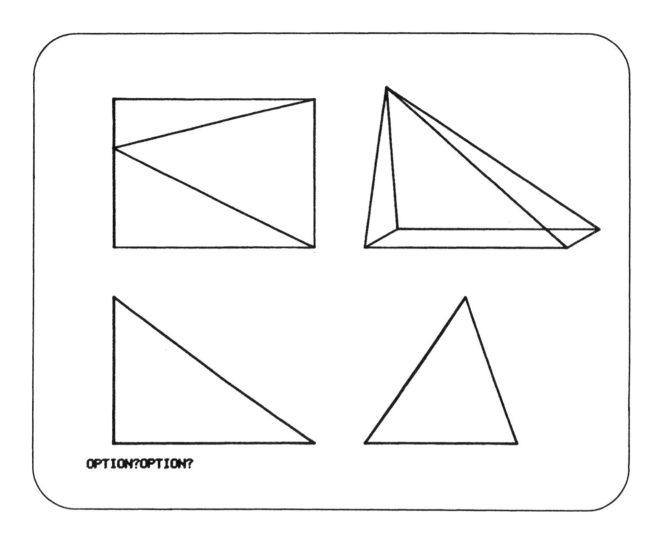

Figure 6.15 Lowest order DVST (4010) screen display of four viewports. (Courtesy DLR Associates and Tektronix Corp.)

Figure 6.15, it asks for the display option
to be used. Several display options exit,
for Figure 6.15 the reference axis shown in
Figure 6.16 were deleted from the X, Y, Z
(pictorial option) data list. Figure 6.16
shows the amount and the order of the data
points used to display the four viewports
in Figure 6.15.

6.4 POINT SPECIFICATION

Point specification is important in a
multiview projection because the pen movement must be taken into consideration. For
example, suppose the five points shown in
Figure 6.16 were:

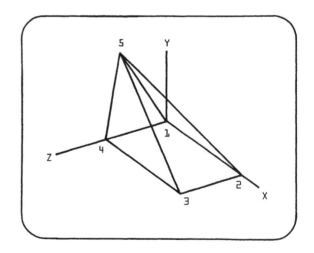

Figure 6.16 Data arrangement.

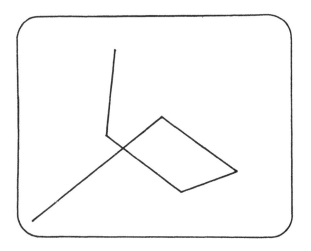

Figure 6.17 Display of data base.

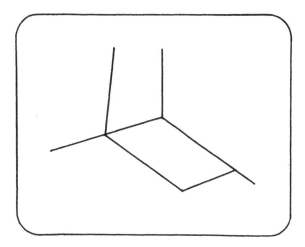

Figure 6.18 Display with AXIS option.

DATAPT	X	Y	Z	IPEN
1	0.0	0.0	0.0	2
2	2.1	0.0	0.0	2
3	2.1	0.0	1.5	2
4	0.0	0.0	1.5	2
5	0.0	1.5	0.5	2

If we entered these points as shown above to the 3-D image process and selected the ALL option, Figure 6.17 would result. We would be less than satisfied if we compared that kind of output to the output shown in Figure 6.15. What went wrong?

First the viewport was set equal to the window (display screen size). This is recommended for small DVST's. So what caused the line from the lower left hand corner of the screen to point 1? The answer is the IPEN data for datapt 1. It is a 2 which means pen down or beam on. If we change the first line of the database to read:

| 1 | 0.0 | 0.0 | 0.0 | 3 |

and we select the ALL and AXIS option, Figure 6.18 is displayed. Better, but not a completed view by any means. Let's see what is missing.

1.   line segment from datapt 5 to 3
2.   line segment from datapt 5 to 1
3.   line segment from datapt 5 to 2
4.   line segment from datapt 4 to 1

Suppose we added these missing lines to the

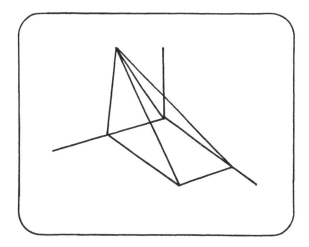

Figure 6.19 Display with AXIS and ICON.

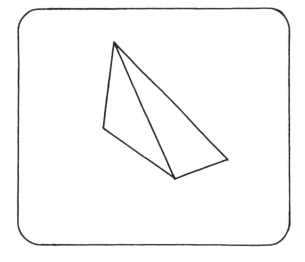

Figure 6.20 Display with AXIS off and ICON on with SOLIDS option. (Courtesy DLR Assoc)

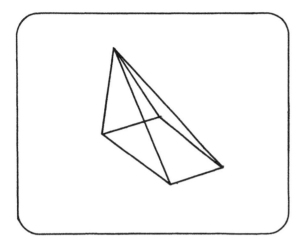

Figure 6.21 Display with ICON only.

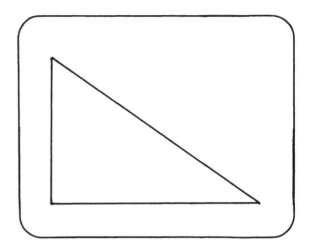

Figure 6.22 Frontal viewport with ICON.

data base as an ICON (point connection system) that would appear as:

ICON = -1,2,3,4,1,5,4,-5,3,-5,2

The ICON data says:

```
GOTO 1 WITH PEN UP........-1
GOTO 2 WITH PEN DOWN..... 2
GOTO 3 WITH PEN DOWN..... 3
GOTO 4 WITH PEN DOWN..... 4
GOTO 1 WITH PEN DOWN..... 1
GOTO 5 WITH PEN DOWN..... 5
GOTO 4 WITH PEN DOWN..... 4
GOTO 5 WITH PEN UP........-5
GOTO 3 WITH PEN DOWN..... 3
GOTO 5 WITH PEN UP........-5
GOTO 2 WITH PEN DOWN..... 2
```

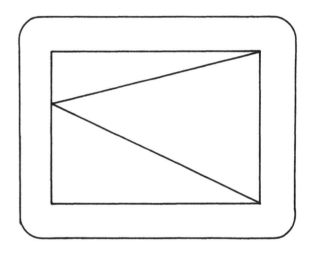

Figure 6.23 Horizontal viewport with ICON.

This is exactly what we need to display
Figure 6.19 if AXIS is on and Figure 6.21
if AXIS is off. Figure 6.20 tests for a
solid object and deletes the hidden lines.

Once the ICON data list has been set,
the other viewports can be plotted by:

1. X,Y,ICON (Figure 6.22)
2. X,Z,ICON (Figure 6.23)
3. Z,Y,ICON (Figure 6.24)

These views should be arranged as shown in
Figure 6.15 and sent to a plotter. The F,
H, and P viewports are required to display
a typical three-view drawing. The ALL viewport is optional, it is shown here for explaination purposes only.

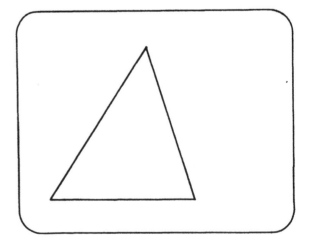

Figure 6.24 Profile viewport with ICON.

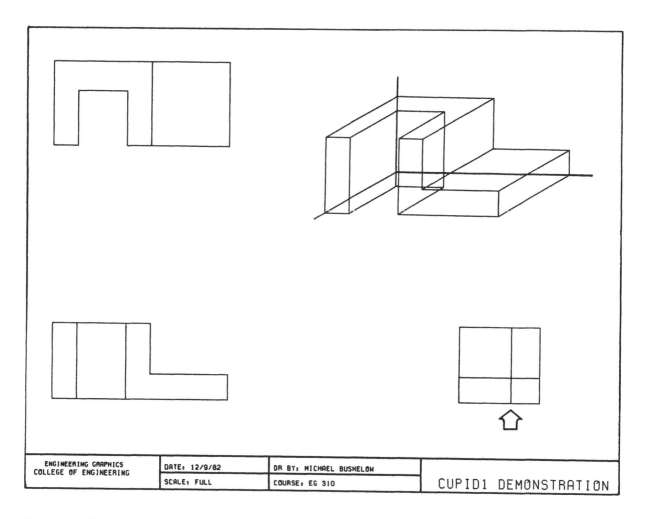

Figure 6.25   Line specification during 3-D image processing.   (Courtesy DLR Associates)

6.5   LINE SPECIFICATION

You will recall from Chapter 2 that a line generator is used to produce a number of different types of lines; solid, dashed, broken, center, cutting plane and so forth. A line specification is needed for multiview projection. In Figure 6.25, it is apparent that not all lines can be moves or draws (open or solids). The arrow locates a case where a hidden line must be used. A draftsperson who understands multiview projection quickly notes the error and may do an image modification on the profile view before it is sent to the plotter. In Chapter 5, this modification was done with the 1 key. In this chapter, it may be done by changing the IPEN code from 2 to 1. A 3-D image processor uses a line generator in-

side it. Therefore, the draftsperson may list the line generator to see if hidden lines are included. This is done by:

LIST 'ACS.USERID.WATLIB(PLOT)'

The line generator would then appear at a DVST screen as shown in Figure 6.26. This is not a complete listing, only the first page is shown in Figure 6.26, but the IPEN 1 is contained and located by the pointing arrow. All parts of the 3-D image processor may be listed, as was the case in 2-D image processing, however, only those that are important to the understanding of multiview projection will be shown in this chapter. The chapter entitled Pictorial Representations will contain a complete discussion of 3-D image processors.

```
SUBROUTINE PLOT(XPAGE, YPAGE,IPEN)
IX=X*130.
IY=Y*130.
IF(IPEN.EQ.2) CALL DRWABS(IX,IY)
IF(IPEN.EQ.3) CALL MOVABS(IX,IY)
IF(IPEN.EQ.-2)CALL DRWREL(IX,IY)
IF(IPEN.EQ.-3)CALL MOVREL(IX,IY)
IF(IPEN.EQ.1)CALL DASHP(IX,IY,.1)
IF(IPEN.EQ.4) CALL DASHP(IX,IY,.3)
IF(IPEN.EQ.5) CALL CENTER(IX,IY,KX,KY
IF(IPEN.EQ.6) CALL CNTRL(IXAR,IYAR,NP
```

Figure 6.26 Listing of line generator.

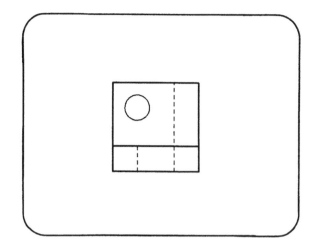

Figure 6.27 Profile viewport with DASHP.

6.6 ARC SPECIFICATION

The use of a line generator becomes a necessary modification tool when other part features are added to multiviews. For example, suppose holes were to be drilled in the object shown in Figure 6.25? An arc generator for 3-D must be provided for this purpose. A circle (type of arc) is shown in Figure 6.27. This circle represents a hole through the entire part. Therefore, four circles are required to illustrate a CAD/CAM function (computer-aided design and computer-aided manufacturing). The draftsperson must first check the 3-D image processor to see if three-dimensional arcs are available. This is done by:

LIST 'ACS.USERID.WATLIB(ZARC)'

A listing for the arc generator in the Z plane of projection is listed at the DVST screen, as shown in Figure 6.28. This is the type of arc that can be generated in the example shown in Figure 6.5. This will not display the circle shown in Figure 6.29, but it will display the circle shown in Figure 6.27. Three arc generators are provided for each of the multiview positions, namely:

1. XARC (see Figure 6.4)
2. YARC (see Figure 6.3)
3. ZARC (see Figure 6.5)

See Figures 6.27, 6.29, and 6.30 for exam-

```
SUBROUTINE ZARC(XP,YP,ZP,R,SANG,EANG,
DIMENSION P(100,3),IC(100),VP(3)
PI=3.14159265
N=IFIX(EANG-SANG)
THETA=(EANG-SANG)/N
THETAR=THETA*PI/180.
SRANG=SANG*PI/180.
DO 1 I=1,N
DX=R*COS(SRANG)
DY=R*SIN(SRANG)
P(I,1)=XP+DX
```

Figure 6.28 Listing of arc generator.

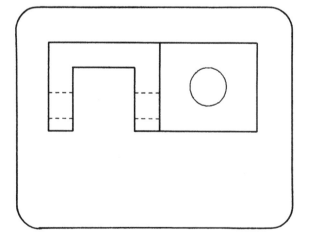

Figure 6.29 Horizontal viewport with XARC.

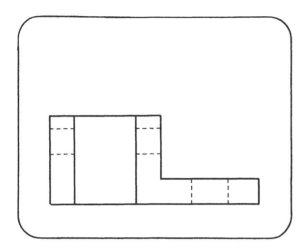

Figure 6.30 Frontal viewport with DASHP.

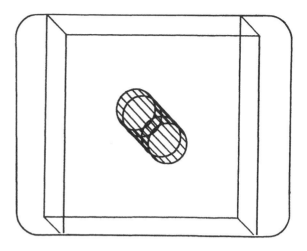

Figure 6.31 CAD display of ARC functions.

ples of ZARC, XARC and DASHP functions as listed in Figure 6.26.

The ARC functions are important in a 3-D image processor because a great deal of graphic information is contained in each arc specification. Figure 6.31 is a display of the ALL viewport, enlarged at the location of the holes shown in each of the other viewports. In viewport shown in Figure 6.27, the hole is displayed as a circle. The hole runs through two sections of the part as shown in Figure 6.29 and Figure 6.30. In both the Horizontal and Frontal viewports, the hole appears as a set of hidden lines. Therefore, four display images are required, and the best view of this is in Figure 6.31. Notice that 4 circles are clearly visible and that the hole has been shaded to represent CAM action.

6.7 VISUALIZATION OF VIEWS

In order to visualize how the position and relationship of views is affected by a line and arc specification, assume that the object shown in Figure 6.32 is to be displayed from the worksheet arrangement shown in Chapter 2. You will remember that this worksheet was arranged so that the frontal viewport represented the single plane into which the remaining viewports are to be displayed. For the purpose of orientation,

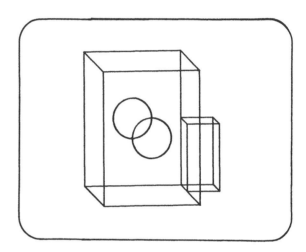

Figure 6.32 Object for visualization.

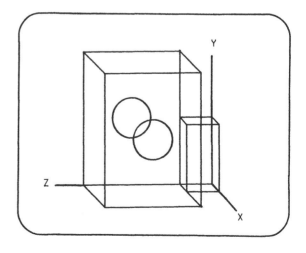

Figure 6.33 Orientation A of the object.

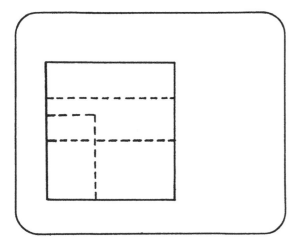

Figure 6.34 Frontal viewport with DASHP.

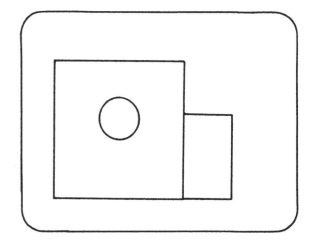

Figure 6.35 Profile viewport with ARC.

imagine the viewports swinging the respec-
tive projection planes shown in Chapter 2
into their correct viweing positions. In
that case, Figure 6.34 would be the correct
orientation of the object shown in Figure
6.33 for the frontal projection. Likewise,
Figure 6.35 represents the profile and Figure 6.36 represents the horizontal or top
view. Each viewport, except the rear, is
considered to be hinged to the frontal viewport. Notice in Figure 6.37 that when the
viewports are swung open, viewport X is
directly above the frontal viewport (Y);
viewport Z is to the right; viewport Z' is
to the left, viewport X' is below and the
rear projection folds so it is to the left
of Z'. Therefore, for any third angle projection, the views of an object are always
located in these positions.

6.8 ORIENTATION OF THE OBJECT

To describe an object completely, an
automated drawing must show all the features
of the object. In many instances certain
edges, intersections, or surfaces cannot be
seen from the position of the digitized input. This is the case for Figure 6.33. A
poor choice of orientation was used in this
example. The correct practice in automated
drafting is to show those hidden features,
as in Figure 6.34, as visible edges if possible. A different orientation of the object

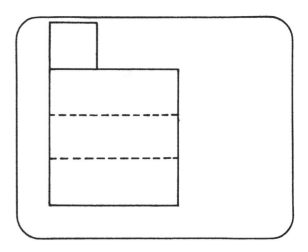

Figure 6.36 Horizontal viewport with DASHP.

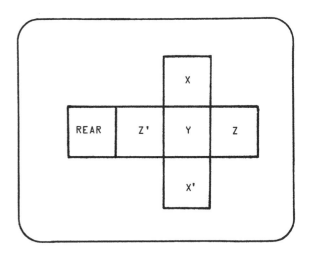

Figure 6.37 Placement of viewports.

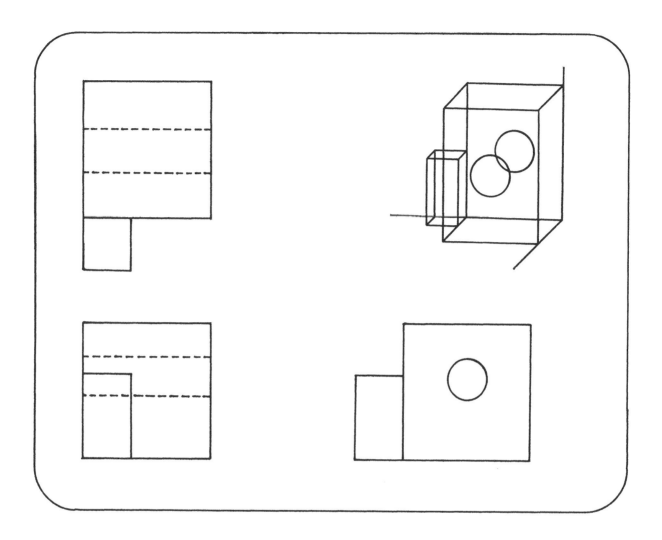

Figure 6.38 Orientation B of same object with 3 viewports showing hidden line elimination.

as shown in Figure 6.38 will result in an elimination of hidden lines. The through hole will remain hidden in at least two of the three views, however. Therefore, dashed lines are shown in the top and front viewports.

Hidden lines consist of evenly spaced dashes with a space between them approximately 1/4 the length of the dash. A hidden line should always begin with a dash in contact with the line from which it starts, except when the dash would form a continuation of a full line. Dashes should touch at corners as shown in Figure 6.34. The amount of hidden lines should always be kept to a minimum in the three principle views (top, front and right side).

6.9 CHANGING OBJECT ORIENTATION

As long as the object to be defined is placed within the third angle projection position, namely a positive X, Y and Z data gathering system; any change of position is acceptable. Figure 6.39 represents just a few of the possible positions prior to data gathering. Once the object is placed at the intersections of the X, Y and Z axes, the first point become 0.0, 0.0, 0.0 automatically. This system is called digitizing. This allows a standard programming approach to be used for automatic view generators. Drafters have access to automatic view generators, they do not write them. Drafters supply the data for automatic view generators.

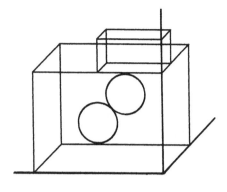

Orientation C

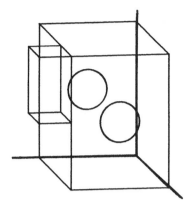

Orientation D

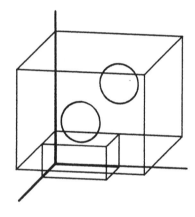

Orientation E

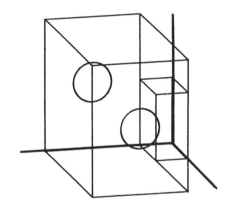

Orientation A

Figure 6.39 Different object orientation affects point data list. (Courtesy DLR Associates)

At least three methods for selecting point data have been presented in this text, they were:

1. Data tablet (see Chapter 4)
2. Image processing (see Chapter 5)
3. Digitizing

When digitizing is selected, two main parts of the display object are broken into separate data elements. The first is the line segments used throughout Figures 6.15 to 6.40. Each starting and ending point of a line has a set of data. These lines are then connected by the ICON data. Arc generators are the second main part of the display object. Each ninety degree arc segment contains 15 straight line data elements which are provided through the use of the ARC functions described earlier. The draftsperson uses a function rather than provide the 15 separate data elements for each 90 degree segment. In the case of a circle, 60 separate pieces of information are needed to display an object. Therefore, Figure 6.38 contains 120 data elements for the two circles plus the straight line data (16 sets of end points), plus the hidden line generated from the function DASHP. It is only natural that computer-assisted techniques are used to help the draftsperson collect all the information needed to automatically display a multiview drawing. One of these techniques is a users manual provided for

```
          SUBROUTINE XARC(XP,YP,ZP,SANG,EANG)
C
C     ****************************************************************
C     * THIS SUBROUTINE GRAPHS CIRCULAR ARCS IN THE X PLANE (IE. THE  *
C     * PLANE PERPENDICULAR TO THE X-AXIS) FROM A THREE DIMENSIONAL    *
C     * DATA BASE IN X, Y, AND Z.*
C     *   XP...THE X COORDINATE OF THE CENTER OF THE ARC              *
C     *   YP...THE Y COORDINATE OF THE CENTER OF THE ARC              *
C     *   ZP...THE Z COORDINATE OF THE CENTER OF THE ARC              *
C     *    R...THE RADIUS OF THE ARC*
C     * SANG...THE STARTING ANGLE OF THE ARC MEASURED IN DEGREES CCW   *
C     *          POSITIVE FROM THE POSITIVE Z AXIS*
C     * EANG...THE ENDING ANGLE OF THE ARC MEASURED IN DEGREES CCW     *
C     *          POSITIVE FROM THE POSITVE Z AXIS                      *
C     ****************************************************************
      DIMENSION P(400,3)
      PI=3.14159265
      N=IFIX(EANG-SANG)
      THETA=(EANG-SANG)/N
      THETAR=THETA*PI/180.
      SRANG=SANG*PI/180.
      DO 1 I=1,N
      DZ=R*COS(SRANG)
      DY=R*SIN(SRANG)
      P(I,1)=XP+DX
      P(I,2)=YP+DY
      P(I,3)=-ZP
      SRANG=SRANG+THETAR
    1 CONTINUE
      CALL VANTAC(IC,VP)
      RETURN
      END
      SUBROUTINE VANTAC
      COMMON /B1/NP,NC,P(750),VP(3)
      DIMENSION PP(750,3),IC(750),VP(3)
      A=ARCTAN(VP(1),VP(3))
      SA=SIN(A)
      CA=COS(A)
      DO 10 J=1,NP
      PP(J,3)=P(J,1)*CA+P(J,1)*SA
   10 PP(J,1)=P(J,1)*CA-P(J,3)*SA
      VPP=VP(3)*CA+VP(1)*SA
      A=ARTAN(VP(2),VPP)
      SA=SIN(A)
      CA=COS(A)
      DO 20 J=1,NP
   20 PP(J,2)=P(J,2)*CA-PP(J,3)*SA
      DO 30 J=1,NC
      IF (C(J).LT.0)GOTO1
      CALL PLOT(PP(IC(J),1),PP(IC(J),2),2)
      GOTO 30
    1 K=-IC(J)
      CALL PLOT(PP(K,1),PP(K,2),3)
   30 CONTINUE
      RETURN
      END
```

Figure 6.40 Typical page from a users manual. All right-facing pages of the manual contain a source listing, in this example it is the subroutine XARC. A single manual is used for both draftspersons and programmers. All right-facing pages are subject to change by the programmer. All left-facing pages, which contain the graphic output of the source programming, are subject to change by the draftsperson. In this manner, working together, the draftsperson and the programmer keep the loose-leaf notebook type manual current with the hardware (items listed throughout Chapters 4, 5 and 6).

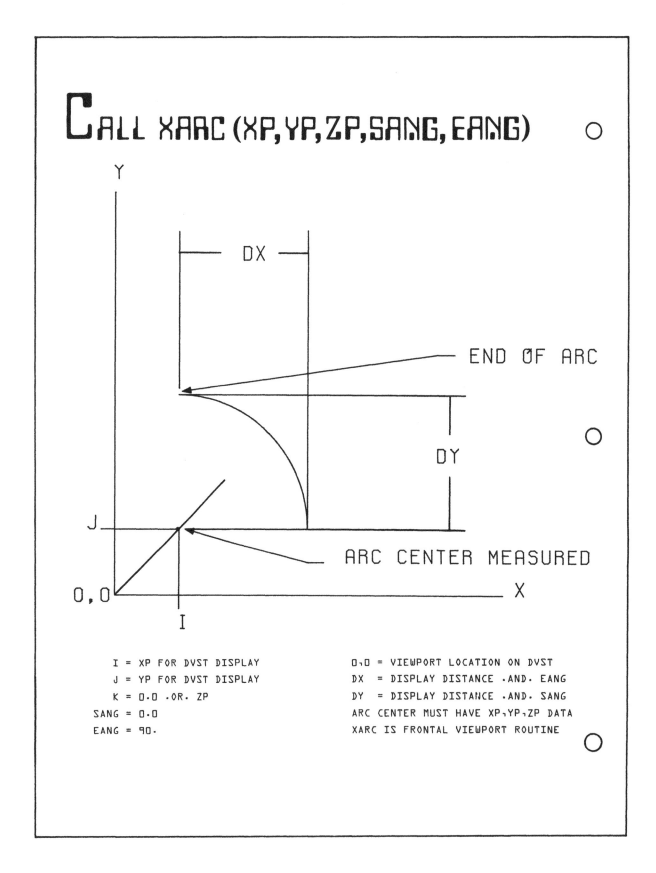

Figure 6.41 Typical left-facing page from a users manual. (Courtesy DLR Associates)

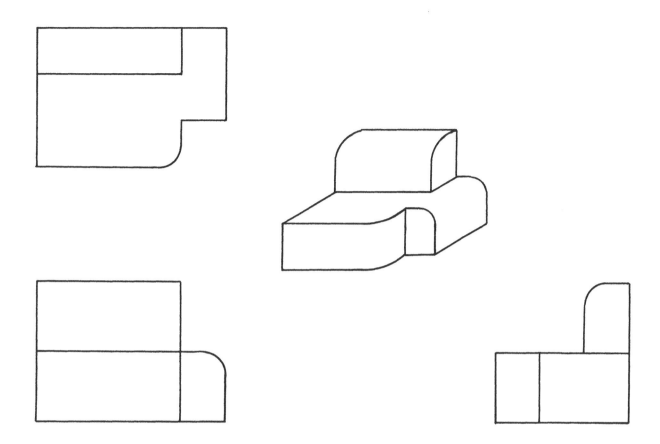

Figure 6.42 Typical multiview drawing produced by CALL XARC, YARC, ZARC and line specification from ICON data list. (Courtesy DLR Associates)

each drafter. The manual is divided into programs, subprograms and certain routines which assist the draftsperson in each of a series of drafting tasks that must be done. One of these tasks is producing a multiview display. Figure 6.40 and 41 introduce the concept of a users manual, if additional information is required, refer to footnote.▲

6.10 NUMBER OF VIEWS

In Figure 6.37 it was shown that an object has six views. This does not mean that all six must be displayed to completely describe an object in a multiview drawing. A view may be left off because it merely duplicates others. This was the case in Figure 6.42. It was decided that just three views were needed and the others did not contribute anything to the display. The principle in selecting the number of views is to include only those views which are absolutely necessary to portray clearly the shape of the part. Therefore, for a cannon ball, one view is enough; where an instrument for brain surgery might require six.

A single view might also be sufficent for certain cylindrical parts if the necessary dimensions are indicated as diameters. Single views may be used for thin flat pieces of uniform thickness such as gaskets, and plates, if a note is given to indicate thickness.

▲   Ryan, D.L. 'Computer Programming for Graphical displays', Text, January 1983. Contains a number of source programs, subprograms, and routines written for use on a micro, mini, and macro computer. Users are introduced to a number of techniques for using hardware and software successfully in an engineering graphics environment.

Symmetrical objects can often be limited to two views. Typical examples were the
Figures 2.10 and 2.11. A drawing of these
types needs only a front and top view, because the profile would be a repeat of the
front.

For most objects three principal views
are all that is necessary as shown in Figures 6.15, 6.25, 6.38 and 6.42. Occasionally a left side might be substituted for
the right profile or the bottom might be
used instead of the top.

6.11 SPACING OF VIEWS

The viewports of a display object are
touching as shown in Figure 6.37. This
does not mean that the representations of
the objects inside the viewports will be
touching when they are routed to a plotter.
Views never are larger than the viewport
which contains them, in fact they are always smaller, therefore a space between
views always occurs during the display.

The viewport placement allows the views
of an object to be arranged so as to present a balanced appearance on the plotter
paper. Ample space must be provided between views to permit the placement of dimensions and notes without crowding.

To provide additional space between
display objects, simply increase the viewport size (see Figure 6.2). A good viewport size to select before data base construction is 6 inches as shown in Figures
6.3 through 6.5. Always make sure that the
viewport is square, otherwise the aspect
ratio will not be equal. For example, X
will not equal Y and circles will be displayed as ellipses.

6.12 COLLECTING THE DATA BASE

The draft person is responsible for
describing the object to be placed in the
3-D image processor for multiview projection. This includes a number of procedures
that must be followed:

1. Select viewport size.
2. Select orientation of the object.

3. Digitize the point data.
4. Add an fillets and rounds by call
    XARC, YARC, or ZARC.
5. Locate holes, slots and other part
    features such as keyways.
6. Determine the order of point con-
    nection (ICON)
7. Submit the data to the 3-D image
    processor.

As mentioned in section 6.11, a good selection for a viewport is 6 inches. Let's use
that for several example problems that we
will work (step 1.).

Step 2 will require us to position an
object, shown in Figure 6.43, so that the
proper orientation will be obtained for
multiview projection. Step 4 does not need
to be considered for this first example because no fillets or rounds of any kind are
present. Likewise no other part features
(step 5) are shown so we progress from the
third step directly to the sixth step. Figure 6.44 represents these two steps (3 & 6).

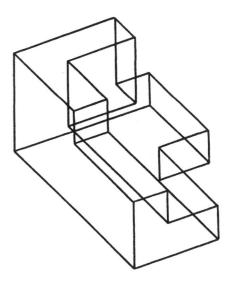

Figure 6.43 Object to be positioned so it
will require the fewest number of hidden
lines. After step 1 (viewport selection),
the object shown is oriented so that none
of the line segments will appear hidden in
any of the three principal views. Step 3
is then completed, shown in the next figure.

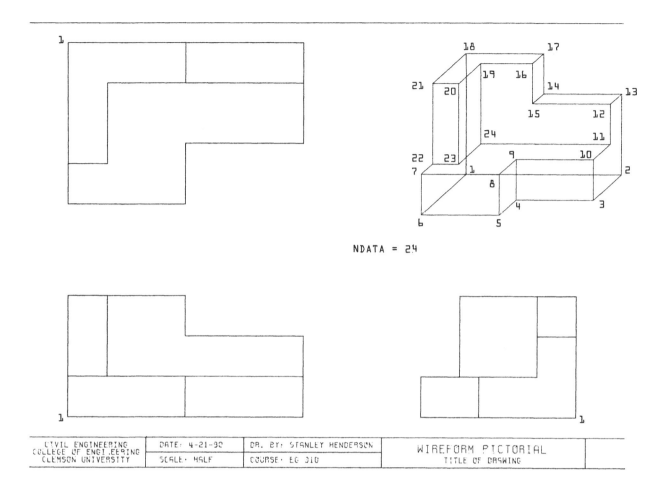

NDATA = 24

CIVIL ENGINEERING COLLEGE OF ENGINEERING CLEMSON UNIVERSITY	DATE: 4-21-80	DR. BY: STANLEY HENDERSON	WIREFORM PICTORIAL	
	SCALE: HALF	COURSE: EG 310	TITLE OF DRAWING	

Figure 6.44 Data base (NDATA = 23) is shown with the display data from ICON =

The order of the point connections (ICON) is as follows:

-1,2,3,4,5,6,7,8,9,10,11,12,15,16,19,
20,21,22,7,-22,23,24,11,-24,19,-16,17,-15,
14,-12,13,-21,18,1,6

These point connections are then read into the program that will display the data base as multiviews shown in Figure 6.44. The program maps the placement of the points (see point 1 in all three views). Every automated, multiview problem is approached in this manner. If arcs are required, the procedure is the same for all straight line data. The arcs are called at their locations as shown in Figure 6.41. This saves a lot of data listing for the draftsperson, 15 data locations for each arc.

6.13 REVIEW PROBLEMS

Each of the review problems is designed to test your ability in the seven steps of multiview projection listed in section 6.12. Work each of the problems assigned by your instructor.

1. SIMPLE BLOCK.

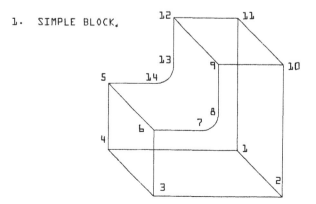

2.   DATA BASE PROBLEM

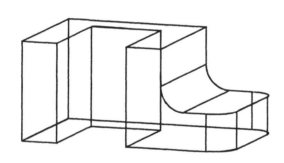

5.   DATA BASE PROBLEM

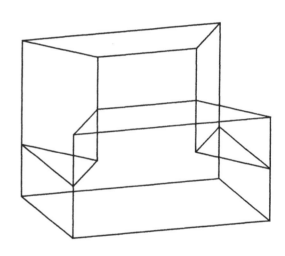

3.   ICON PROBLEM

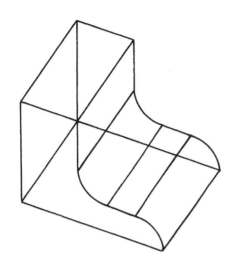

6.   ICON PROBLEM

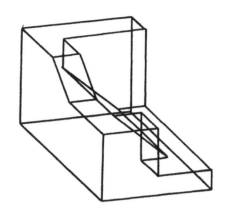

4.   DIGITIZING PROBLEM

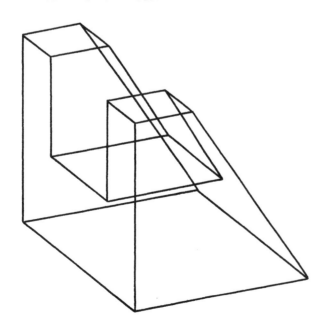

7.   DIGITIZING PROBLEM

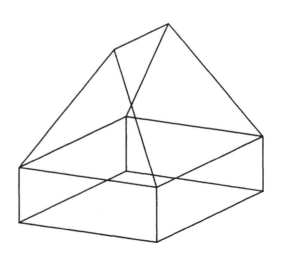

8. Enter the following program and save.

```
//MSJHN1 JOB (0923-1-001-03-  ,:07,7),'JOHNSON'
//STEP1 EXEC FTG1CLG,PLOTTER=VERSATEC
//C.SYSIN DD *
C
C
C     **************************************************
C     *                                                *
C     *    PROGRAM TO DISPLAY MORE THAN ONE DRAWING     *
C     *                                                *
C     **************************************************
C
      CALL PLOTS
      CALL DIGIT(0.,1.)
      CALL CIRCUT(0.,12.)
      CALL DIGIT(9.,-12.0)
      CALL VIEW4(0.0,12.0)
      CALL CUPID1(9.0,-12.0)
      CALL FEATUR(0.,12.0,1)
      CALL FEATUR(9.0,-12.0,0)
      CALL CUPID3(0.0,12.0)
      CALL PLOT(0.,0.,999)
      STOP
      END
```

9. Enter the following subprogram.

```
      SUBROUTINE DIGIT(TRANX,TRANY)
C     **********************************************
C     *                                            *
C     *   DIGITIZING ROUTINE FOR PREVIEW USING      *
C     *   MSJ SYSTEM OF DATA POINTS                 *
C     *                                            *
C     **********************************************
C
C
      DIMENSION P(100,3)
      REAL RADIUS,STARTA,ENDA,CODE
      READ (1,*) NPTS,NPTR
      WRITE(3,*) NPTS,NPTR
C
      DO 30 I=1,NPTS
      READ (1,*) N,P(I,1),P(I,2)
C     WRITE(3,*) N,P(I,1),P(I,2)
 30   CONTINUE
C     CALL PLOTS
      CALL PLOT(TRANX,TRANY,-3)
      CALL TITLE
      CALL PLOT(-.5,-.5,-3)
      CALL NEWPEN(4)
      DO 20 I=1,NPTR
      READ (1,*) IPEN,N
      IF ((IPEN.EQ.0).OR.(IPEN.EQ.-24)) CALL SMOOT(P(N,1),P(N,2),IPEN)
      IF (IPEN.EQ.-2) CALL SMOOT(P(N,1),P(N,2),IPEN)
      IF (IPEN.EQ.1) READ(1,*) RADIUS,STARTA,ENDA,CODE
      IF (IPEN.EQ.1) CALL CIRCL(P(N,1),P(N,2),STARTA,ENDA,RADIUS,
     *RADIUS,CODE)
      IF ((IPEN.EQ.2).OR.(IPEN.EQ.3)) CALL PLOT(P(N,1),P(N,2),IPEN)
 20   CONTINUE
C     CALL PLOT(0.,0.,999)
      CALL NEWPEN(1)
      CALL PLOT(.5,.5,-3)
      RETURN
      END
C
C
      SUBROUTINE VIEW4(TRANX,TRANY)
      CALL PLOT(TRANX,TRANY,-3)
      CALL TITLE
      CALL FRAME
      CALL S1(2.07,8.)
      CALL S2(5.94,7.5)
      CALL S3(2.07,3.25)
      CALL S4(5.94,3.)
C     CALL PLOT(0.,0.,999)
      RETURN
      END
C
C
      SUBROUTINE FRAME
C     CALL NEWPEN(4)
      CALL RECT(.25,1.,4.5,3.63,0.,3)
      CALL RECT(4.13,1.,4.5,3.62,0.,3)
      CALL RECT(.25,5.75,4.49,3.63,0.,3)
      CALL RECT(4.13,5.75,4.49,3.62,0.,3)
C     CALL NEWPEN(1)
      RETURN
      END
C
C
      SUBROUTINE S1(X,Y)
      CALL PLOT(X,Y,-3)
C     CALL NEWPEN(3)
      CALL CIRCL(.433,.75,30.,150.,.5,.5,0.0)
      CALL CIRCL(-1.,-1.,270.,150.,.5,.5,0.0)
      CALL CIRCL(1.,-1.,270.,390.,.5,.5,0.)
      CALL CIRCL(.3,0.,0.,360.,.3,.3,0.)
      CALL MARK(0.,0.,.3)
      CALL PLOT(-1.,-1.,3)
      CALL PLOT(1.,-1.,2)
      CALL PLOT(.433,.75,3)
      CALL PLOT(1.433,-.25,2)
```

```
      CALL PLOT(-.433,.75,3)
      CALL PLOT(-1.433,-.25,2)
      CALL PLOT(-X,-Y,-3)
C     CALL NEWPEN(1)
      RETURN
      END
C
C
      SUBROUTINE S2(X,Y)
      CALL PLOT(X,Y,-3)
      CALL NEWPEN(3)
      CALL PLOT(-1.,-.25,3)
      CALL PLOT(.433,-.25,2)
      CALL CIRCL(.433,-.25,270.,390.,.5,.5,0.)
      CALL CIRCL(0.0,1.0,90.,30.,1.,1.,0.0)
      CALL CIRCL(0.3,0.3,0.0,360.,.3,.3,0.)
      CALL MARK(0.,0.3,.3)
      CALL FIT(-1.0,-.25,-.3,1.0,0.0,1.0)
      CALL PLOT(-X,-Y,-3)
C     CALL NEWPEN(1)
      RETURN
      END
C
C
      SUBROUTINE S3(X,Y)
      CALL PLOT(X,Y,-3)
C     CALL NEWPEN(3)
      CALL CIRCL(.65,1.125,30.,150.,.75,.75,0.0)
      CALL CIRCL(-.25,-.933,230.,310.,.39,.39,0.0)
      CALL CIRCL(0.3,0.0,0.0,360.,.3,.3,0.)
      CALL PLOT(-.65,1.125,3)
      CALL PLOT(-1.5,0.0,2)
      CALL PLOT(.65,1.125,3)
      CALL PLOT(1.5,0.0,2)
      CALL PLOT(-.25,-.933,3)
      CALL PLOT(-1.5,0.0,2)
      CALL PLOT(.25,-.933,3)
      CALL PLOT(1.5,0.0,2)
      CALL MARK(0.,0.0,.3)
      CALL PLOT(-X,-Y,-3)
C     CALL NEWPEN(1)
      RETURN
      END
C
C
      SUBROUTINE S4(X,Y)
      DIMENSION DX(3),DY(3),DR(3),DSANG(3),NN(3),
     DTHETA(3)
      CALL PLOT(X,Y,-3)
C     CALL NEWPEN(3)
      CALL FACTOR(.5)
      DATA DX/-1.5,0.,1.5/
      DATA DY/0.,1.5,0./
      DATA DR/.875,1.245,.875/
      DATA DSANG/270.,225.,45./
      DATA NN/45,18,45/
      DATA DTHETA/3*5./
      DO 1 I=1,3
      CALL CIRCLE
     (DX(I),DY(I),DR(I),DSANG(I),NN(I),DTHETA(I))
 1    CONTINUE
      CALL CIRCL(-1.2,0.,0.,360.,.3,.3,0.)
      CALL CIRCL(1.8,0.,0.,360.,.3,.3,0.)
      CALL MARK(-1.5,0.,.3)
      CALL MARK(1.5,0.,.3)
      CALL PLOT(-1.5,-.875,3)
      CALL PLOT(1.5,-.875,2)
C     CALL NEWPEN(1)
      CALL FACTOR(1.)
      CALL PLOT(-X,-Y,-3)
      RETURN
      END
C
C
      SUBROUTINE MARK(X,Y,R)
      DIMENSION XRAY(4),YRAY(4)
      DIMENSION CXRAY(4),CYRAY(4)
      R=R*1.25
      XRAY(1)=X+R
      YRAY(1)=Y
      XRAY(2)=X-R
      YRAY(2)=Y
      XRAY(3)=0.0
      YRAY(3)=0.0
      XRAY(4)=1.0
      YRAY(4)=1.0
      CALL CNTRL(XRAY,YRAY,2,1)
      CXRAY(1)=X
      CYRAY(1)=Y-R
      CXRAY(2)=X
      CYRAY(2)=Y+R
      CXRAY(3)=0.0
      CYRAY(3)=0.0
      CXRAY(4)=1.0
      CYRAY(4)=1.0
      CALL CNTRL(CXRAY,CYRAY,2,1)
      R=R-.25
      RETURN
      END
C
C
```

10. Enter the following subprogram.

```
      SUBROUTINE CUPID1(TRANX,TRANY)
C     *************************************************
C     *                                               *
C     *       VIEW3 -- PROGRAM USING XARC,YARC,ZARC,AND  *
C     *       FORVEW TO PLOT THREE VIEWS AND ISOMETRIC  *
C     *                                               *
C     *************************************************
C
C
      DIMENSION X1(10),Y1(10),Z1(10),RAD(10),SNG(10),ENG(10)
      DIMENSION X(50),Y(50),Z(50),IPEN(50)
      DATA X1/4*2.0,4*4.0/
      DATA Y1/2*2.0,2*3.2,2*2.0,2*1.0/
      DATA Z1/1.0,4.0,1.0,4.0,2.0,3.0,2.0,3.0/
      DATA RAD/2*1.0,2*.8,4*1.0/
      DATA SNG/2*180.,2*0.,270.,0.,270.,0./
      DATA ENG/2*270.,2*90.,360.,90.,360.,90./
C     *************************************************
      READ(1,*) NPTS
      DO 1 I=1,NPTS
      READ(1,*) X(I),Y(I),Z(I),IPEN(I)
C     WRITE(3,9) X(I),Y(I),Z(I),IPEN(I)
C9    FORMAT(1X,3F10.3,I3)
1     CONTINUE
C     CALL PLOTS
      CALL PLOT(TRANX,TRANY,-3)
      CALL TITLE
      CALL NEWPEN(4)
      CALL PLOT(1.5,2.,-3)
      CALL FORVEW(X,Y,Z,IPEN,NPTS,.5)
      DO 100 L=1,4
         CALL ZARC(X1(L),Y1(L),Z1(L),RAD(L),SNG(L),ENG(L))
100   CONTINUE
      DO 200 L=5,8
         CALL YARC(X1(L),Y1(L),Z1(L),RAD(L),SNG(L),ENG(L))
200   CONTINUE
C     CALL PLOT(0.,0.,999)
      CALL FACTOR(1.)
      CALL PLOT(-1.5,-2.0,-3)
      CALL NEWPEN(1)
      RETURN
      END
C
C
      SUBROUTINE YARC(XP,YP,ZP,R,SANG,EANG)
C     *************************************************
C     *                                               *
C     *       THIS SUBROUTINE PLOTS ARCS IN THE Y PLANE  *
C     *                                               *
C     *************************************************
C
C
      DIMENSION XX(400),YY(400),ZZ(400),IPEN(400)
      DATA IPEN/3,399*2/
      PI=3.14159265
      N=IFIX(EANG - SANG)
      THETA=(EANG-SANG)/N
      THETAR=THETA*PI/180.
      SRANG=SANG*PI/180.
      DO 1 I=1,N
         DX=R*COS(SRANG)
         DZ=R*SIN(SRANG)
         XX(I)=XP+DX
         YY(I)=YP
         ZZ(I)=(ZP+DZ)
         SRANG=SRANG+THETAR
1     CONTINUE
      CALL FORVEW(XX,YY,ZZ,IPEN,N,.5)
      RETURN
      END
C
      SUBROUTINE ZARC(XP,YP,ZP,R,SANG,EANG)
C
C     *************************************************
C     *                                               *
C     *       SUBROUTINE PLOTS ARCS IN THE Z PLANE  IE.  *
C     *       THE PLANE PERPINDICULAR TO THE X AXIS  *
C     *                                               *
C     *************************************************
C
C
      DIMENSION XX(400),YY(400),ZZ(400),IPEN(400)
      DATA IPEN/3,399*2/
      PI=3.14159265
      N=IFIX(EANG-SANG)
      THETA=(EANG-SANG)/N
      THETAR=THETA*PI/180.
      SRANG=SANG*PI/180.
      DO 1 I=1,N
         DX=R*COS(SRANG)
         DY=R*SIN(SRANG)
         XX(I)=XP+DX
         YY(I)=YP+DY
         ZZ(I)=ZP
         SRANG=SRANG+THETAR
1     CONTINUE
      CALL FORVEW(XX,YY,ZZ,IPEN,N,.5)
      RETURN
      END
C
C
      SUBROUTINE CUPID3(TRANX,TRANY)
      DIMENSION X(500),Y(500),Z(500)
C     PROGRAM TO TEST SURF
```

```
      ICNT=0
      DO 200 J=1,10
      DO 300 K=1,10
      ICNT=ICNT+1
      X(ICNT)=J
      Y(ICNT)=K
C     READ(10,*)Z(ICNT)
C 100 FORMAT(F10.2)
300   CONTINUE
200   CONTINUE
      IC = 0
      DO 400 I = 1,10
      READ(1,11) (Z(IC+J),J=1,10)
C     WRITE(3,11) (Z(J),J=1,10)
      IC = IC + 10
400   CONTINUE
11    FORMAT(10F4.2)
      NAX=2
C     CALL PLOTS
      CALL PLOT(TRANX,TRANY,-3)
      CALL TITLE
      CALL FACTOR(1.0)
      CALL PLOT(-6.,2.,-3)
      CALL NEWPEN(4)
      CALL SURF(X,Y,Z,ICNT,NAX)
C     CALL PLOT(0.,0.,999)
      CALL PLOT(6.0,-2.,-3)
      CALL NEWPEN(1)
      RETURN
      END
C
C
      SUBROUTINE SURF(A,B,C,N,NAX)
      DIMENSION A(500),B(500),C(500),P(20,20)
      DATA AXLN/3.0/
      AXLN=5.
      AXLN1=AXLN+0.001
      XH=0.0
      YH=0.0
      ZH=0.0
      XL=100.
      YL=100.
      ZL=100.
      NN=10
      DO 1 I=1,N
         IF(XH.LE.A(I))XH=A(I)
         IF(YH.LE.B(I))YH=B(I)
         IF(ZH.LE.C(I))ZH=C(I)
         IF(XL.GE.A(I))XL=A(I)
         IF(YL.GE.B(I))YL=B(I)
         IF(ZL.GE.C(I))ZL=C(I)
1     CONTINUE
      DO 2 I=1,N
         A(I)=(A(I)-XL)/(XH-XL)*AXLN
         B(I)=(B(I)-YL)/(YH-YL)*AXLN
         C(I)=(C(I)-ZL)/(ZH-ZL)*AXLN
2     CONTINUE
      DO 3 I=1,NN
         DO 4 J=1,NN
         P(I,J)=0.
4        CONTINUE
3     CONTINUE
      DO 5 I=1,N
         IX=A(I)/AXLN1*NN+1.
         IY=B(I)/AXLN1*NN+1.
         IF(C(I).GT.P(IX,IY))P(IX,IY)=C(I)
5     CONTINUE
      IF(NAX.NE.1) GO TO 10
      X=AXLN
      Y=-YL/(YH-YL)*AXLN
      Z=-ZL/(ZH-ZL)*AXLN
      CALL ROTATE(X,Y,Z)
      CALL PLOT(X,Y,3)
      X=-XL/(XH-XL)*AXLN
      Y=-YL/(YH-YL)*AXLN
      CALL ROTATE(X,Y,Z)
      CALL PLOT(X,Y,2)
      Z=AXLN
      X=-XL/(XH-XL)*AXLN
      Y=-YL/(YH-YL)*AXLN
      CALL ROTATE(X,Y,1)
      X=-XL/(XH-XL)*AXLN
      Z=-ZL/(ZH-ZL)*AXLN
      Y=-YL/(YH-YL)*AXLN
      CALL ROTATE(X,Y,Z)
      CALL PLOT(X,Y,3)
      Y=AXLN
      X=-XL/(XH-XL)*AXLN
      Z=-ZL/(ZH-ZL)*AXLN
      CALL ROTATE(X,Y,Z)
      CALL PLOT(X,Y,2)
10    CONTINUE
      DO 6 I=1,NN
         X=I*AXLN/NN
         SX=X
         IP=3
         DO 7 J=1,NN
            X=SX
            Y=J*AXLN/NN
            Z=P(I,J)
            CALL ROTATE(X,Y,Z)
            CALL PLOT(X,Y,IP)
            IP=2
7        CONTINUE
6     CONTINUE
      DO 8 I=1,NN
         Y=I*AXLN/NN
         SY=Y
         IP=3
```

10.   Continued

```
      DO 9 J=1,NN
        Y=SY
        X=J*AXLN/NN
        Z=P(J,I)
        CALL ROTATE(X,Y,Z)
        CALL PLOT(X,Y,IP)
        IP=2
 9      CONTINUE
 8    CONTINUE
    RETURN
    END
C
C
    SUBROUTINE ROTATE(X,Y,Z)
      DATA C1/0.707/,C2/0.808/,S1/0.707/,S2/-0.587/
      SY=Y
      Y=Z*C2+X*C1*S2+Y*S1*S2+5.
      X=SY*C1-X*S1+10.
      RETURN
      END
C
C
    SUBROUTINE TITLE
C   CALL PLOTS
    CALL NEWPEN(5)
    CALL RECT(0.,0.,10.5,8.,0.,3)
    CALL PLOT(0.,.75,3)
    CALL PLOT(8.,.75,2)
    CALL PLOT(6.25,0.,3)
    CALL PLOT(6.25,.75,2)
    CALL PLOT(6.25,.375,3)
    CALL PLOT(8.,.375,2)
    CALL NEWPEN(2)
    CALL SYMBOL(.25,.25,.25,'COMPUTER-AIDED GRAPHICS',0.,23)
    CALL SYMBOL(6.25,.5244,.07608,' DR BY MATTHEW JOHNSON ',0.,23)
    CALL SYMBOL(6.75,.125,.125,'EG 310',0.,6)
    CALL NEWPEN(1)
C   CALL PLOT(0.,0.,999)
    RETURN
    END
C
C
    SUBROUTINE CIRCLE(X,Y,R,SANG,N,THETA)
      X=X-R
      SANG=(3.14/180.)*SANG
      XX=R*(1.-COS(SANG))
      YY=R*(SIN(SANG))
      DX=X+XX
      EY=Y+YY
C   WRITE(3,1) DX,EY
    CALL PLOT(DX,EY,3)
C1  FORMAT(1H ,2F10.5)
      THETA=(3.14/180.)*THETA
      THETA1=THETA
C   WRITE(3,7)
C7  FORMAT('0',1H ,25HPOINTS PRODUCED BY CIRCLE)
      DO 2 I=1,N
      FEE=SANG+THETA
      PX=R*(1.-COS(FEE))
      SY=R*(SIN(FEE))
      DX=X+PX
      EY=Y+SY
C   PRINT POINTS PRODUCED BY CIRCLE
C   WRITE(3,3) PX,SY,DX,EY
C3  FORMAT(1H ,4F10.5)
      CALL PLOT(DX,EY,2)
 2    THETA=THETA+THETA1
    RETURN
    END
```

11.   Enter the following subprogram.

```
C
    SUBROUTINE CIRCUT(TRANX,TRANY)
    DIMENSION P(100,2)
    DIMENSION X(10),Y(10),CHT(10),ROT(10),ILEN(10),MESAG(10)
    DATA Y/8.0,7.8,7.6,5.3,7.725,6.825,5.925,6.825,6.025,6.125/
    DATA X/1.0,1.0,1.0,1.0,2.2,2.5,2.2,4.1,5.4,6.5/
    DATA CHT/10*.15/
    DATA ROT/10*0./
    DATA ILEN/4*1,3,2,3,2,3,1/
    DATA MESAG /1HA,1HB,1HC,1HD,3HAND,2HOR,3HAND,2HOR,3HAND,1HF/
    CALL PLOT(TRANX,TRANY,-3)
    CALL TITLE
    CALL NEWPEN(4)
    READ(1,*) NPTS,NPTR
    DO 100 I=1,NPTS
100 READ(1,*) N,P(I,1),P(I,2)
    DO 200 I=1,NPTR
    READ(1,*) IPEN,N1
    IF ((IPEN.EQ.0).OR.(IPEN.EQ.-24)) CALL SMOOT(P(N1,1),P(N1,2),IPEN)
    IF (IPEN.EQ.-2) CALL SMOOT(P(N1,1),P(N1,2),IPEN)
    IF (IPEN.EQ.1) READ(1,*) RADIUS,STARTA,ENDA,CODE
    IF (IPEN.EQ.1) CALL CIRCL(P(N1,1),P(N1,2),STARTA,ENDA,RADIUS,
   *RADIUS,CODE)
    IF ((IPEN.EQ.2).OR.(IPEN.EQ.3)) CALL PLOT(P(N1,1),P(N1,2),IPEN)
200 CONTINUE
    CALL NEWPEN(2)
    DO 300 I=1,10
    CALL SYMBOL(X(I),Y(I),CHT(I),MESAG(I),ROT(I),ILEN(I))
300 CONTINUE
    CALL NEWPEN(1)
    RETURN
    END
```

12.   Enter the following subprogram.

```
C
    SUBROUTINE FEATUR(TRANX,TRANY,IFLAG)
    CALL PLOT(TRANX,TRANY,-3)
    CALL TITLE
    CALL PLOT(0.,-1.,-3)
    CALL BLOCK
    IF (IFLAG.EQ.1) CALL NOCUT
    IF (IFLAG.EQ.0) CALL CUT
    CALL PLOT(0.,1.0,-3)
    RETURN
    END
C
C
    SUBROUTINE BLOCK
    DIMENSION X(50),Y(50),IPEN(50),SX1(50),SY1(50),
   SX2(50),SY2(50)
C   CALL NEWPEN(4)
    CALL FACTOR(.6)
    READ(1,*) ICONT
    DO 1 I=1,ICONT
    READ(1,6) X(I),Y(I),IPEN(I)
 1  CALL PLOT(X(I),Y(I),IPEN(I))
C   CALL NEWPEN(2)
    CALL SYMBOL(4.5,11.,0.25,'PLAN',0.0,4)
C   CALL NEWPEN(1)
    CALL DIMEN(2.0,7.50,.75,0.0,1.)
    CALL DIMEN(2.75,10.50,1.75,0.0,1.)
    CALL DIMEN(4.5,7.50,1.563,0.0,1.)
    CALL DIMEN(6.063,8.75,1.125,0.0,1.)
    CALL DIMEN(7.188,7.50,.813,0.0,1.)
    CALL DIMEN(9.5,7.50,2.625,0.0,1.)
    CALL DIMEN(2.0,11.50,3.25,0.0,1.)
    CALL DIMEN(5.25,11.50,2.75,0.0,1.)
    CALL DIMEN(1.75,9.875,.75,270.0,1.)
    CALL DIMEN(9.25,10.25,2.625,270.0,1.)
    CALL DIMEN(8.25,7.75,.75,90.0,1.)
    CALL DIMEN(3.5,13.625,1.875,270.0,1.)
    CALL NEWPEN(4)
    CALL RECT(9.5,7.75,.75,2.625,0.,3)
    CALL RECT(9.5,7.75,2.625,2.625,0.,3)
    CALL CIRCL(7.1875,13.0625,0.,360.,.5625,.5625,0.)
    CALL CNTL(6.6250,13.0625,0.65)
    CALL CIRCL(10.625,9.5,0.,360.,.375,.375,0.)
    CALL CNTL(10.250,9.5,0.43)
    DIMENSION A(50),B(50),C(50),E(50),F(50)
    DATA E/26*.06/
    DATA F/8*0.,5*90./
    READ(1,*) ICONT
    DO 2 I=1,ICONT
    READ(1,7) A(I),B(I),C(I)
 2  CALL DASH(A(I),B(I),C(I),E(I),F(I))
C   CALL NEWPEN(1)
C 2 WRITE(10,6) A(I),B(I),C(I),E(I),F(I)
 6  FORMAT(2(F7.4,3X),1I1)
 7  FORMAT(3(F7.4,3X))
    CALL FACTOR(1.)
    RETURN
    END
C
C
    SUBROUTINE NOCUT
C   CALL NEWPEN(2)
    CALL FACTOR(.6)
    CALL SYMBOL(3.0,7.,0.25,'FRONT ELEVATION',0.0,15)
    CALL FACTOR(1.)
C   CALL NEWPEN(1)
    RETURN
    END
C
C
    SUBROUTINE CUT
    DIMENSION SX1(25),SY1(25),SX2(25),SY2(25)
C   CALL NEWPEN(2)
    CALL FACTOR(.6)
    CALL KNIFE
    CALL SYMBOL(4.5,7.0,0.25,'SECTION A-A',0.0,11)
    CALL SYMBOL(3.0,15.5,0.5,' SECTIONED CUT ',0.0,16)
    DATA SX1/4.5,5.25,5.25,6.0625,6.0625,0.0,1.0/
    DATA SY1/9.125,9.125,8.5,8.5,7.75,0.0,1.0/
    DATA SX2/4.5,4.5,6.0625,0.0,1.0/
    DATA SY2/9.125,7.75,7.75,0.0,1.0/
    CALL NEWPEN(4)
    CALL SHADE(SX1,SY1,SX2,SY2,.1,150.,.5,1,3,1)
    CALL BAR(7.1875,7.75,0.0,0.75,0.813,0.750,3,8)
    CALL BAR(4.5,9.875,0.0,.375,.75,.375,3,8)
    CALL BAR(2.0,7.75,0.0,1.375,.750,1.375,3,8)
    CALL BAR(2.0,9.875,0.0,.375,.75,.375,3,8)
    CALL PLOT(4.5,9.125,3)
    CALL PLOT(5.25,9.125,2)
    CALL PLOT(6.0625,8.5,3)
    CALL PLOT(6.0625,7.75,2)
    CALL FACTOR(1.)
    CALL NEWPEN(1)
    RETURN
    END

    SUBROUTINE DASH(X,Y,TL,DL,THETAL)
    CALL NEWPEN(4)
    CALL PLOT(X,Y,3)
    X1=X
    Y1=Y
    THETA=THETAL*.017453
    M=(TL/DL)/3
    DO 100 I=1,M
    TCOS=DL*COS(THETA)
    TSIN=DL*SIN(THETA)
```

12. Continued

```
        X1=TCOS+X1
        Y1=TSIN+Y1
        CALL PLOT(X1,Y1,2)
        X1=TCOS+X1
        Y1=TSIN+Y1
        CALL PLOT(X1,Y1,3)
        X1=TCOS+X1
        Y1=TSIN+Y1
100     CALL PLOT(X1,Y1,2)
        X1=X
        Y1=Y
C       CALL NEWPEN(1)
        RETURN
        END
C
C
        SUBROUTINE KNIFE
C       CALL NEWPEN(5)
        CALL AROHD(1.25,12.5,1.25,13.25,.25,.125,12)
        CALL AROHD(8.75,13.0625,8.75,14.125,.25,.125,12)
        CALL PLOT(1.25,12.5,3)
        CALL PLOT(2.5,12.5,2)
        CALL PLOT(5.0,12.5,3)
        CALL PLOT(5.5,12.5,2)
        CALL PLOT(5.5,13.0625,2)
        CALL PLOT(6.0,13.0625,2)
        CALL PLOT(7.5,13.0625,3)
        CALL PLOT(8.75,13.0625,2)
C       CALL NEWPEN(1)
        RETURN
        END
C
C
        SUBROUTINE CNTL(X,Y,RAD)
        DIMENSION XX(4),YY(4)
        CALL NEWPEN(2)
        XX(1)=X-RAD
        XX(2)=X+RAD
        XX(3)=0.0
        XX(4)=1.0
        YY(1)=Y
        YY(2)=Y
        YY(3)=0.0
        YY(4)=1.0
        CALL CNTRL(XX,YY,2,1)
        XX(1)=X
        XX(2)=X
        YY(1)=Y-RAD
        YY(2)=Y+RAD
        CALL CNTRL(XX,YY,2,1)
        CALL NEWPEN(1)
        RETURN
        END
```

13. Enter the following data for the program shown in problem 8.

```
/*
//G.PLOTPARM DD *
 &PLOT MODEL = 8236 &END
//G.SYSIN DD *
16 24
 1 4.0 6.0
 2 4.0 5.0
 3 5.0 5.0
 4 5.0 6.0
 5 5.0 6.5
 6 3.5 6.0
 7 4.0 4.5
 8 5.5 5.0
 9 5.5 7.0
10 3.0 6.5
11 3.5 4.0
12 6.0 4.5
13 6.0 6.5
14 3.5 7.0
15 3.0 4.5
16 5.5 4.0
 3  1
 2  2
 2  3
 2  5
 2 10
 2 15
 2 11
 2 16
 2 12
 2 13
 2  9
 2 14
 2 10
 3  1
 2  6
 2 11
 3  2
 2  7
 2 12
 3  3
 2  8
 2  9
 3  4
 2  1
```

```
62 68
 1 2.0 8.1
 2 2.8 8.1
 3 1.2 8.0
 4 1.2 7.8
 5 1.2 7.6
 6 2.0 8.0
 7 2.0 7.8
 8 2.0 7.6
 9 2.0 7.5
10 2.8 7.5
11 3.1 7.8
12 3.4 7.8
13 3.4 7.1
14 3.808 7.1
15 3.89 6.9
16 3.808 6.7
17 3.6 7.2
18 4.4 7.2
19 4.7 6.9
20 4.4 6.6
21 3.6 6.6
22 3.808 6.7
23 3.4 6.7
24 3.4 6.0
25 3.1 6.0
26 2.8 6.3
27 2.8 5.7
28 2.0 5.7
29 2.0 6.3
30 2.0 6.1
31 2.0 6.0
32 2.0 5.9
33 2.0 5.8
34 1.8 5.8
35 1.4 5.9
36 1.6 6.0
37 1.8 6.1
38 1.6 6.8
39 1.4 7.0
40 2.0 7.2
41 2.28 7.0
42 2.28 6.8
43 2.0 6.6
44 3.1 6.9
45 5.0 5.3
46 5.0 6.0
47 5.0 6.2
48 5.2 6.2
49 5.2 6.0
50 5.2 6.4
51 5.2 5.8
52 6.0 6.4
53 5.2 5.8
54 6.3 6.1
55 1.4 8.0
56 1.6 7.8
57 1.8 7.6
58 1.3 5.3
59 1.8 5.3
60 5.0 6.9
61 2.8 7.2
62 6.5 6.1
 3  1
 2  2
 1  2
.3 90. -90. 0.
 2  9
 2  1
 3  3
 2  6
 3  4
 2  7
 3  5
 2  8
 3 55
 2 35
 2 32
 3 56
 2 36
 2 31
 3 57
 2 37
 2 30
 3 29
 2 26
 1 26
.3 90. -90. 0.
 2 28
 2 29
 3 58
 2 45
 3 59
 2 34
 2 33
 3 45
 2 46
 2 49
 2 50
 2 52
 1 52
.3 90. -90. 0.
 2 51
 2 49
 3 48
 2 47
 2 60
 2 19
 3 21
```

13. Continued

```
  1 21
  .3 270. 450. 0.
  2 18
  1 18
  .3 90.  -90. 0.
  2 21
  3 22
  2 23
  2 24
  2 25
  3 44
  2 15
  3 14
  2 13
  2 12
  2 11
  3 40
  2 61
  1 61
  .3 90.  -90. 0.
  2 43
  1 43
  .3 270. 450. 0.
  3 41
  2 39
  3 42
  2 38
  3 54
  2 62
 34 38
  1 2.24 6.5
  2 2.8 6.5
  3 3.02 6.9
  4 3.28 7.0
  5 3.5 6.97
  6 3.7 6.8
  7 3.76 6.5
  8 5.8 6.5
  9 5.8 6.7
 10 6.0 6.9
 11 6.4 6.91
 12 6.7 6.6
 13 7.05 6.89
 14 7.12 7.12
 15 6.33 7.39
 16 5.69 7.49
 17 5.1 7.5
 18 4.39 7.5
 19 3.34 7.16
 20 2.23 6.73
 21 2.33 6.6
 22 3.59 7.17
 23 3.78 7.16
 24 4.22 7.14
 25 4.61 7.17
 26 5.0 7.21
 27 5.48 7.24
 28 5.86 7.22
 29 3.9 6.61
 30 4.9 6.61
 31 3.3 6.3
 32 6.24 6.3
 33 4.41 7.15
 34 4.81 7.19
  3 20
  2 21
  2 1
  2 2
  2 3
  2 5
  2 7
  2 8
  2 9
  2 10
  2 11
  2 12
  2 13
  2 14
  2 15
  2 16
  2 17
  2 18
  2 19
  0 20
 -2 19
 -2 22
 -2 23
 -2 24
 -2 33
 -2 25
 -2 34
 -2 26
 -2 27
 -2 28
-24 14
  3 17
  2 26
  3 18
  2 22
  2 29
  2 30
  2 26
 26
  1.0 2.0 1.0 3
  1.0 4.0 1.0 2
```

```
  1.0 4.0 4.0 2
  1.0 2.0 4.0 2
  2.0 1.0 1.0 3
  4.0 1.0 1.0 2
  2.0 1.0 4.0 3
  4.0 1.0 4.0 2
  2.8 2.0 1.0 3
  4.0 2.0 1.0 2
  2.8 2.0 4.0 3
  4.0 2.0 4.0 2
  2.8 2.0 1.0 3
  2.8 3.2 1.0 2
  2.8 2.0 4.0 3
  2.8 3.2 4.0 2
  1.0 4.0 1.0 3
  2.0 4.0 1.0 2
  1.0 4.0 4.0 3
  2.0 4.0 4.0 2
  5.0 1.0 2.0 3
  5.0 1.0 3.0 2
  5.0 2.0 2.0 3
  5.0 2.0 3.0 2
  2.8 2.0 1.0 3
  2.8 2.0 4.0 2
 24
  2.7500      7.7500      3
  2.7500     10.2500      2
  2.0000     10.2500      2
  2.0000      7.7500      2
  8.0000      7.7500      2
  8.0000      8.5000      2
  5.2500      8.5000      2
  5.2500     10.2500      2
  4.5000     10.2500      2
  4.5000      7.7500      2
  2.7500      7.7500      2
  5.2500     14.3750      3
  5.2500     11.7500      2
  8.0000     11.7500      2
  8.0000     14.3750      2
  2.0000     14.3750      2
  2.0000     11.7500      2
  2.7500     11.7500      2
  2.7500     13.6250      2
  4.5000     13.6250      2
  4.5000     11.7500      2
  5.2500     11.7500      2
  2.7500     10.2500      3
  4.5000     10.2500      2
 13
  2.0000      9.1250      0.7500
  4.5000      9.1250      0.7500
  2.0000      9.8750      0.7500
  4.5000      9.8750      0.7500
  2.0000     12.1250      0.7500
  4.5000     12.1250      0.7500
  2.0000     12.8750      0.7500
  4.5000     12.8750      0.7500
 10.1875      7.7500      0.7500
 11.3125      7.7500      0.7500
 11.3750      7.7500      2.6250
  7.1875      7.7500      0.7500
  6.0625      7.7500      0.7500
 24
  2.7500      7.7500      3
  2.7500     10.2500      2
  2.0000     10.2500      2
  2.0000      7.7500      2
  8.0000      7.7500      2
  8.0000      8.5000      2
  5.2500      8.5000      2
  5.2500     10.2500      2
  4.5000     10.2500      2
  4.5000      7.7500      2
  2.7500      7.7500      2
  5.2500     14.3750      3
  5.2500     11.7500      2
  8.0000     11.7500      2
  8.0000     14.3750      2
  2.0000     14.3750      2
  2.0000     11.7500      2
  2.7500     11.7500      2
  2.7500     13.6250      2
  4.5000     13.6250      2
  4.5000     11.7500      2
  5.2500     11.7500      2
  2.7500     10.2500      3
  4.5000     10.2500      2
 13
  2.0000      9.1250      0.7500
  4.5000      9.1250      0.7500
  2.0000      9.8750      0.7500
  4.5000      9.8750      0.7500
  2.0000     12.1250      0.7500
  4.5000     12.1250      0.7500
  2.0000     12.8750      0.7500
  4.5000     12.8750      0.7500
 10.1875      7.7500      0.7500
 11.3125      7.7500      0.7500
 11.3750      7.7500      2.6250
  7.1875      7.7500      0.7500
  6.0625      7.7500      0.7500
 0.0 0.0 0.0 0.0 0.0 0.0 0.0 0.0 0.0 0.0 0.0
 0.0 0.2 0.4 0.6 0.8 0.8 0.6 0.4 0.2 0.0
 0.0 0.4 0.6 0.8 1.0 1.0 0.8 0.6 0.4 0.0
 0.0 0.6 0.8 1.0 1.2 1.2 1.0 0.8 0.6 0.0
 0.0 0.4 0.6 0.8 1.0 1.0 0.8 0.6 0.4 0.0
 0.0 0.2 0.4 0.6 0.8 0.8 0.6 0.4 0.2 0.0
 0.0 0.0 0.0 0.0 0.0 0.0 0.0 0.0 0.0 0.0
```

13. Continued

```
0.0 0.0 0.0 0.0 0.0 0.0 0.0 0.0 0.0 0.0
0.0 0.0 0.0 0.0 0.0 0.0 0.0 0.0 0.0 0.0
0.0 0.0 0.0 0.0 0.0 0.0 0.0 0.0 0.0 0.0
/*
//STEP2 EXEC VTECP,DESC=VER36
//
```

14. Store problems 8 through 13 as one large data set and submit the job for processing.

15. Route the computed job to the plotter.

6.14 CHAPTER SUMMARY

Multiview projection is a technique to display three-dimensional objects in two-dimensional space, namely the screen of the DVST or the plotter as shown in Figure 6.1. This type of representation is sometimes called orthographic projection as shown in Figure 6.15. Engineers, designers, and detailers all use the multiview approach to describe the views of an object that will become part of an automated, computer generated working drawing as shown in Chapter 2. These drawings are produced by draftsperson who describe the objects as data. A data base is used inside a computer program to produce the final drawing.

Projection theory was introduced from a childs block (see Figure 6.2). From this the concept of viewports was introduced as shown in Figures 6.3, 6.4 and 6.5. Next, view construction was described using the equipment introduced in Chapters 4 and 5. This equipment is common throughout the text and different items are featured in each of the chapters.

View manipulation depends upon the type of equipment present for the draftspersons use. Figures 6.10 through 6.14 are the recommended types. Once the equipment is present, points for the object may be specified as shown in Figure 6.16. A detailed example was then given through Figures 6.17, 6.18 and 6.19. To avoid repeating points, an ICON command was introduced. This allowed the draftsperson to connect the points as desired.

Line specification is also important because some lines will need to be dashed, which represent a hidden feature. The line generator from Chapter 5 was reviewed for this purpose. Many of the 2-D techniques shown in Chapter 5 can be adopted to multiview (3-D image processing), the arc is an example. Figure 6.27 shows an arc generator which displays data in 3-D data. The best way to obtain this type of data is through a subprogram as shown in Figure 6.28. Of course the draftsperson must know how to use these types of subprograms - so the concept of a users manual was introduced in Figure 6.40 and 6.41. The use of a display program and subprograms produces the visualization of the views. Figures 6.29 through 6.39 show the concepts of visualization of 3-D objects through the use of viewports. The object is oriented inside these viewports and displayed as shown in Figure 6.37. If changes are needed then the object is reoriented through an automatic view generator.

Once the object is displayed on the face of the DVST it is still subject to modifications before routing to a plotter. One modification is the deletion of object lines that should appear as hidden, another is the addition of fillets and rounds as shown in Figure 6.40 and 6.41. The output to the plotter will then appear as shown in Figure 6.42.

The number of views, spacing of views, and the collecting of data for the object is left to the draftsperson. How to do this was then explained in sections 6.10 through 6.12. These concepts are better illustrated by the review problems shown next. No previous computer experience is needed to work problems 1 through 7. Each of these can be worked with pencil and paper, by drawing in the display axis and measuring along it to determine the data for each of the points making up the object. Once the data base is found, the ICON array which contains the order of the point connections can be written and given to a typist. The typist can then enter these items to the program CUPID which automatically displays the multiviews.

7

DIMENSIONING PRACTICES

An automated drawing is expected to show exact information regarding every detail of of the object represented in Chapter 6. A definite specification pattern expressed by dimensions will be introduced in this chapter. This will indicate clearly any engineering intent for fabrication, construction or production of the object. An object that is correctly dimensioned will permit ease of production and interchangeability of its various subparts and pieces.

7.1 ANSI STANDARDS AND CORE PRACTICES

According to ANSI X3/H3 and the CORE extensions defined in section 1.3 of this text, dimensioning may be done at either the 2-D level or the 3-D level. Both methods will be used inside this chapter. A 2-D system is useful for adding a single size notation, for example, but would be worthless to automatically size a multiview display. So each has its own application areas. We will begin with 2-D, simple examples and work our way through the more advanced usages.

The following are the 2-D basic guidelines which should be observed in image processing:

1. Show only enough dimensions so that the intent can be determined without calculating or assuming any distances. (See Figure 7.1).

2. Display each dimension clearly, so that it can be interpreted in only one way by the viewer. (See Figure 7.2)

3. Display the dimension (sizes) between points, lines or surfaces which have a necessary and specific relation to each other or which control the location of other components or mating parts.(See Fig. 7.3)

4. Always display the dimensions to avoid accumulations of tolerances which may permit various sizes of mating parts. (See Figure 7.4)

5. Display each basic size; X,Y,Z only once. (See Figure 7.5).

6. Always dimension a feature of an object where it appears in profile. (See Figure 7.6)

7. When subparts are used, always select modular (standard) pieces.

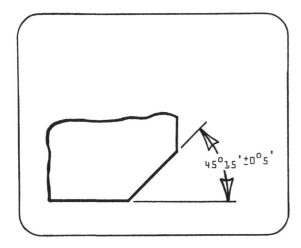

Figure 7.1   ANSI guideline 1.

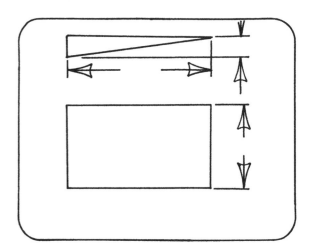

Figure 7.4   ANSI guideline 4.

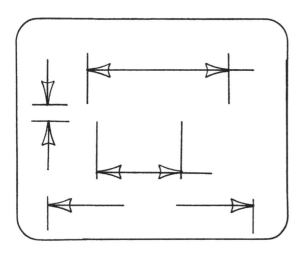

Figure 7.2   ANSI guideline 2.

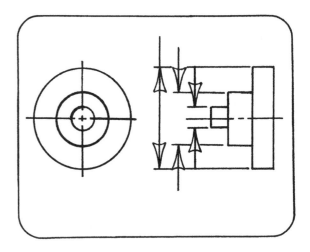

Figure 7.5   ANSI guideline 5.

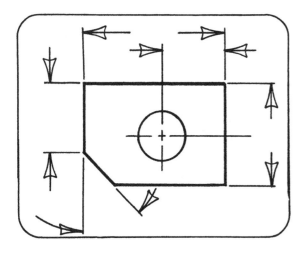

Figure 7.3   ANSI guideline 3.

Figure 7.6   ANSI guideline 6.

Six simple guidelines. Sounds easy
doesn't it? While they sound rather simple
they are difficult to practice. The design
engineer must work with a large number of
different objects, two are never exactly
the same. The treatment of even similar
objects can be different depending upon the
manufacturing and production methods. The
drafter tries to describe the multiview object so the guidelines are followed in the
spirit, if not the letter of the guideline.

For example, Figure 7.1 assumes that
degrees, minutes and seconds are required.
And more than just angles apply underguide-
line 1 of the ANSI X3/H3 standards, all the
Figures shown on page 122 apply under guideline 1. Likewise, all the figures apply
under guideline 2. No one ever attempts to
confuse the drawing reader, but confusion
results many times.

Guideline 3 applies to all figures on
page 122, as do guidelines 4, 5 and 6. So
we might say, learning dimensioning skills
is a little like learning English; there
are rules, but the rules are broken.

7.2 HOW TO DISPLAY DIMENSIONS

Nearly ever text written for drafting
assumes that a reader knows what a dimension
looks like. Dimensions take a variety of
forms, if we are talking about a linear description, then most people have seen this.
For those who have not, look at Figure 7.7.
Here we see a computer display of the command issued and the result. Notice that a
users manual, left-facing page was used to
produce this display. The draftsperson
located the section of the manual about how
to specify linear dimensions and found the
call to:

CALL DIMEN(X,Y,DIME,THETA,SCALE)

The sheet of instructions which explain a
subroutine like DIMEN are shown in Figure
7.8. Using both Figures 7.7 and 7.8, we
can begin to understand how this works. A
call to a storage location contains all of
the instructions necessary to display the

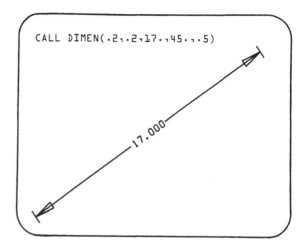

Figure 7.7 Display of DIMEN routine.

linear dimension shown in Figure 7.7. Now
a check of the users manual will tell us how
the information was placed inside the routine.
First the X and Y screen or plotter locations
were given as:

X = .2 inches from origin
Y = .2 inches from origin

This location is the lower left hand corner
of the dimension line shown in Figure 7.7.
The length of the line and its rotation on
the screen are given next as:

DIME = 17. inches in length
THETA = 45. degrees off the X axis

The angle is clearly shown in the figure,
but its length looks short. That is because
a scale of .5 or half size was requested for:

SCALE = .5 or half size

Therefore, the dimension was displayed at
8.5 inches in length on the screen. The
arrowheads are placed automatically, the
dimension line is divided automatically, the
value of the dimension is inserted automatically and two short extension lines are
then provided. These extension lines are
never long enough and draw commands are used
to extend them to their proper length.

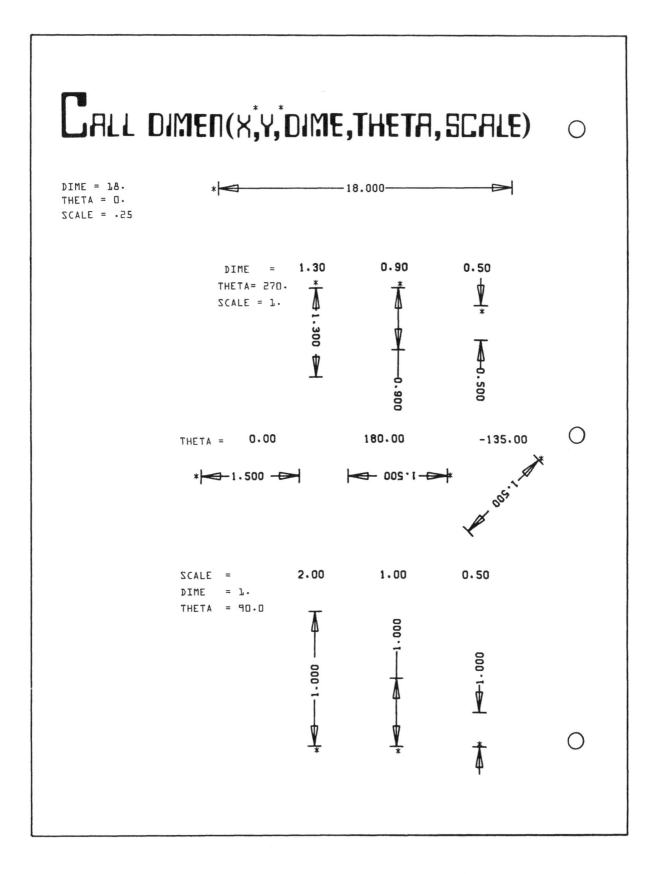

Figure 7.8 Typical left-facing page from a users manual. (courtesy DLR Associates)

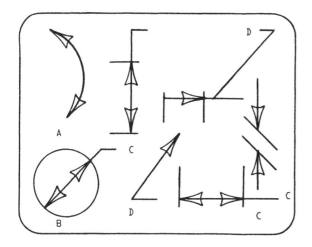

Figure 7.9 Non-linear type dimension lines.

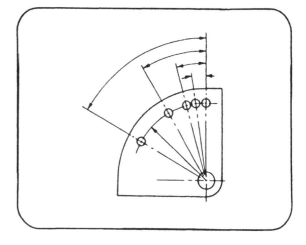

Figure 7.10 A type non-linear dimensions.

7.3 DISPLAY UNITS

Dimensions on an automated drawing are
expressed in display units. If the display
is a digital plotter, the common display
unit is an inch. If a DVST is used the unit
may be either English or Metric. If dimensions are to be shown in SI, a general note
must be included stating this fact. When
both inch and millimeter dimensions are
used, the inch symbol should be included on
figures representing inches and the abbre-
viation MM on figures representing millimeters. For linear type displays of dimensions this is quite clear, however, not all
dimension lines are shown in a linear fashion as shown in Figure 7.9. In this figure,
beginning with:

A. Angles should be dimensioned by an
arc drawn with the vertex of the angle as a
center and the angular dimension inserted
in a break in the arc as shown in Figure 7.10

B. Angular and linear distances are
shown on a drawing by dimension lines as
shown in Figure 7.11.

C. Dimension lines are thin (NEWPEN=
1 or 2), solid lines terminated with arrowheads. A break is left in the line for all
sizes over 1.2 inches, under 1.2 the line
is left closed and the size figure is placed
outside the arrowhead placement. See 7.12.

D. Single arrowheads may point to a
feature in order to size, describe, or loc-

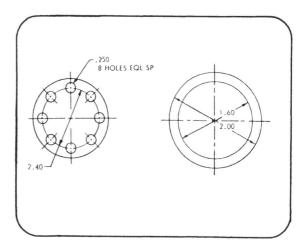

Figure 7.11 B type non-linear dimensions.

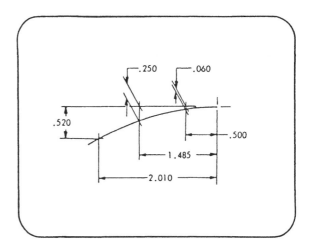

Figure 7.12 C type non-linear dimensions.

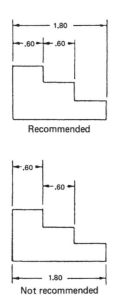

Figure 7.13 Placement of linear dimensions for simple objects.

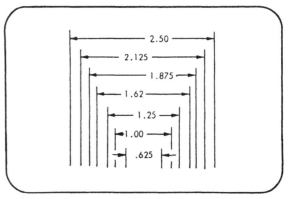

Figure 7.14 Location of linear dimensions.

ate an important part feature.

7.4 DIMENSIONAL CAPABILITIES

Automated dimensioning techniques are aligned whenever it is possible and grouped uniformly as shown in Figure 7.13. This is always possible with simple objects like Figure 7.13, it is more difficult to do when the object becomes more detailed as shown in Figure 7.14.

All parallel dimension lines should be spaced about 3/8 of a display unit apart and should not be closer than 1/2 of a display unit to the outline of the object. If several parallel dimensions lines are necessary, the numerals should be staggered as shown in Figure 7.15. Center lines, hidden line, or object lines should never be used as dimension lines. These are all used as reference lines as shown in Figure 7.16. Every effort should be taken to avoid crossing dimension lines.

In addition to the proper use of the dimension line, the following dimensional capabilities should be understood:

1. Extension lines
2. Arrowheads
3. Numeric annotation
4. Dimensional expressions

Figure 7.15   Parallel placement of dimensions.

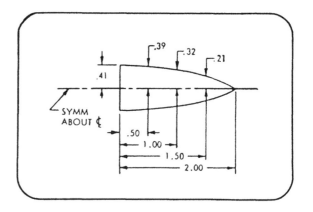

Figure 7.16   Ref. dimensions about centerline.

7.5 EXTENSION LINES

The pointers for a dimension line are called **extension lines**. Examples of these are shown in figures used thus far in this chapter. A detailed explanation is shown in Figures 7.20 and 22. First, in Figure 7.20 you will notice that a line weight is assigned by the NEWPEN command. This is set at 1 so these pointers will not over-power the display of the dimension line. In Figure 7.22 extension lines are shown. They do not touch the object shown in a larger newpen. They start approximately 1/16 display unit form the object. They should always extend about 1/8 display unit beyond the outermost arrowhead of the dim-ension line as shown in Figure 7.21. You will notice that dimension lines are dis-played as indicated in Figure 7.21. A short pointer is included, but this is not an extension line, it is the location for the extension line.

Extension lines should be displayed so they do not cross one another or cross dim-ension lines. While this is desired, it can not always be displayed as shown in Figure 7.17. In this example the extension lines are going to cross, no matter how they are placed in the output. Crossing lines can be kept to a minimum if the shortest dimension lines are displayed nearest the outline of the object. Therefore, parallel dimensions follow in order of their length as shown in Figure 7.18.

When it is impossible to avoid cross-ing other extension lines, dimension lines, or object lines, the extension lines should be displayed solid, not broken, as shown in Figure 7.19. The exception is when an ex-tension line crosses a dimension line close to arrowheads, as shown in Figure 7.19, in this case the extension line may be broken.

If a reference point is located just by extension lines as shown in Figure 7.24, the extension lines should pass through the reference point making a small cross. Variations like the crossing extension lines may occasionally be necessary depending on the type of object to be dimensioned.

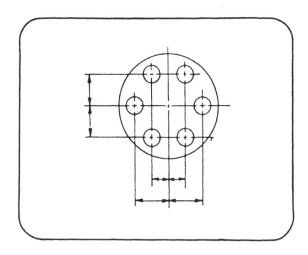

Figure 7.17   Placement of extension lines.

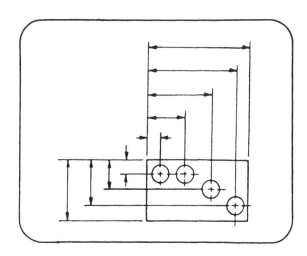

Figure 7.18   Parallel placement is correct.

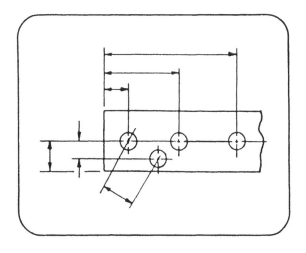

Figure 7.19   Breaking and crossing lines.

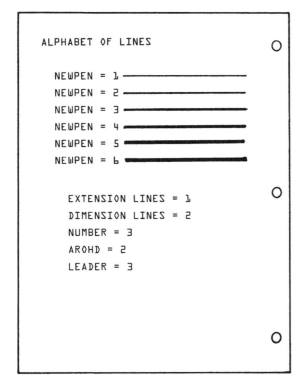

Figure 7.20 Alphabet of lines (manual page)

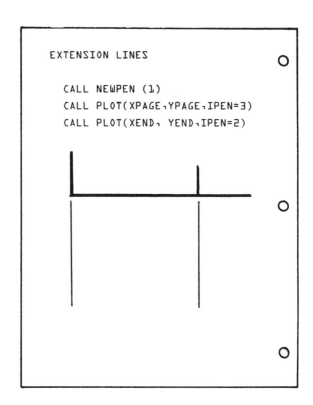

Figure 7.22 Extension lines (users manual)

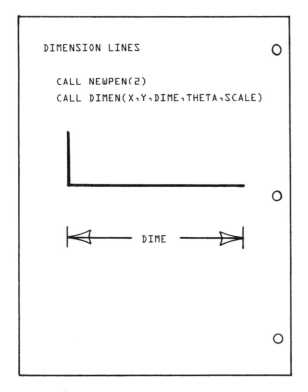

Figure 7.21 Dimension lines (users manual)

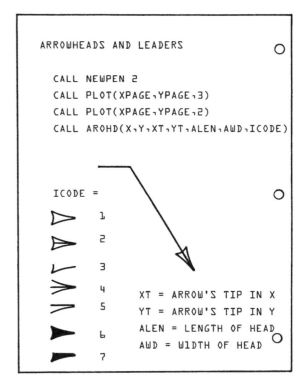

Figure 7.23 Arrowheads from a users manual.

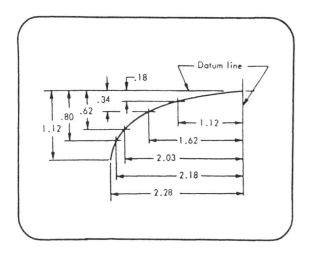

Figure 7.24 Points by extension lines.

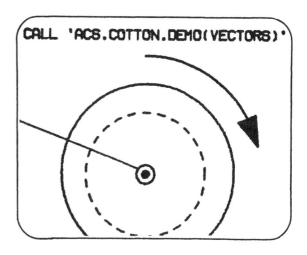

Figure 7.25 Larger arrowheads for meaning.

7.6 ARROW HEADS

The arrowheads which are part of the

CALL DIMEN(X,Y,DIME,THETA, SCALE)

routine shown in Figure 7.8 use the ICODE
of 2 as indicated in Figure 7.23. Most
drafters prefer the solid type arrowhead
(ICODE = 6) instead of the ICODE = 2. In
these cases, the programmer may change the
routine DIMEN. It is changed by locating

CALL AROHD(X,Y,XT,YT,ALEN,AWD, 12)

inside the DIMEN routine and changing it to

CALL AROHD(X,Y,XT,YT,ALEN,AWD,16).

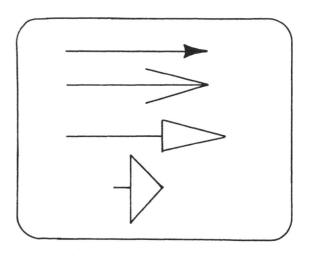

Figure 7.26 Changing ALEN and AWD.

Now two types of arrowheads may be displayed from DIMEN with the addition of five additional types if needed. Therefore, many
types of arrowheads are shown in the figure
illustrations shown in this chapter. The
type may be changed within a drawing if it
is felt that the change will add meaning.

The length of the arrowhead should be
approximately 15 display units on small
drawings and up to twenty   display units
on large drawings. Both are demonstrated
in Figure 7.24 and 7.25 (larger). The width
of the arrowhead (AWD) may be changed independent of the length (ALEN). Again most

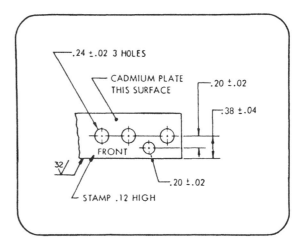

Figure 7.27 Correct use of a leader.

drafters like to keep the width to length in proportion. Figure 7.26, however, indicates that any combination of length to width may be used with any type of arrowhead.

If the drafter does want to change from the normal, then ALEN is set at .125 and AWD is set at 0. This does not mean that the arrowhead has zero thickness, it means that a standard proportion is presented on the DVST.

7.7 NUMERIC ANNOTATION

Two systems are used for numeric annotation inside dimension lines - the system used in this chapter called aligned and another called unidirectional. The unidirectional system is often preferred for certain special applications such as the automotive industry. It is not considered to be a general application use, however. And for this reason, the aligned system was chosen for this chapter. Review Figure 7.8 for a detailed explanation of the aligned system.

7.8 DIMENSIONAL EXPRESSIONS

The total terminology, except for the area of limits and tolerances (covered in a separate chapter), consists of the following terms and their meanings.

1. SIZE. This is an indication of magnitude. When a value is assigned to a dimension line it is referred to as the 'size'. Sizes are given as nominal (reference only), basic (engineering intent), actual (sized produced by manufacturing), and design (sizes which will be used to determine limits and tolerances later).

2. OVERALL DIMENSIONS. These dimensions appear outside the string dimensions which indicate the size of an object.

3. LOCATION. Dimensions should be placed outside the outline of the object and between the views whenever possible.

4. DATUMS. A datum is a point, line or plane which establishes the location of certain sizes. See Figure 7.16.

5. HIDDEN LINES. A line which indicates a surface which is behind another. Dimensions should not orinate or terminate at hidden lines.

6. OUT OF SCALE. Dimensions which are not to scale should be avoided as much as possible. However, when the original display is changed or a dimension cannot be made to scale, that dimension should be underlined and a note: NOT TO SCALE should be indicated with a leader.

7. LEADER. A leader is a fine oblique line used to direct attention to a note or give the sizes of part features. See Figure 7.27.

7.9 DIMENSIONING PART FEATURES

In automated drafting, objects for multiview projection are made up of parts. The ANSI standards include over 60 different dimensioning techniques for part features. Figure 7.28 represents a sample or a compression of those shown in the ANSI standard. The reader should refer to the current CORE or ANSI standards for more information. Any conventional drafting book will also include detailed information on dimensioning part features.

For our purposes we will study Figure 7.28 and its many examples. They are:

1. External diameters (parts A and B)
2. Holes (parts C thru F)
3. Hole placement (parts G thru M)
4. Slots (part P)
5. Hole location (parts N,O,Q, & R)

7.10 REVIEW PROBLEMS

Practice in the use of CALL DIMEN is required to complete the chapter on dimensioning. Work those problems assigned by your instructor.

1. Use a 2-D image processor and a CALL DIMEN to display Figure 7.1 on your DVST.

2. Practice the CALL DIMEN inside a display program using Figure 7.2 as a guide.

3. Use the 2-D image processor and

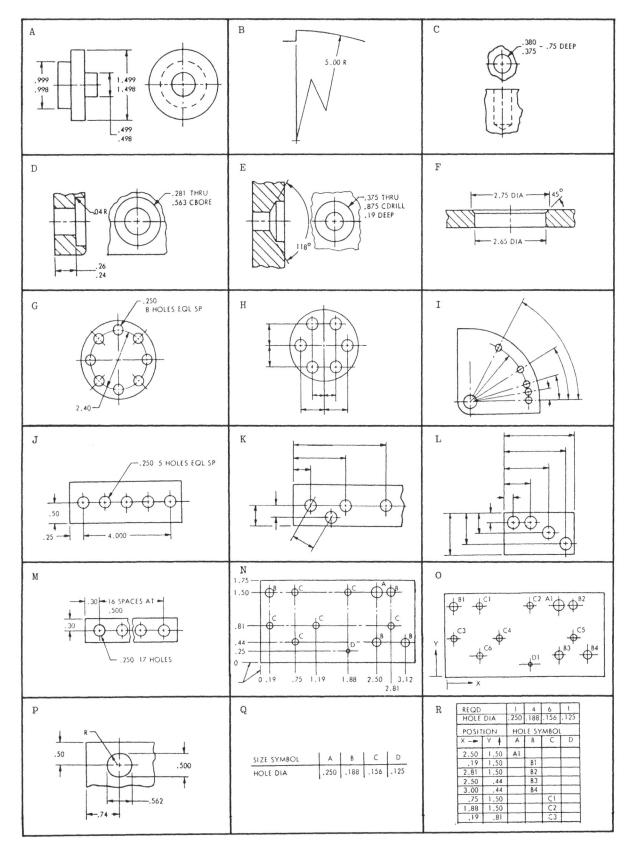

Figure 7.28   Summary of ANSI part features.

Figure 7.3 with the correct dimensions displayed inside the dimension lines.

4.   Write a short display program which
will repeat Figure 7.5 in detail.

5.   Use the multiview generator from
Chapter 6 and CALL DIMEN from this chapter
to display Figure 7.6 on your DVST.

6.   Practice dimensioning part features
as shown in Figure 7.10, hole placement.

7.   Practice dimensioning part features
as shown in Figure 7.11, external diameters.

8.   Write a short display procedure for
the Figure 7.12.

9.   Use the example from Chapter 6,
shown in Figure 7.14, rotate it, and dimension it properly.

10.   Practice the use of stacked dimensions to reproduce Figure 7.15 at your
workstation.

11.   Practice the use of extension line
work by reproducing Figure 7.16 at your
DVST terminal.

12.   Route Figures 7.17 through 19 to
a pen plotter.

13.   List the subroutine DIMEN at your
workstation. Make the necessary changes
so that different type arrowheads can be
displayed as shown in Figure 7.26.

14.   Using problem 13 as a guide, reproduce Figure 7.25 at your workstation.

15.   List the subroutine AROHD at your
workstation and identify the parts shown
in Figure 7.21 through 23. Change it for 16.

16.   Practice the use of CALL LEADER as
shown in Figure 7.27

17.   Use the 2-D image processor and
produce the part shown in Figure 7.28A.

18.   Use the 2-D image processor and
produce the part shown in Figure 7.28B.

19.   Use the 2-D image processor and
produce the part shown in Figure 7.28C.

20.   Use the 3-D image processor and
produce the part shown in Figure 7.28D.

21.   Use the 3-D image processor and
produce the part shown in Figure 7.28E.

22.   Use the 2-D image processor and
produce the part shown in Figure 7.28F.

23.   Use the 2-D image processor and
reproduce the parts shown in G.H, and I.

24.   Use the 2-D image processor and
reproduce the parts shown in J,K,L, and M.

25.   Develop a display program which
uses either N and Q or O and R in its final
presentation.

7.11   CHAPTER SUMMARY

According to ANSI X3/H3 and the CORE
extensions defined in section 1.3 of this
book, dimensioning may be done at either
the 2-D or the 3-D level. As several of the
review problems pointed out, this 2-D or 3-D
depends upon the image processor chosen to
display the object to be dimensioned. It
does not change the routine for placing that
dimension after the object has been displayed.

Seven basic rules for doing this were
introduced on page 121, examples of these
were shown on page 122. While dimensioning
sounds simple, it took the next 12 pages to
fully explain there usage. The rest of the
chapter presented how to display simple sets
of dimensions, what a display unit was, and
the various capabilities of the subroutine
DIMEN.

Other routines are necessary in order
to size, locate and describe objects. These
were AROHD, LEADER. NUMBER, SYMBOL, and certain image processors like rectangle, circle
and the like. Together these techniques are
used to describe a part, object or assembly
of parts to be manufactured. While it is
very important that as many review problems
be worked as possible; it should be kept in
mind that this chapter represents an activity
that a draftsperson uses througout the automated drafting process. It is not something
that can be learned and then not used in a
later application. Dimensioning is basic to
all forms of automated drafting output. Even
in the early introduction chapters you will
find objects that contain dimensions. Now
you know how to do it!

8

SECTION VIEWS

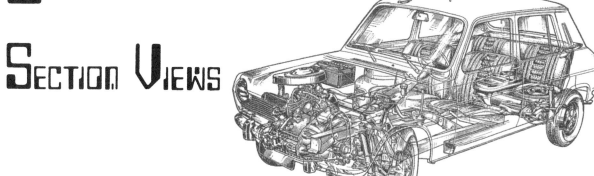

Automated drawings for objects which are
simple in design are created by the techniques shown in Chapter 6 Multiview Projection. Many objects, however, have internal
shapes which are so complicated in nature
that it is impossible to show their true
shape without using a number of hidden features. For example, Figure 8.1 is a situation where intermediate or interior construction cannot be clearly shown in multi-
exterior views. In these cases, the draftsperson resorts to the use of one or more
section views.

8.1 DISPLAYING SECTION VIEWS

A computer display is where a cross-
section of an object is obtained by passing
an imaginary cutting plane through the object. This imaginary cutting plane passes
through the solid object as shown in Figure 8.2 (upper right corner). Once the
plane has been located, a portion of the
object is removed. To indicate which portion has been removed, arrows are used as
shown in the top view of Figure 8.2.

Figure 8.1 Section views show the internal
construction of an object. (Courtesy Simca)

The arrows point to the direction of
sight used by an observer. In this case,
the sectioned view is the front view. In
Figure 8.2, the entire surface is shown filled. A number of different filling lines
may be used depending upon the material of
the object.

In addition to the filling lines, there
are various types of sectional views, such as:

1. Full section (Figure 8.2)
2. Half section (Figure 8.5)
3. Offset section (Figure 8.3)
4. Broken-out section (Figure 8. 7)
5. Revolved section (Figure 8.8)
6. Removed section (Figure 8.9)
7. Auxiliary section (Figure 8.10)
8. Thin section (Figure 8.10)

In each of the next several sections of this
chapter we will study the eight types. In
each of the eight cases, however, the view

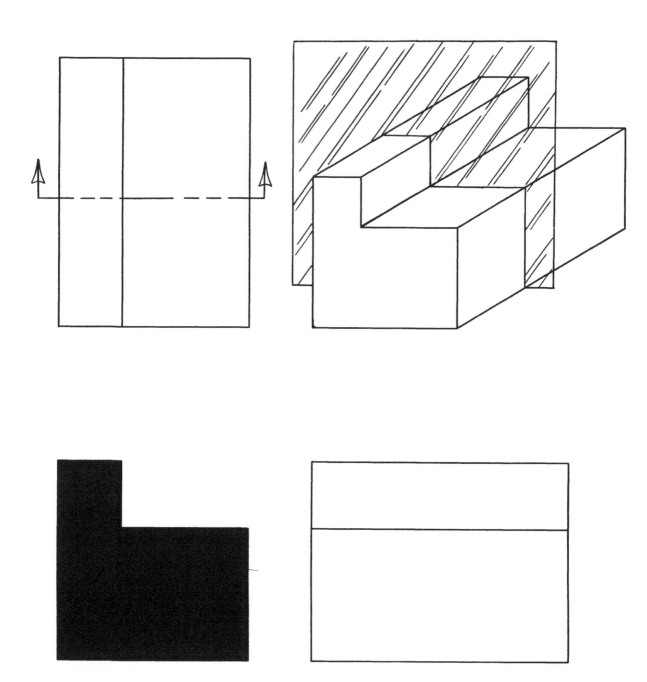

Figure 8.2 Full section of a simple object to show cutting plane theory.

produced is called the **section view**.

8.2 FULL SECTION VIEWS

The position of the cutting plane determines the type of section view produced. In Figure 8.2, above, the cutting plane is passed through the full object. Hence the term **full section**. The view which contains the cutting plane line is always adjacent to the section view. In this manner a subroutine can be used to calculate and place the type of section lines desired. A good example of this is shown in Figure 8.3.

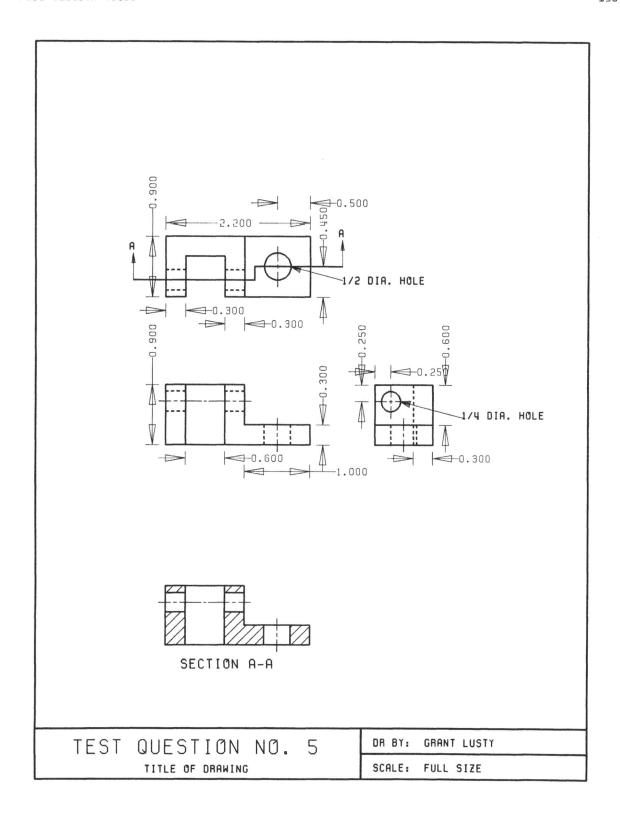

TEST QUESTION NO. 5

TITLE OF DRAWING

DR BY: GRANT LUSTY

SCALE: FULL SIZE

Figure 8.3 Example of full section displayed on the same output as a simple object introduced in Chapter 1, Figure 1.14. It has progressed through the various chapters being displayed as a multiview project, used as a dimensioning example and now a cutting plane is
added in the top view. The front view already existed, so an adjacent section view is shown.

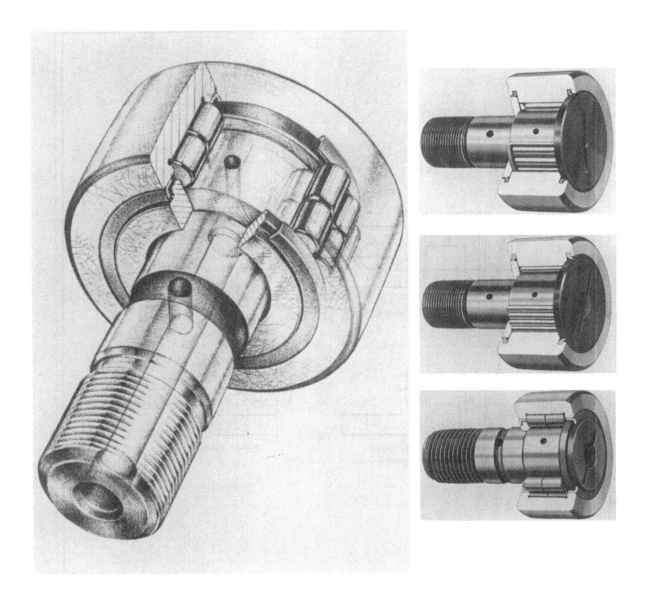

Figure 8.4 Half section. (Courtesy INA Bearing Company Inc.; Bensalem, PA)

8.3 HALF SECTION VIEWS

A half section view shown in Figure 8.
4 results when a cutting plane shaped like
that shown in Figure 8.5 is passed through
an object. As shown in Figure 8.5, the 90
degree cutting plane is positioned along a
centerline or symmetrical axes of an object.
As indicated in Figure 8.4, this permits
the removal of one-fourth the object and
therefore, only one-half of the object is
shown sectioned. For most symmetrical objects this is preferred to full sectioning.

A half section has the advantage over
the full section of showing the interior of
an object, while at the same time maintaining the shape of the exterior. While it is
the recommended section view for nearly all
symmetrical objects, it is difficult to use
for non-symmetrical objects. Linear dimensions are difficult to place on half section
iews, while they are easy to place on full
section views as shown in Figure 8.3. For
this reason, a half section is rarely seen
on a dimensioned drawing (detail drawing),
but is often seen on assembly details such

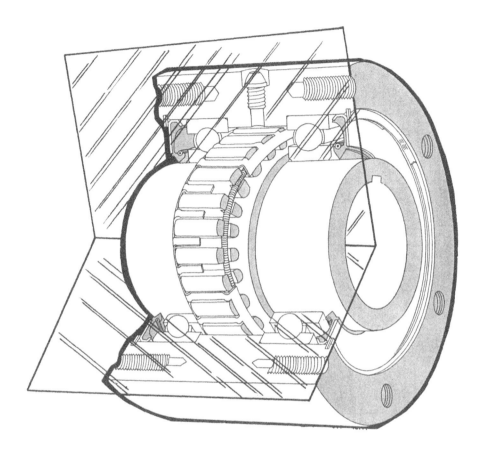

Figure 8.5 Cutting plane position for a half section of a symmetrical object. Usually the entire quarter section is removed, but in this example only the outer section is removed.

as Figure 8.6 illustrates. In this example the cutting plane position is the same as shown in Figure 8.5. Dimensions are limited to notes placed with leaders. Figure 8.6's greatest value is in this type of assembly presentation where it is necessary to show both internal and external construction on the same view.

When indicating a cutting plane line on a multiview drawing as shown in Figure 8.3. a half section cutting line is shown as a center line with the same type of arrowheads. A half section could be taken along any centerline, so imagine a half section taken through the hole in Figure 8.3. In this example only the right arrowhead is left where it is. The left arrowhead is removed from the left side of the top view, the cutting plane ends at the

middle of the hole and a center line continues at 90 degrees straight down the page.

On objects having only one major center line in which the cutting plane is assumed to pass through the axis of symmetry, the practice is to omit that part of the cutting plane line since its position is already clear that the section is taken along that center line. In Figure 8.3 the cutting plane line is bent or offset to go through both sets of holes. This too is a common practice to show the object more clearly for production.

8.4 OFFSET SECTION VIEWS

With the use of the offset cutting plane line as shown in Figure 8.3, features which are off the main center line can be

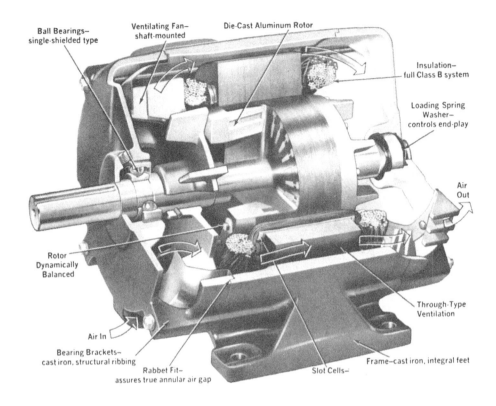

Figure 8.6 Half section used in assembly drawing. (Courtesy Industrial Fasteners Institute)

shown in full section view. Note that off-section views do not look different from full section views, only the cutting plane line is changed in the opposite multiview.

8.5 BROKEN-OUT SECTION VIEWS

While offset sections are considered to be variations of full and half sections, broken-out sections can be used for either detail or assembly drawings with good results. Figure 8.7 represents two examples of broken-out section views. The one on the left is used inside a multiview projection and clearly is useful in linear dimensioning. The one on the right illustrates a broken-out section used in a pictorial (assembly) drawing. In both of these examples of broken-outs, neither has a cutting plane line to represent the section. The sectioned area is outlined by a short break line and the resulting part is known

as a broken-out section.

8.6 REVOLVED SECTION VIEWS

Revolved section views are used to show the cross-sectional shape of such objects as tool handles, spokes, ribs, and forged parts. Forged parts of all types make good use of revolved sectional views. The cutting plane is passed perpendicular to the axis of the forged part and then reveolved 90 degrees into the same view as the orthographic projection. Figure 8.8 represents a revolved sectional view, notice that no cutting plane line is required, like the broken-out view in Figure 8.7.

8.7 REMOVED SECTION VIEWS

Greater explanation is possible if a revolved section is placed off the orthographic view. In these cases, the cross-

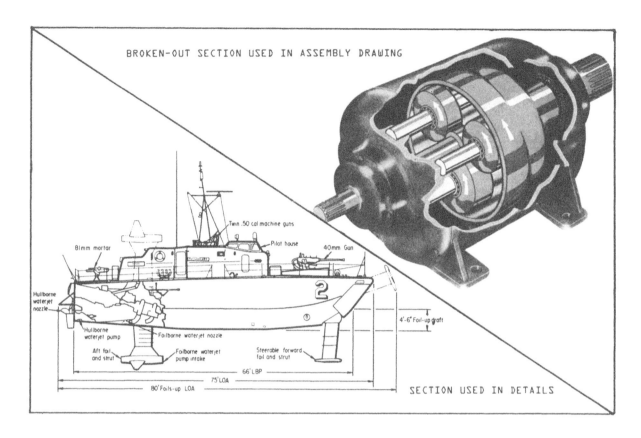

Figure 8.7   Broken-out sections.   (Courtesy Industrial Fasteners Institute)

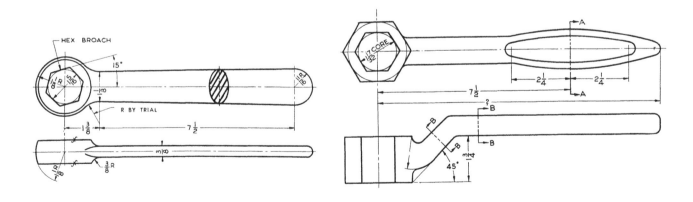

Figure 8.8   Revolved sectional view.   Figure 8.9   Removed sections to be placed.

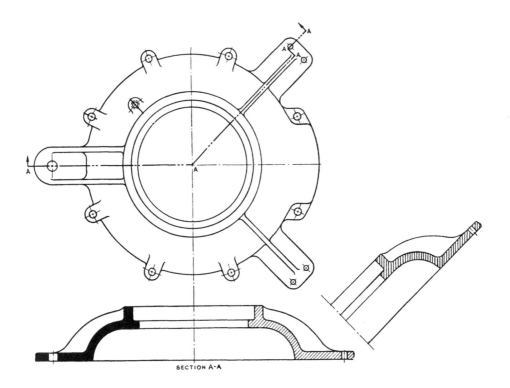

Figure 8.10 Auxiliary and thin section views used on same object.

section shape is removed from the orthographic view (see Figure 8.8) and placed
at another convenient location (see Figure
8.9).

8.8 AUXILIARY AND THIN SECTION VIEWS

Many objects provide for two other
types of views, they are auxiliary and thin.
In an ausiliary view, the vieport has been
rotated as shown in Figure 8.10 above. In
this example, the right side of SECTION A-A
is projected into an auxiliary position. A
view of this nature may be sectioned as in
Figure 8.9. or because it represents a thin
section of material, it could be represented as shown in Figure 8.2. The left side
of the object shown in Figure 8.10 is represented as a thin section.

8.9 COMPUTER-AIDED SECTIONING TECHNIQUES

Each of the eight types of sectional
views can be displayed on the equipment in-

troduced in this book. The draftsperson uses
the automated drafting manual and finds how
to display those elements required in sectioning an object. The most common is shown
in Figure 8.11. All section views, except
thin sections use CALL SHADE to automatically
reproduce any type of section lines required.
Thin sections require Figure 8.12 to present
a solid section. Some types of sections require cutting plane lines, others don't. If
required, Figure 8.13 shows how to call these
from computer memory. One type of section,
revolved, requires a centerline display. In
Figure 8.14 this technique is demonstrated.

8.10 REVIEW PROBLEMS

To really understand automated generation of section views, one must follow a
good example. Figure 8.3 is such a case,
suppose we review how to automate each part
of that example.

1. A main computer program is organized
to display Figure 8.3. Input the following:

```
CALL SHADE(XARRAY,YARRAY,X2,Y2,DLIN,
           ANGLE,NPT1,I1,NPT2,I2)

XARRAY,YARRAY = DATA FOR TOP BOUNDARY
X2,Y2 = DATA FOR BOTTOM BOUNDARY OF VIEW
DLIN = DISTANCE BETWEEN SECTION LINES
ANGLE = ANGLE OF SECTION LINES
NPT1 = NUMBER OF POINTS IN XARRAY
I1 = INDEX OF DATA IN XARRAY
NPT2 = NUMBER OF POINTS IN X2
I2 = INDEX OF DATA IN X2
```

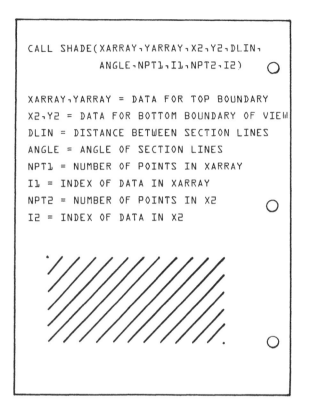

Figure 8.11 Routine for section lines.

```
CALL ARROW(XARRAY,YARRAY,NPT,I,ITYPE)

XARRAY,YARRAY = DATA POINTS FOR C'PLANE
NPT = NUMBER OF DATA POINTS IN C'PLANE
INC = STYLE OF C'PLANE(NORMAL= 1)
ITYPE = 1 then a single arrowhead
        2 then half section c'plane
        3 then full arrowheads
        4 then offset section
        5 then auxiliary section
```

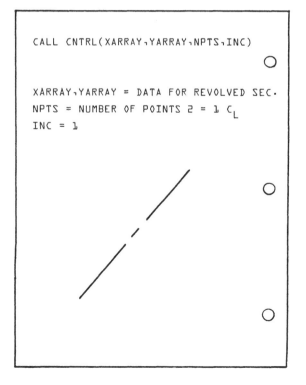

Figure 8.13 Routine for cutting plane line.

```
CALL BAR(X,Y,THETA,HT,WD,SH,IAT,NP)

X,Y = LOWER LEFT OF THIN SECTION
THETA = ROTATION ABOUT X,Y OF THIN SEC.
HT = HEIGHT OF THIN SECTION
WD = WIDTH OF THIN SECTION
SH = WD
IAT = 4
NP = 100
```

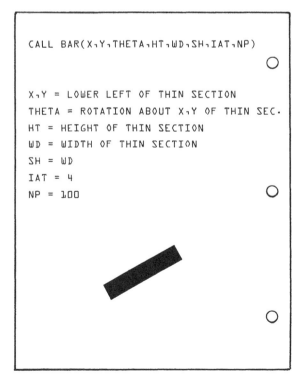

Figure 8.12 Routine for thin sections.

```
CALL CNTRL(XARRAY,YARRAY,NPTS,INC)

XARRAY,YARRAY = DATA FOR REVOLVED SEC.
NPTS = NUMBER OF POINTS 2 = 1 C_L
INC = 1
```

Figure 8.14 Routine for center lines.

```
FORTRAN IV G1  RELEASE 2.0            MAIN            DATE = 82344        11/15/24           PAGE 0001
      0001                    CALL PLOTS                                              00000040
      0002                    CALL FACTOR(.9)                                         00000042
      0003                    CALL TITLE                                              00000045
      0004                    DIMENSION X(34),Y(34),Z(34),IPEN(34)                    00000060
      0005                    DATA X/0.,.3,.3,.9,.9,2.2,2.2,0.,0.,0.,.3,.3,.9,.9,1.2,1.2 00000070
                             C,2.2,2.2,1.2,1.2,0.,0.,0.,0.,.3,.3,.9,.9,2.2,2.2        00000080
                             C,2.2,2.2,1.2,1.2/                                       00000090
      0006                    DATA Y/0.,0.,0.,0.,0.,0.,0.,0.,0.,.9,.9,.9,.9,.9,.3,.3,.3,.3 00000100
                             C,.9,.9,.9,.9,0.,.9,0.,.9,0.,.3,0.,.3,0.,.9,.9/          00000110
      0007                    DATA Z/0.,0.,.6,.6,0.,0.,.9,.9,0.,0.,.6,.6,0.,0.,.6,.0.0. 00000120
                             C,.9,.9,.9,.9,0.,.9,0.,.9,0.,0.,0.,0.,0.,.0.,.9,.9,.9,0./ 00000130
      0008                    DATA IPEN/ 3,21*2, 3,2, 3,2, 3,2, 3,2, 3,2, 3,2/        00000140
      0009                    XTRANS = 3.2                                            00000150
      0010                    YTRANS = 2.5                                            00000160
      0011                    CALL PLOT(4.,1.25,-3)                                   00000170
      0012                    CALL SECT(X,Y,IPEN)                                     00000180
      0013                    CALL PLOT(0.,2.,-3)                                     00000190
      0014                    DO 10 I=1,34                                            00000200
      0015                 10 CALL PLOT(X(I),Y(I),IPEN(I))                            00000210
      0016                    DO 20 J=1,34                                            00000220
      0017                    Z2 = Z(J) + YTRANS                                      00000230
      0018                 20 CALL PLOT(X(J),Z2,IPEN(J))                              00000240
      0019                    DO 30 K=1,34                                            00000250
      0020                    Z3 = Z(K) + XTRANS                                      00000260
      0021                 30 CALL PLOT(Z3,Y(K),IPEN(K))                              00000270
      0022                    X1 = 1.7                                                00000280
      0023                    Y1 = .25                                                00000290
      0024                    Y1 = Y1 + YTRANS                                        00000300
      0025                    CALL CIRCL(X1,Y1,-90.,270.,.2,.2,0.5)                   00000310
      0026                    CALL CIRCL(X1,Y1+.02,-90.,270.,.18,.18,0.0)             00000315
      0027                    X2 = .25                                                00000320
      0028                    Y2=.5                                                   00000330
      0029                    X2 = X2 + XTRANS                                        00000340
      0030                    CALL CIRCL(X2,Y2,-90.,270.,.15,.15,0.0)                 00000350
      0031                    CALL CNTRLS                                             00000360
      0032                    CALL DASHES                                             00000370
      0033                    CALL DIMENS                                             00000380
                      C                                                              00000390
      0034                    CALL PLOT(-0.5,.25+YTRANS,3)                            00000400
      0035                    CALL PLOT(1.35,.25+YTRANS,2)                            00000410
      0036                    CALL PLOT(1.35,.45+YTRANS,2)                            00000420
      0037                    CALL PLOT(2.7,.45+YTRANS,2)                             00000430
      0038                    CALL AROHD(-0.5,.25+YTRANS,-.5,.65+YTRANS,.15,0.05,16)  00000440
      0039                    CALL AROHD(2.7,.45+YTRANS,2.7,.85+YTRANS,.15,0.05,16)   00000450
      0040                    CALL AROHD(2.7,.25+YTRANS,1.9,.45+YTRANS,.15,0.05,16)   00000480
      0041                    CALL AROHD(1.4+XTRANS,.4,.4+XTRANS,.65,.15,0.05,16)     00000490
      0042                    CALL SYMBOL(-.55,.7+YTRANS,.1,'A',0.0,1)               00000496
      0043                    CALL SYMBOL(2.65,.9+YTRANS,.1,'A',0.0,1)               00000497
      0044                    CALL SYMBOL(2.7,.2+YTRANS,.1,'1/2 DIA. HOLE',0.0,13)   00000500
      0045                    CALL SYMBOL(1.35+XTRANS,.4,.1,'1/4 DIA. HOLE',0.0,13)  00000510
                      C                                                              00000520
      0046                    CALL PLOT(0.,0.,999)                                   00000530
      0047                    STOP                                                   00000540
      0048                    END                                                    00000550
```

2. The above program requires several sub-programs in order to display Figure 8.3. Input subroutine TITLE shown below.

```
FORTRAN IV G1  RELEASE 2.0            MAIN            DATE = 82344        11/15/24           PAGE 0001
                      C                                                              00002271
      0001                    SUBROUTINE TITLE                                       00002281
      0002                    CALL RECT(1.,0.,.75,3.5,0.,3)                           00002282
      0003                    CALL RECT(0.,0.,11.25,17.,0.,3)                         00002283
      0004                    CALL RECT(0.,0.,.75,1.,0.,3)                            00002284
      0005                    CALL RECT(4.5,0.,.375,2.5,0.,3)                         00002285
      0006                    CALL RECT(4.5,.375,.375,2.5,0.,3)                       00002286
      0007                    CALL RECT(7.0,0.,.375,3.5,0.,3)                         00002287
      0008                    CALL RECT(7.0,.375,.375,3.5,0.,3)                       00002288
      0009                    CALL RECT(10.5,0.,.75,5.25,0.,3)                        00002289
      0010                    CALL RECT(15.75,0.,.75,1.25,0.,3)                       00002290
      0011                    CALL SYMBOL(.1,.1,.08,11HDRAWING NO.,0.,11)             00002291
      0012                    CALL SYMBOL(1.875,.5,.09,20HENGINEERING GRAPHICS,0.,20) 00002292
      0013                    CALL SYMBOL(1.625,.3125,.1,22HCOLLEGE OF ENGINEERING,0.,22) 00002293
      0014                    CALL SYMBOL(1.750,.1,.12,18HCLEMSON UNIVERSITY,0.,18)   00002294
      0015                    CALL SYMBOL(4.625,.12,.1,6HSCALE:,0.,6)                 00002295
      0016                    CALL SYMBOL(4.625,.45,.1,5HDATE:,0.,5)                  00002296
      0017                    CALL SYMBOL(7.125,.12,.1,7HCOURSE:,0.,7)                00002297
      0018                    CALL SYMBOL(7.125,.45,.1,6HDR BY:,0.,6)                 00002298
      0019                    CALL SYMBOL(12.625,.1,.1,16HTITLE OF DRAWING,0.,16)     00002299
      0020                    CALL SYMBOL(16.25,.1,.1,5HGRADE,0.,5)                   00002300
      0021                    CALL SYMBOL(8.,.12,.1,6HEG 310,0.,6)                    00002301
      0022                    CALL SYMBOL(8.,.45,.1,8HG. LUSTY,0.,8)                  00002302
      0023                    CALL SYMBOL(5.25,.12,.1,7H1.4"=1",0.,7)                 00002303
      0024                    CALL SYMBOL(.375,.25,.25,1H5,0.,1)                      00002304
      0025                    CALL SYMBOL(11.525,.375,.2,19HTEST QUESTION NO. 5,0.0,19) 00002305
      0026                    CALL SYMBOL(5.25,.45,.1,8H12-06-82,0.,8)               00002306
      0027                    RETURN                                                 00002307
      0028                    END                                                    00002308
```

3. The second subprogram required is SECT.

```
FORTRAN IV G1  RELEASE 2.0              MAIN              DATE = 82344        11/15/24           PAGE 0001

                   C                                                                               00001640
       0001                 SUBROUTINE SECT(X,Y,IPEN)                                              00001650
       0002                 DIMENSION X(34),Y(34),IPEN(34),X2(4),Y2(6),X3(6),Y3(8),X4(6),Y4(8)00001660
                   C,X5(4),Y5(6),X6(6),Y6(8),X7(6),Y7(8),X8(4),Y8(4),X9(8),Y9(8)                  00001670
                   C,X10(4),Y10(6),X11(6),Y11(8),XX(6),YY(6),XXX(4),YYY(4)                         00001680
       0003                 DATA X2/0.,.3,0.,1./                                                   00001690
       0004                 DATA X3/0.,0.,.3,.3,0.,1./                                             00001700
       0005                 DATA X6/.9,1.2,0.,1./                                                  00001710
       0006                 DATA X7/.9,.9,1.2,1.2,0.,1./                                           00001720
       0007                 DATA X8/.9,1.2,0.,1./                                                  00001730
       0008                 DATA X9/.9,.9,1.5,1.5,1.2,1.2,0.,1./                                   00001740
       0009                 DATA X10/1.9,2.2,0.,1./                                                00001750
       0010                 DATA X11/1.9,1.9,2.2,2.2,0.,1./                                        00001760
       0011                 DATA Y2/.9,.9,0.,1./                                                   00001770
       0012                 DATA Y3/.9,.8,.8,.9,0.,1./                                             00001780
       0013                 DATA Y5/.5,0.,0.,.5,0.,1./                                             00001790
       0014                 DATA Y6/.5,.5,0.,1./                                                   00001800
       0015                 DATA Y8/.5,.5,0.,1./                                                   00001810
       0016                 DATA Y9/.5,0.,0.,.3,.3,.5,0.,1./                                       00001820
       0017                 DATA Y10/.3,.3,0.,1./                                                  00001830
       0018                 DATA Y11/.3,0.,0.,.3,0.,1./                                            00001840
       0019                 DO 10 I=1,34                                                           00001850
       0020              10 CALL PLOT(X(I),Y(I),IPEN(I))                                           00001860
       0021                 CALL SHADE(X2,Y2,X3,Y3,0.1,45.,2,1,4,1)                                00001870
       0022                 CALL SHADE(X2,Y4,X3,Y5,0.1,45.,2,1,4,1)                                00001880
       0023                 CALL SHADE(X6,Y2,X7,Y3,0.1,45.,2,1,4,1)                                00001890
       0024                 CALL SHADE(X8,Y8,X9,Y9,0.1,45.,2,1,6,1)                                00001900
       0025                 CALL SHADE(X10,Y10,X11,Y11,0.1,45.,2,1,4,1)                            00001910
       0026                 CALL PLOT(0.,0.5,3)                                                    00001920
       0027                 CALL PLOT(.3,.5,2)                                                     00001930
       0028                 CALL PLOT(.9,.5,3)                                                     00001940
       0029                 CALL PLOT(1.2,.5,2)                                                    00001950
       0030                 CALL PLOT(0.,.8,3)                                                     00001960
       0031                 CALL PLOT(.3,.8,2)                                                     00001970
       0032                 CALL PLOT(.9,.8,3)                                                     00001980
       0033                 CALL PLOT(1.2,.8,2)                                                    00001990
       0034                 CALL PLOT(1.5,0.,3)                                                    00002000
       0035                 CALL PLOT(1.5,.3,2)                                                    00002010
       0036                 CALL PLOT(1.9,0.,3)                                                    00002020
       0037                 CALL PLOT(1.9,.3,2)                                                    00002030
       0038                 CALL PLOT(1.52,0.,3)                                                   00002031
       0039                 CALL DASHP(1.52,.3,.05)                                                00002032
       0040                 CALL PLOT(1.88,0.,3)                                                   00002033
       0041                 CALL DASHP(1.88,.3,.05)                                                00002034
       0042                 XX(1)=-0.1                                                             00002040
       0043                 XX(2)=0.4                                                              00002050
       0044                 XX(3)=0.8                                                              00002060
       0045                 XX(4)=1.3                                                              00002070
       0046                 XX(5)=0.0                                                              00002080
       0047                 XX(6)=1.0                                                              00002090
       0048                 YY(1)=.65                                                              00002100
       0049                 YY(2)=.65                                                              00002110
       0050                 YY(3)=.65                                                              00002120
       0051                 YY(4)=.65                                                              00002130
       0052                 YY(5)=0.0                                                              00002140
       0053                 YY(6)=1.0                                                              00002150
       0054                 CALL CNTRL(XX,YY,4,1)                                                  00002160
       0055                 XXX(1)=1.7                                                             00002170
```

```
FORTRAN IV G1  RELEASE 2.0              SECT              DATE = 82344        11/15/24           PAGE 0002

       0056                 XXX(2)=1.7                                                             00002180
       0057                 XXX(3)=0.0                                                             00002190
       0058                 XXX(4)=1.0                                                             00002200
       0059                 YYY(1)=-0.1                                                            00002210
       0060                 YYY(2)=0.4                                                             00002220
       0061                 YYY(3)=0.0                                                             00002230
       0062                 YYY(4)=1.0                                                             00002240
       0063                 CALL CNTRL(XXX,YYY,2,1)                                                00002250
       0064                 CALL SYMBOL(.25,-0.35,.13,'SECTION A-A',0.0,11)                        00002252
       0065                 RETURN                                                                 00002260
       0066                 END                                                                    00002270
```

4. Input the following subprogram:

```
FORTRAN IV G1  RELEASE 2.0              CNTL              DATE = 82344        11/15/24           PAGE 0001

       0001                 SUBROUTINE CNTL(X,Y,RAD)                                               00001270
       0002                 DIMENSION XX(4),YY(4)                                                  00001280
       0003                 XX(1)=X-RAD                                                            00001290
       0004                 XX(2)=X+RAD                                                            00001300
       0005                 XX(3)=0.0                                                              00001310
       0006                 XX(4)=1.0                                                              00001320
       0007                 YY(1)=Y                                                                00001330
       0008                 YY(2)=Y                                                                00001340
       0009                 YY(3)=0.0                                                              00001350
       0010                 YY(4)=1.0                                                              00001360
       0011                 CALL CNTRL(XX,YY,2,1)                                                  00001370
       0012                 XX(1)=X                                                                00001380
       0013                 XX(2)=X                                                                00001390
       0014                 YY(1)=Y-RAD                                                            00001400
       0015                 YY(2)=Y+RAD                                                            00001410
       0016                 CALL CNTRL(XX,YY,2,1)                                                  00001420
       0017                 RETURN                                                                 00001430
       0018                 END                                                                    00001440
```

5. Input the CNTRLS routine:

```
FORTRAN IV G1  RELEASE 2.0              CNTRLS           DATE = 82344      11/15/24              PAGE 0001

        0001              SUBROUTINE CNTRLS                                             00000560
        0002              DIMENSION XX(6),YY(6),XXX(4),YYY(4)                           00000570
        0003              XTRANS=3.2                                                    00000580
        0004              YTRANS=2.5                                                    00000590
        0005              CALL CNTL(1.7,.45+YTRANS,.3)                                  00000600
        0006              CALL CNTL(.25+XTRANS,.65,.25)                                 00000610
        0007              XX(1)=-0.1                                                    00000620
        0008              XX(2)=0.4                                                     00000630
        0009              XX(3)=0.8                                                     00000640
        0010              XX(4)=1.3                                                     00000650
        0011              XX(5)=0.0                                                     00000660
        0012              XX(6)=1.0                                                     00000670
        0013              YY(1)=0.25+YTRANS                                             00000680
        0014              YY(2)=0.25+YTRANS                                             00000690
        0015              YY(3)=0.25+YTRANS                                             00000700
        0016              YY(4)=0.25+YTRANS                                             00000710
        0017              YY(5)=0.0                                                     00000720
        0018              YY(6)=1.0                                                     00000730
        0019              CALL CNTRL(XX,YY,4,1)                                         00000740
        0020              YY(1)=.65                                                     00000750
        0021              YY(2)=.65                                                     00000760
        0022              YY(3)=.65                                                     00000770
        0023              YY(4)=.65                                                     00000780
        0024              CALL CNTRL(XX,YY,4,1)                                         00000790
        0025              XXX(1)=1.7                                                    00000800
        0026              XXX(2)=1.7                                                    00000810
        0027              XXX(3)=0.0                                                    00000820
        0028              XXX(4)=1.0                                                    00000830
        0029              YYY(1)=-0.1                                                   00000840
        0030              YYY(2)=0.4                                                    00000850
        0031              YYY(3)=0.0                                                    00000860
        0032              YYY(4)=1.0                                                    00000870
        0033              CALL CNTRL(XXX,YYY,2,1)                                       00000880
        0034              XXX(1)=0.45+XTRANS                                            00000890
        0035              XXX(2)=0.45+XTRANS                                            00000900
        0036              CALL CNTRL(XXX,YYY,2,1)                                       00000910
        0037              RETURN                                                        00000920
        0038              END                                                           00000930
```

6. Input the DASHES routine:

```
FORTRAN IV G1  RELEASE 2.0              DASHES           DATE = 82344      11/15/24              PAGE 0001

        0001              SUBROUTINE DASHES                                             00000940
        0002              XTRANS=3.2                                                    00000950
        0003              YTRANS=2.5                                                    00000960
        0004              CALL PLOT(0.,0.5,3)                                           00000970
        0005              CALL DASHP(.3,.5,.043)                                        00000980
        0006              CALL PLOT(.9,.5,3)                                            00000990
        0007              CALL DASHP(1.2,.5,.043)                                       00001000
        0008              CALL PLOT(0.,.8,3)                                            00001010
        0009              CALL DASHP(.3,.8,.043)                                        00001020
        0010              CALL PLOT(.9,.8,3)                                            00001030
        0011              CALL DASHP(1.2,.8,.043)                                       00001040
        0012              CALL PLOT(1.5,0.,3)                                           00001050
        0013              CALL DASHP(1.5,.3,.043)                                       00001060
        0014              CALL PLOT(1.9,0.,3)                                           00001070
        0015              CALL DASHP(1.9,.3,.043)                                       00001080
                C                                                                       00001090
        0016              CALL PLOT(.25+XTRANS,0.,3)                                    00001100
        0017              CALL DASHP(.25+XTRANS,.3,.043)                                00001110
        0018              CALL PLOT(.65+XTRANS,0.,3)                                    00001120
        0019              CALL DASHP(.65+XTRANS,.3,.043)                                00001130
        0020              CALL PLOT(.6+XTRANS,0.,3)                                     00001140
        0021              CALL DASHP(.6+XTRANS,.9,.043)                                 00001150
                C                                                                       00001160
        0022              CALL PLOT(0.,.1+YTRANS,3)                                     00001170
        0023              CALL DASHP(.3,.1+YTRANS,.043)                                 00001180
        0024              CALL PLOT(.9,.1+YTRANS,3)                                     00001190
        0025              CALL DASHP(1.2,.1+YTRANS,.043)                                00001200
        0026              CALL PLOT(0.,.4+YTRANS,3)                                     00001210
        0027              CALL DASHP(.3,.4+YTRANS,.043)                                 00001220
        0028              CALL PLOT(.9,.4+YTRANS,3)                                     00001230
        0029              CALL DASHP(1.2,.4+YTRANS,.043)                                00001240
        0030              RETURN                                                        00001250
        0031              END                                                          00001260
```

7. Input the DIMENS routine:

```
FORTRAN IV G1  RELEASE 2.0              DIMENS           DATE = 82344      11/15/24              PAGE 0001

        0001              SUBROUTINE DIMENS                                             00001450
        0002              XTRANS=3.2                                                    00001460
        0003              YTRANS=2.5                                                    00001470
        0004              CALL DIMEN(.3,-.2,.6,0.,1.)                                   00001480
        0005              CALL DIMEN(1.2,-.4,1.,0.,1.)                                  00001490
        0006              CALL DIMEN(2.4,0.,.3,90.,1.)                                  00001500
        0007              CALL DIMEN(-.2,0.,.9,90.,1.)                                  00001510
        0008              CALL DIMEN(0.,-.2+YTRANS,.3,0.,1.)                            00001520
        0009              CALL DIMEN(0.9,-.4+YTRANS,.3,0.,1.)                           00001530
        0010              CALL DIMEN(1.7,1.4+YTRANS,.5,0.,1.)                           00001540
        0011              CALL DIMEN(2.4,YTRANS,.45,90.,1.)                             00001550
        0012              CALL DIMEN(0.,1.1+YTRANS,2.2,0.,1.)                           00001560
        0013              CALL DIMEN(-.2,YTRANS,.9,90.,1.)                              00001570
        0014              CALL DIMEN(-.2+XTRANS,.65,.25,90.,1.)                         00001580
        0015              CALL DIMEN(XTRANS,1.1,.25,0.,1.)                              00001590
        0016              CALL DIMEN(.6+XTRANS,-.2,.3,0.,1.)                            00001600
        0017              CALL DIMEN(1.1+XTRANS,.3,.6,90.,1.)                           00001610
        0018              RETURN                                                        00001620
        0019              END                                                          00001630
```

8.12 CHAPTER SUMMARY

Automated section drawing is most interesting because of its many different types
of sectional views. This chapter introduces
eight different types for you to study:

1. Full
2. Half
3. Offset
4. Broken-out
5. Revolved
6. Removed
7. Auxiliary
8. Thin

Each different type was shown in an
example figure, but only one type was completely automated. That was the offset
shown in Figure 8.3 and solved in the review problems section. Each of the other
types may be presented in a similar manner
by your instructor. In the interest of
time and difficulty for this type of book,
other examples would simply be a repeat of
the procedures shown in Figures 8.11 through
8.14. These left facing pages of a users
manual are very familiar by this time in
your study of automated drafting. In the
earlier chapters, many more examples were
given to give you practice in dealing with
the newness of the computer graphics equipment and procedures. At this point you are
probably feeling good about the use of the
equipment and need as many different applications for drafting as possible. In this
spirit, the next several chapters will present traditional materials in an automated
mode.

9

Auxiliary Views

A view of an object that is not one of the six primary views shown in Chapter 6 is an additional or auxiliary view. When views are planned for computer output, it is often convenient to locate these views on a planning sheet introduced in Chapter 2, Figure 2.14. The gridded section represented the locations for the six primary views from Chapter 6, while the open space is used for auxiliary (additional) views. It should be pointed out that six orthographic views are rarely used in an automated drawing, therefore, the gridded space not used, may also be a convenient space to locate a combination auxiliary and section view, and this was shown in Chapter 8.

9.1 PRIMARY AUXILIARY VIEWPORTS

A primary auxiliary viewport is a viewport which is rotated perpendicular to one of the six principle viewports and inclined to the other two. Figure 9.1 represents a frontal viewport as described in Chapter 6. Figure 9.2 represents how primary auxiliary viewports can be located in relationship to

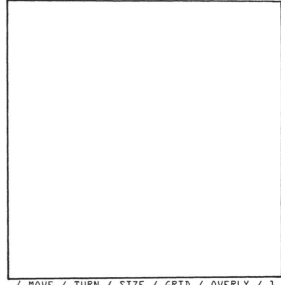

/ MOVE / TURN / SIZE /.GRID / OVERLY / 1

Figure 9.1 Viewport with menu bar for constructing auxiliary views. A pointing device is placed over the first command MOVE and any object within the viewport may be moved (shifted). Likewise TURN will rotate and SIZE will scale the object. GRID & OVERLY produce the planning sheet reference.

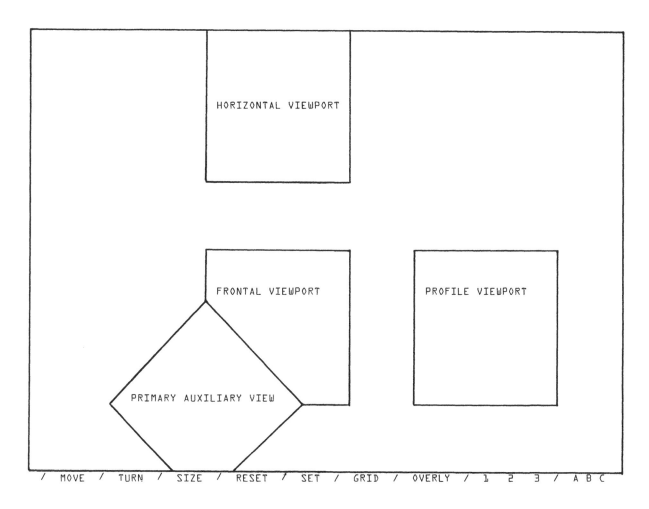

Figure 9.2 Drawing window and primary auxiliary viewport locations.   (Courtesy DLR Assoc.)

a frontal viewport for example. In Figure
9.2 a window containing all three primary
viewports; horizontal, frontal, and profile
is shown. Notice that the menu bar is now
located at the bottom of the window. Therefore, in Figure 9.1 the frontal viewport was
the same size as the display window. Many
combinations of primary auxiliary viewport
locations exist. A primary auxiliary view
may be next to any of the six primary viewports and at any rotation angle and position
to it. It must represent a perpendicular
'folding' plane relationship as explained
in Chapter 6, once this relationship is
found (normally parallel to an inclined
surface on the object), the viewport is
established.
     Figure 9.3 is an example of how this
process begins. First the drafter selects

at least two views. In this case the top
(horizontal) and frontal of an object. In
Figure 9.3 both viewports fill the entire
screen, so the menu bar item SIZE is selected to expand the object in the frontal
viewport until the inclined surface is located. Figure 9.4 shows the frontal viewport after the zooming has taken place. An
operator next begins the auxiliary viewport
process. He selects a location that is perpendicular to the frontal viewport and parallel to the inclined feature that is to be
displayed in the auxiliary viewport. His
left hand is located over a bank of function
buttons programmed for this purpose.

9.2 DISPLAYING THE AUXILIARY VIEW

     When displaying the auxiliary view, it

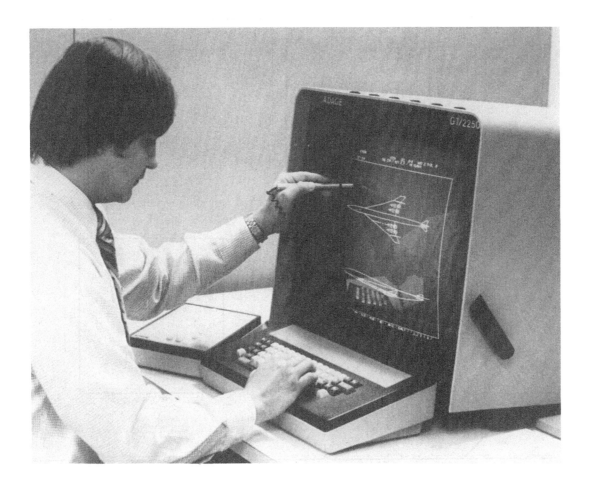

Figure 9.3 Drafter begins the auxiliary view process. (Courtesy ADAGE Corp.)

considered good practice to show the actual contour of only the inclined surface - so by using the SIZE command data which is not part of the inclined surface can be clipped (removed from the auxiliary viewport). The projection of the entire view data usually adds very little to the shape description and is at such a small scale that little useful information is obtained. Notice in Figure 9.4 how much more detail is shown about the object as compared to Figure 9.3 which is the same object - only at a much smaller scale. Always take auxiliary views from the largest display data available.

9.3 LOCATIONS FOR PRIMARY AUXILIARY VIEWS

In our examples shown so far, the auxiliary views have all been located off the frontal viewport. Generally, auxiliary views are located off the horizontal, frontal, or profile; whichever is most convenient for the draftsperson. The view's location depends upon the placement of the inclined surface within the three primary views. An auxiliary view is constructed to show things like true length, true shape, true angle and so forth. Therefore, the parallel relationship is important to the location of the auxiliary view. Any convenient location which is inside the display window can be used to display a primary auxiliary view. The draftsperson decides where the view will be located. In Figure 9.5, she decides that an auxiliary view will be located below the frontal-expanded view from Figure 9.4. Because of the angle of the inclined surface to be represented in the auxiliary view, she

Figure 9.4   Drafter enlarges frontal view to locate inclined surface.   (Courtesy ADAGE Corp)

elects not to place the auxiliary view above
the front view, as this will fold over the
horizontal viewport. A hardcopy is returned to her for varification of auxiliary view
location.

9.4 DATA BASE CONSTRUCTION

Once the location of the auxiliary viewport is established, the object must be defined relative to that viewport. The data
for typical auxiliaries are grouped by:
1.   Symmetrical objects
2.   Random data base objects
3.   Curved surfaces
4.   Multiple view objects
The object represented in Figures 9.3 - 9.6
contain all of the above data base types.
In Figure 9.6 the drafter is providing for

the data base construction. This procedure
involves the following steps:
1.   The position of the auxiliary viewport is placed on the display area. If the
object to be represented is symmetrical, the
top edge of the viewport, called a **reference**
line, can be placed along the centerline.
One-half the object will be projected as an
auxiliary view and this is called a **unilat-
eral auxiliary view**. If the object is a
random data base view than it is not symmetrical and this type of auxiliary view is
called a **bilaterial**.
2.   The data describing the auxiliary
view is then plotted or displayed from the
reference line into the auxiliary view. A
data base is established, you will remember
for Chapter 6 from the viewport edges for
each view. Therefore, this same data can

Figure 9.5 Drafter selects location parallel and away from inclined surface. (Courtesy of ADAGE Corp.)

be used again to display the object feature in the auxiliary view. Look at Figure 9.7 for an understanding of this concept. This is a very simple example so that you can understand how the data base construction is used in a primary auxiliary viewport. Two viewports are represented in Figure 9.7, the top and front and a third viewport is located parallel to the inclined surface labeled Q. This third viewport is an auxiliary viewport. Notice that a true shape of the surface marked Q appears inside this viewport. Also notice that the location inside this viewport is related to the data which described the location of the other two views. Check this by taking a pair of dividers or a ruler. Measure up from the edge of the viewports into the top viewport

where the inclined surface looks like a square. Is it a square? No it is a rectangle as shown in the auxiliary view. Does the distance up into the top viewport equal the distance over in the auxiliary viewport? It should if the same data base was used.

3. Surfaces which appear in true shape are labeled and unnessary views are deleted. Notice that the profile view was deleted in Figure 9.7.

4. Hidden lines are not used in the auxiliary projection.

9.5 SYMMETRICAL OBJECTS

To save computer processing time, the practice is to include only half of a view for all symmetrical objects as shown in the

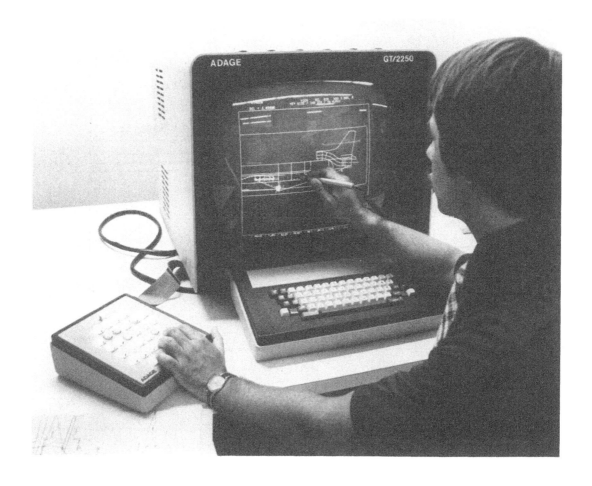

Figure 9.6   Drafter positions viewport parallel to inclined feature.   (Courtesy ADAGE Corp.)

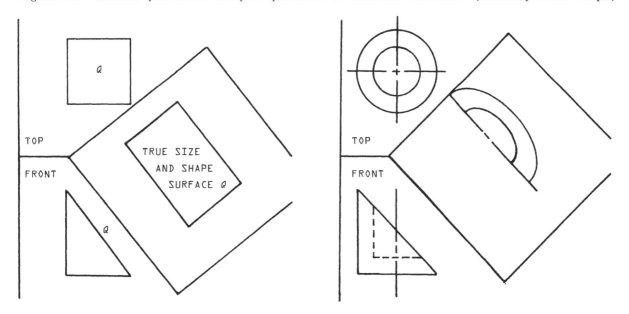

Figure 9.7 Random data base auxiliary view.   Figure 9.8   Symmetrical auxiliary view.

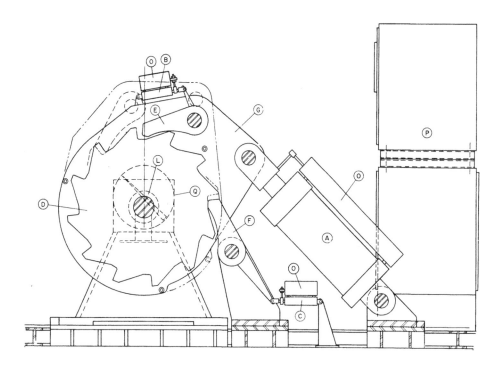

Figure 9.9 Curved surface auxiliary view problem.

Figure 9.8.

9.6 RANDOM DATA BASE OBJECTS

This type of auxiliary projection is
the most common found in automated drafting.
Often symmetrical parts and intermixed with
other types, and it is easier to handle the
auxiliary view as a random data base construction. A good example of this type of
problem is shown in Figure 9.9. Here some
parts are curved, symmetrical and random in
nature. They are grouped by part letter as:

Random	Symmetric	Curved Surfaces
	A	A
	B	B
	C	C
D		D
E		E
F		F
		G
		L
	O	O
P		
Q		

9.7 MULTIAUXILIARY VIEWS

The clear majority of automated drawings usually involve a single, primary auxiliary view. However, there are instances
where an auxiliary viewport is constructed
from another auxiliary viewport. For these
cases a multiauxiliary called a **double auxiliary** is required. The rules and steps to
display a double auxiliary are the same as
those listed in section 9.4 and are shown
in Figure 9.10. Like primary auxiliaries,
double auxiliaries may be taken from any of
the six primary views. Once a primary view
is choosen the multiauxiliary process will
begin.

9.8 AUXILIARY AND SECTION VIEWS

As described in the last chapter, a
special relationship exists between auxiliary and section views. Many of the rules
and procedures are the same for both type
of automated drawing. Figure 9.11 is an
example of this special relationship. Some
types of engineering graphics lend themselves

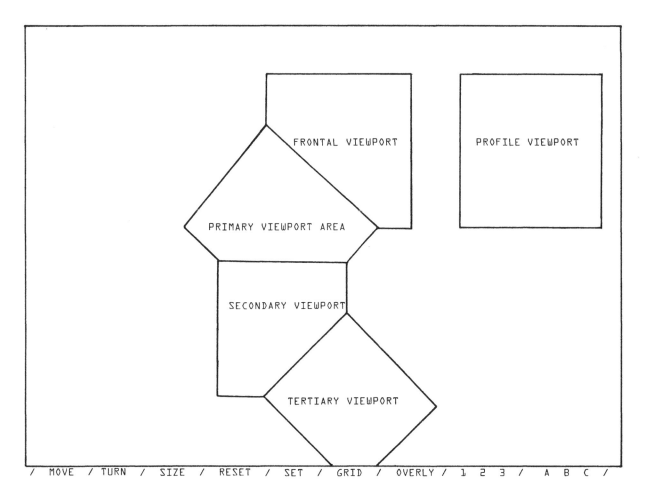

Figure 9.10 Multiauxiliary viewports taken from front view. (Courtesy DLR Associates)

to combination of section and auxiliary. A good example is the Civil Engineering drawing shown in example Figure 9.11. Mechanical Engineering drawings like that shown in Figure 9.9 tend to lend themselves to a multiview approach as shown in Figure 9.10. If the object shown in Figure 9.9 were modeled inside the format shown in Figure 9.10, the drafter would use a pointer to select the primary viewing space. This is done by pressing the 1 2 or 3 shown along the menu bar. These numbers coorespond to the three principle views of 1 = horizontal, 2 = frontal, and 3 = profile. To output a Figure like 9.10, the drafter would select 2 and 3 to display the front and profile views of an object. Next a primary auxiliary viewport is selected by pressing the A from the menu bar. Where A = primary, B=

secondary auxiliary viewport, and C = tertiary auxiliary viewport. In this manner and combination, multi viewports can be created to solve most any graphics view problem.

9.9 REVIEW PROBLEMS

Practice in the creation of auxiliary views is necessary for you to understand a concept of additional views. Begin your practice of auxiliary views by preparing several manual drawn examples assigned by your instructor. Then review the objects that you completed for Chapter 6 and 8. Do any of these objects require an auxiliary? This is the question that all draftspersons ask themselves when preparing multiview objects in a manual or automated mode.

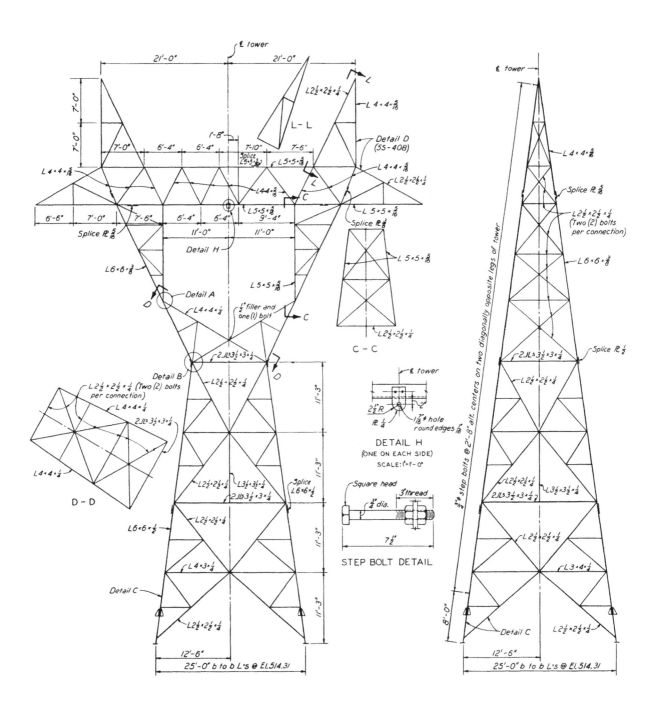

Figure 9.11 Typical Civil Engineering drawing using a combination of section and auxiliary view construction. (Courtesy DLR Associates)

1.

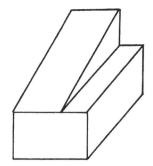

2.

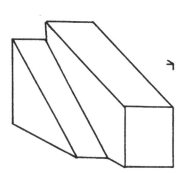

3.

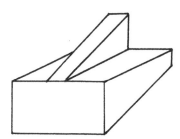

4.

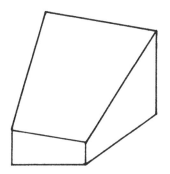

5.

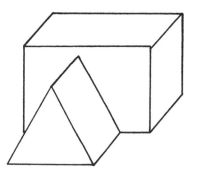

6.

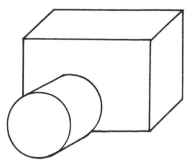

7.

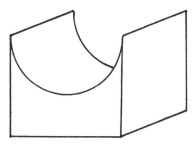

8.

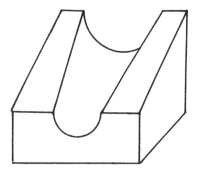

9.

13.

10.

11.

14.

12.

15.

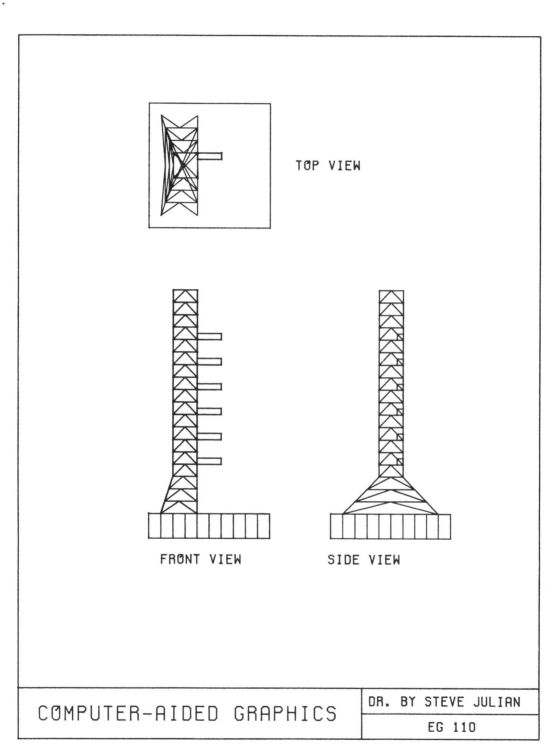

TOP VIEW

FRONT VIEW   SIDE VIEW

COMPUTER-AIDED GRAPHICS

DR. BY STEVE JULIAN

EG 110

Figure 9.12 Data tablet used with planning sheet Figure 2.14 to produce auxiliary views in
a static mode for DVST display. (Courtesy Tektronix Inc., Information Display Division)

9.10 CHAPTER SUMMARY

This chapter has presented the ideal
manner for auxiliary view construction. It
is typical of most industrial approaches
and students should be aware that this is
the correct technique. However, education
is not industry and a much more economical
method exists for presenting auxiliary views
in a learning situation. The use of a DVST
can greatly reduce the cost of providing a
CRT and light pen. However, a pointing device must be used. An excellent choice is
a small data tablet as shown in Figure 9.12
and introduced in Chapter 4 of this book.
The data tablet is used connected to the
terminal so that the stylus can locate the
points necessary to locate a viewport for

any type of auxiliary view required.

It is common practice to use a clear
plastic overlay on top of the data tablet.
This overlay would contain those items shown
in Figure 9.1. When the stylus is placed
over the menu bar and pressed, the proper
function is relayed to the face of the DVST
and in this manner the light pen and CRT
can be simulated. Of course, dynamic graphic
display is not possible on the lower order
DVST terminal models. So the move command
is really an ERASE & RELOCATE command as
there is no provision for shifting an image
on the DVST. Likewise, the other menu bar
items are modified to represent the DVST
limitations of direct display. In other
words the ERASE function is performed first
and then TURN, SIZE, or GRID functions.

10

Advanced Theory

At about this point in any textbook a reader begins to ask, " What good is all this stuff, and how does it differ from the manual drafting I have been doing ?" This is where you get your answer.

Automated drafting is not an end in itself, if it is properly done it is a beginning. The beginning to a sequence which includes graphics, design and production or manufacture. The whole idea behind drafting is communication of design intent. With an automated drafting system we can include all the information for design analysis,(testing) prototype construction,(experimentation) and manufacture of a product.

10.1 DESIGN AND MANUFACTURING NETWORK

If we assume that the drafting automation discussed in the first nine chapters is only one part of an automated network, what are the other parts, and what do they look like? To study this, examine Figure 10.1, it is a design/manufacturing network diagram. It indicates that drafting is a function that can be automated and follows the automation of engineering design. Manufacturing planning, production control, tool design and manufacturing operations follow.

Notice that a stream of information connects all of these functions, it is the data base for the entire network. While this book is particular in its treatment of the drafting function, the other functions can not be ignored. All drafting personnel must understand how the network is organized so that they can get information from and put information into the data base. Figure 10.2 is a better illustration of the total data base for the network. Notice that it is CORE divided into 2-D, 3-D and non-graphic. When Figures 10.1 and 10.2 are compared, an understanding of function is clear. In Figure 10.1, starting at the left of the diagram, engineering design has two main data types:

1.   Geometric modeling (input)

2.   Finite element modeling (input/utput). The Core provides for I/O, 2-D/3-D data base types.

Moving to the right, drafting has five

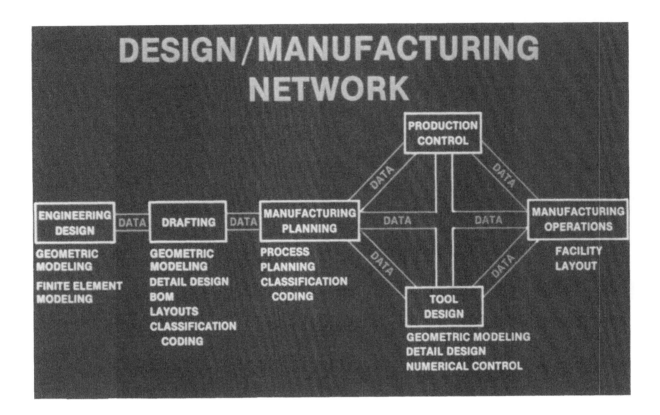

Figure 10.1 Design/manufacturing network. (Courtesy Gerber Systems Technology, Inc.)

main data types:

1. Geometric modeling (shared with engineering design and used as basic input for all drawing--output)

2. Detail design (output)

3. Bill of materials (input to manufacturing planning)

4. Layout drawings (input to production control and manufacturing operations)

5. Classification coding (input to tool design and manufacturing operations)

Moving along the network, manufacturing planning has two main data types:

1. Process planning (input from drafting, output to production control and tool design)

2. Classification coding

The network branches into production control, tool design and manufacturing operations which have four main data types:

1. Geometric modeling (input from both drafting and engineering design)

2. Detail design

3. Numerical control (automated machine tool operation)

4. Facility Layout

10.2 DATA BASE COMPONENTS

Figure 10.2 gives a good representation of the types of data bases and how they are used. The CORE representation is:

1. 2-D, schematic diagram representations of all types.

2. 3-D, geometric modeling, detail drafting, finite element modeling and numerical control.

3. Non-graphic, bill of materials and schedules of all types.

These data base components are used to produce the following items in the remainder of this chapter:

- Parts
- Patterns
- Symbols
- Output files(data sets)

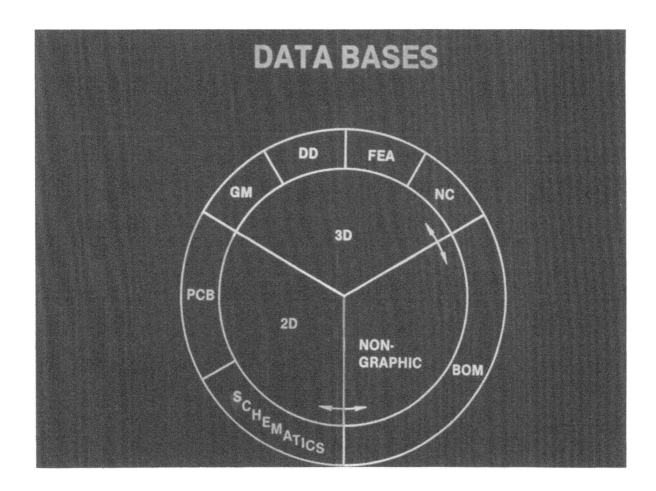

Figure 10.2 Data base components. (Courtesy Gerber Systems Technology, Inc.)

10.3 PARTS OF THE SYSTEM

An operating system for the entire network is shown in Figure 10.3. This is known
as the software architecture, and it is in
two main parts. One part is the application
control and the other support of these applications. The main functions listed earlier
are manipulated through the graphic subrou-
time package described in the first nine
chapters of this text. The subroutines become useful for each application through the
skillful use of data base access routines.
These routines are:

    ARC .... variation of XARC, YARC, ZARC
    ANGLE .. modification of rotation
    ATANGL . display other that THETA
    BEARG .. bearing of a line
    CENTER . indicates tool travel --CNTRL

    CYLIND .. circular cylinder tool form
    DATAPT .. space location in 3-D
    DOT...... tool point -solid object
    GO ...... tool control
    HOME .... origin of tool position
    IN ...... inside tolerance for tool
    LINES ... tool path -- displacement
    OUT ..... outside tolerance for tool
    PLANE ... three or more DATAPT's
    PT ...... 2-D space location
    R ...... radius of part feature
    RTHETA .. arc segments along radius
    SLOPE.... angle in degrees of feature
    TANTO .. tangent to
    UNIT ... display element size
    VECTOR .. magnitude and direction
    XYPLAN .. surface on a planning model
    XZPLAN .. surface on a planning model
    YZPLAN .. surface on a planning model

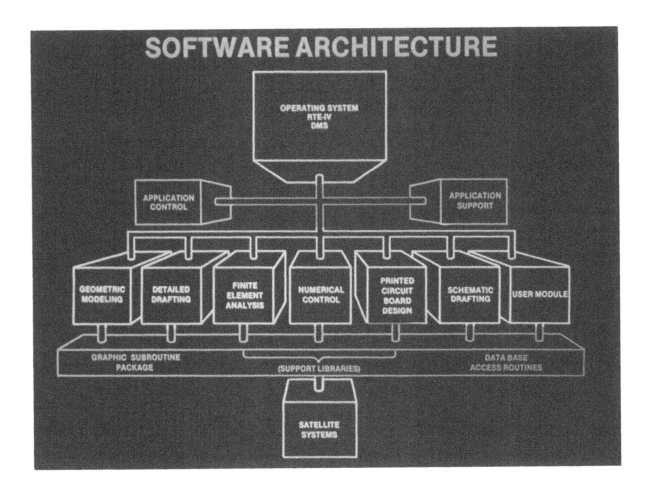

Figure 10.3 Parts of the software architecture. (Courtesy Gerber Systems Technology, Inc.)

The listing of data base access routines are used by the processing unit of each satellite system as shown in Figure 10.4. Notice that the communications controller keeps us in touch with the network described in Figure 10.1 and the various parts shown in Figure 10.3. The items shown in Figure 10.4 have been described in detail throughout the first nine chapters, but now you can see how they fit into the overall network approach to advanced theory. The main concern here is to understand these access routines and the applications of these to network operations.

10.4 ARC

The ARC routine generates a pen path, CRT beam, or tool path which is composed of tangential slope lines to the desired or true arc. As indicated in Chapter 6, these lines cover minute distances which change very rapidly for drafting output, or tool control. When displaying arcs the center of arc radius must be given to the graphics routine. The I and J dimensions in Figure 10.5 provide the information for the arc center offset in 3-D along the Z axis. I represents the distance along the X axis between the beginning of the arc and the center of the arc. J, in turn, is the Y distance offset. The end point of the arc is given as DX and DY in notation and as X and Y distances in an access routine. This differs from the XARC, YARC, ZARC subroutines shown in Chapter 6. Here the entire arc was always in the same plane. In access routines the starting and ending

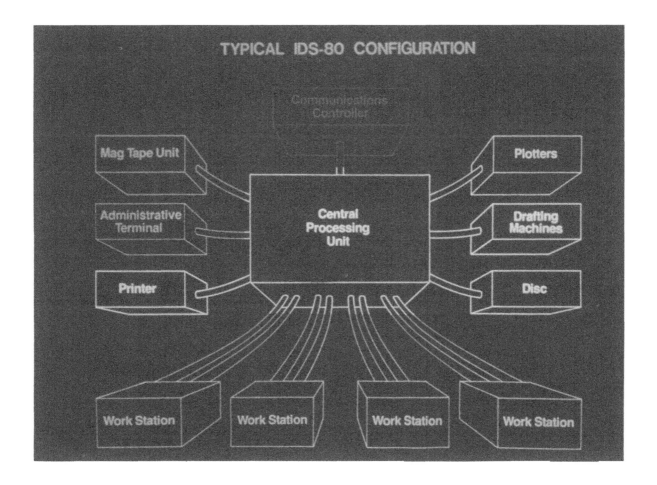

Figure 10.4 Relationship between drafting user and system network. (Courtesy Gerber Systems Technology, Inc.)

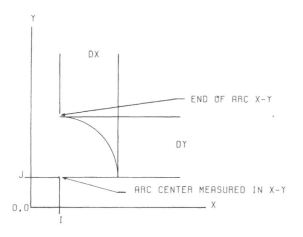

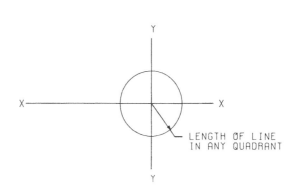

Figure 10.5 Arc access routine.   Figure 10.6 Angle access routine.

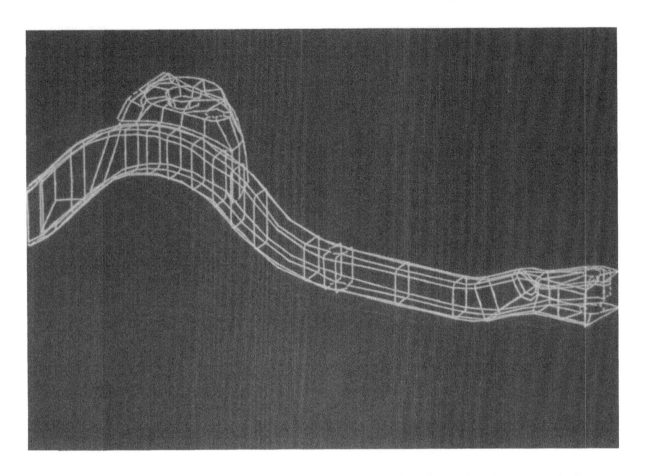

Figure 10.7 Example of access routines ARC, ANGLE, ATANGL and BEARG. (Courtesy Gerber
Systems Technology, Inc.)

points do not have to be in the same plane.
For example, an arc can be produced from a
different display (Z axis) depth as shown
in Figure 10.7.

10.5 ANGLE

This access routine is used to describe
any engineering design angle less than 90
degrees that is measured from the intersection of two axes on a 2-dimensional plane
surface within a 3-D model. Two examples
are shown in Figure 10.7; X-Y plane and
Y-Z plane. In Figure 10.6, a 2-D example
of the intersection of the X and Y display
axis is demonstrated. Using the ANGLE routine, an engineer can send a 3-D model of
a machine part to a drafters terminal for
inspection and detail drawing development.

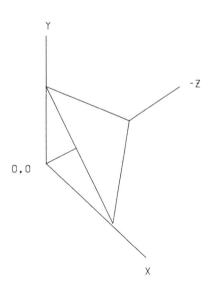

Figure 10.8 Atangle access routine.

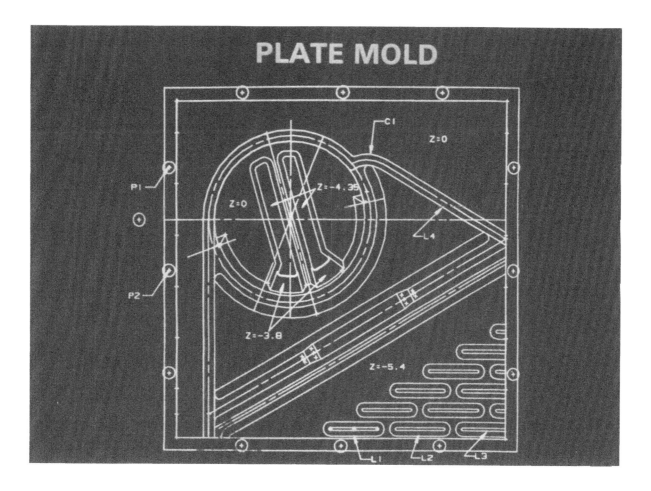

Figure 10.9 Example of ATANGL, BEARG, and CENTER access routines. (Courtesy Gerber Systems Technology, Inc.)

10.6 ATANGL

All engineering design angles greater than 90 degrees are accessed through ATANGL. Figure 10.9 demonstrates this concept which generates from the center of a circle by the rotation of the cutter. This then produces a tool path, direction of cut, depth of cut, and profile of the finished part. The plane of the cut may be at any convenient angle as shown in Figure 10.8. Atangl is used to describe the location of a manufactured plane surface. It is referenced by the angle between the reference axes and the surface as shown in Figure 10.8. Examples of manu-factured plane surfaces are shown in Figure 10.9 above. The plate mold is a flat sur-face plate before the cutting tool removes the material to form the geometric profiles.

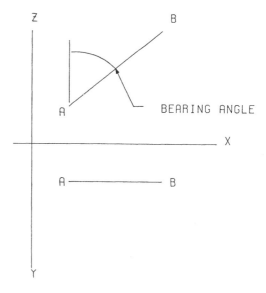

Figure 10.10 Bearg access routine.

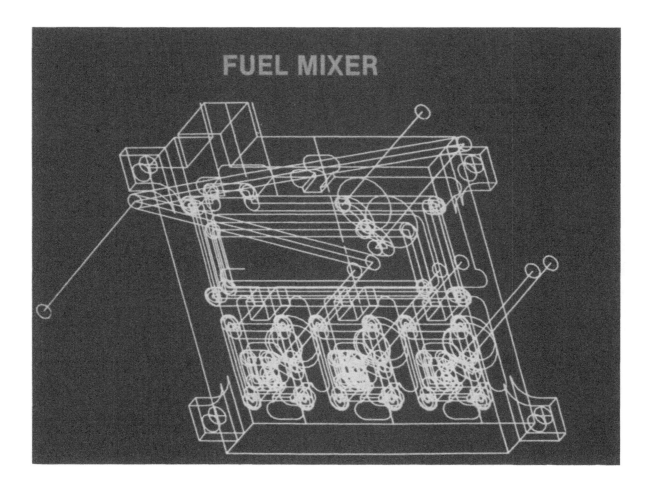

Figure 10.11   Example of CENTER, CYLIND, and DATAPT. (Courtesy Gerber Systems Technology, Inc.

10.7   BEARG

The flat piece part shown in Figure 10.9 is typical of the use for the access routine shown in Figure 10.10. A bearing angle is read from the the horizontal viewport and helps in the specification of the tooling.

10.8   CENTER

Figure 10.11 clearly demonstrates that a CENTER location need not be in two dimensions with a Z-axis reading of zero. If the center of a circle which represents the diameter of a tool has coordinate readings of 4 in X, 3 in Y, and 5 in Z, the circle image would appear smaller than the actual size of a tool in two dimensions or could appear as an ellipse as shown in Figure 10.11. Of course this is an optical illusion but necessary if a draftsperson is to enter this into the data base so that manufacturing operations can produce the require part as shown in Figure 10.12. Compare the design intent shown in Figure 10.11 with the finished product shown in Figure 10.12. Another reference to Figure 10.1 shows the left to right movement of design information called automated data base. Figure 10.3 indicates the flow of information from the access routines to the drafting phase and finally into the data base for the automated manufacture of the finished part. Drafting becomes the link between the engineer's idea and the machine tool operations necessary to control the final form of the product. Automated drafting now becomes more than the production of drawings.

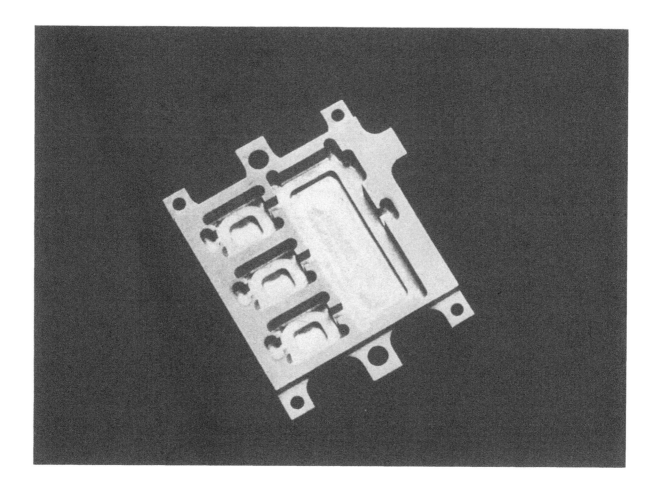

Figure 10.12 Finished part. (Courtesy Gerber Systems Technology, Inc.)

10.9 CYLIND

A circular cylinder is produced by a generation of a plane surface, denoted as a circle in Figure 10.13, through space. Access routines like CYLIND are useful in displaying the many positions that a tool may take during a machining operation.

The finished part shown in Figure 10.12 is a result of the proper use of CYLIND in Figure 10.11. Of course CYLIND is used with other access routines to define the part outline, tool travel, material removal rates and depth of cuts used to machine a finished part. A CYLIND routine can be displayed with a top and bottom circle and centerline as shown in Figure 10.12 or a pair of tangent lines may be displayed by the use of TANTO as shown in Figure 10.13.

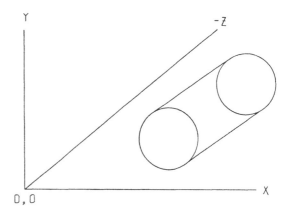

Figure 10.13 Cylind access routine.

Figure 10.14 Example of DATAPT, DOT, and GO access routines. (Courtesy Gerber Systems Technology, Inc.)

10.10 DATAPT

A data point is a 3-dimensional location for a point, expressed in display units
and direction of travel. Data points are
more easily used when stored in a computer
on a space model as shown in Figure 10.15.
Points are used to specify graphic location
for other access routines as well, Figure
10.14 is a good example of this.

10.11 DOT

The access routine DOT is an example
of a graphic shape that can be located by
DATAPT as shown in Figure 10.14 above. A
dot pattern is more common than the use of
a single dot. The electronics fabrication
industry makes excellent use of DOT.

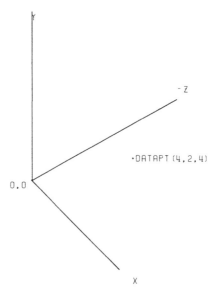

Figure 10.15 Datapt access routine.

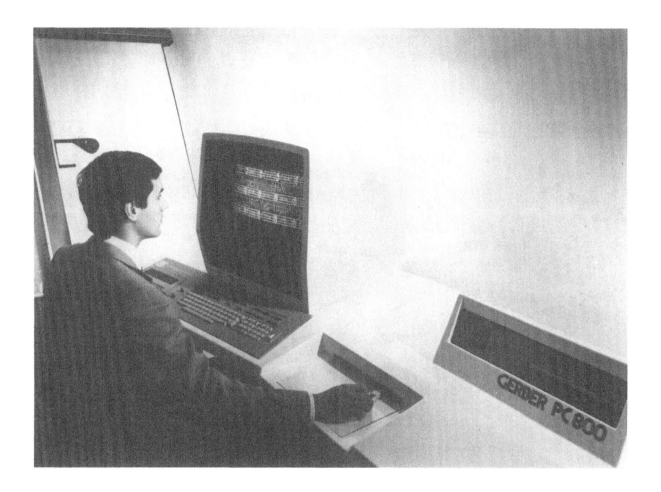

Figure 10.16 Draftsman uses access routines GO, HOME, IN and OUT along with LINES to produce data base information. (Courtesy Gerber Systems Technology, Inc.)

10.12 GO

A GO access routine is used to indicate either pen motion or tool motion. It
can be shown directly as the MOVE graphics
command learned in earlier chapters or the

    CALL PLOT(XPAGE,YPAGE,3)

computer program statement.

10.13 HOME

A home access routine returns the pen
to the origin or returns the tool to the
starting position. The HOME position is
clearly shown in Figure 10.11 and will appear again in many of the later diagrams.

10.14 IN

When 3-dimensional objects are displayed
in a 2-dimensional display area, two important concepts are used. They are inside
and outside control boundaries. IN is used
to control the boundary closest to HOME.
Figure 10.17 illustrates the control necessary for manufacturing operations, while
Figure 10.18 illustrates the concept for
drafting output.

10.15 LINES - VECTORS

The access routine LINES is probably
the most used routine in the examples shown
in this chapter. It can be directly compared to DRAW in the graphics commands or

Figure 10.17 Control from IN and OUT access routines during manufacturing operations. (Courtesy Gerber Systems Technology, Inc.)

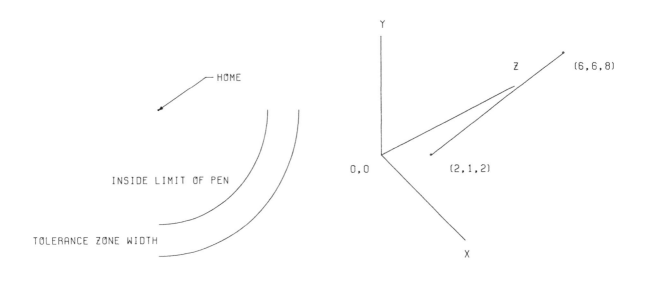

Figure 10.18 In and Out access routines.   Figure 10.19 Lines access routine.

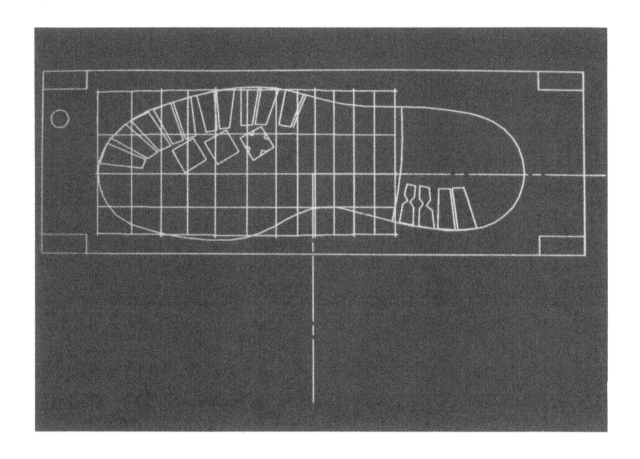

Figure 10.20 Examples of OUT, PLANE, PT, R, RTHETA and TANTO. (Courtesy Gerber Systems Technology, Inc.)

CALL PLOT(XPAGE,YPAGE,2)

computer program statement.

10.16 OUT

This access routine controls the outside of the tool path farthest from HOME. Figures 10.17 and 18 demonstrated this.

10.17 PLANE

Shown in Figures 10.20 and 21, a PLANE has 3-dimensional data and at least three data points.

10.18 PT

Shown in Figure 10.10, a PT is a two-dimensional representation of DATAPT.

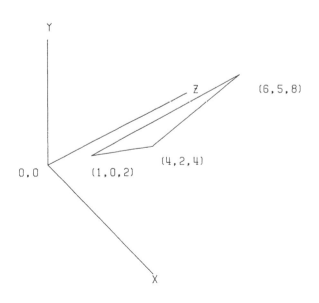

Figure 10.21 Plane access routine.

Figure 10.22 Manufacturing operation from data base supplied from Figure 10.20. (Courtesy Gerber Systems Technology, Inc.)

10.19 R

Without the R access routine the operations shown in Figure 10.22 could not
be prgrammed. R is the notation or storage
symbol for the radius of a circle or the
arc of a surface.

10.20 RTHETA

This access routine describes the tool
path taken in Figure 10.22. It is illustrated in Figure 10.23 and demonstrated for
modeling purposes in Figure 10.24.

10.21 SLOPE

Any line that can not be read from a

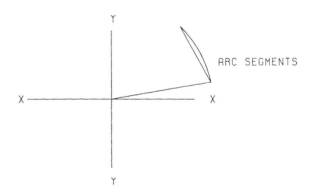

Figure 10.23 Rtheta access routine.

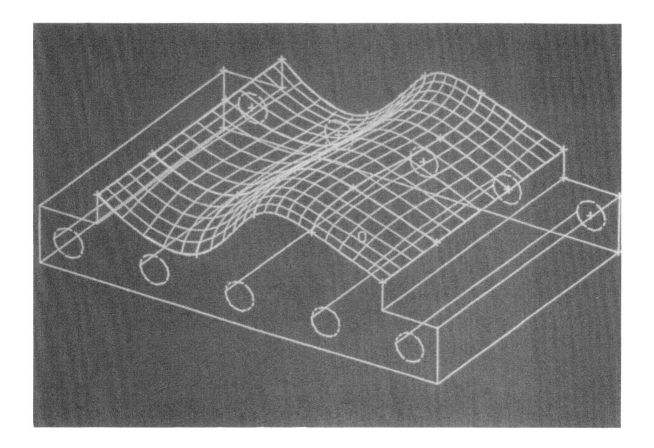

Figure 10.24   Example of RTHETA, SLOPE, TANTO and PLANE; continuation of Figures 10.20 and 10.22.   (Courtesy Gerber Systems Technology, Inc.)

data base location must be calculated and then stored for later use.   One form that a line can take is SLOPE.

10.22   TANTO-UNIT

Tangent to is a symbol for computer use and describes the exact point at which two graphic shapes are tangent. This enables the access routines to adjust to the grid units being used for display or manufacture.

10.23   XY,XZ,YZ PLAN

Shown in Figure 10.25 as XYPLAN it is composed of a grid matrix of units.   This allows plane surface features to be added to a planning, detail, or manufacturing display as shown in Figure 10.26.

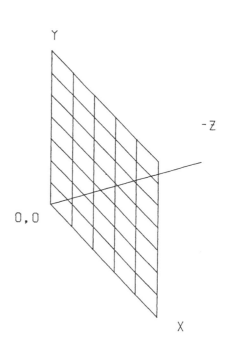

Figure 10.25   Xyplan access routine.

Figure 10.26 Addition of a plane surface; continuation of figure sequence 20-26. (Courtesy Gerber Systems Technology, Inc.)

10.24 REVIEW PROBLEMS

The following 12 problems are keyed to Figures 10.27 through 10.39. These Figures do not have captions, but represent a typical output that each problem requires.

1. Prepare a detailed drawing as it might come from the DRAFTING function shown in Figure 10.1. A typical output is shown is Figure 10.27.

2. Prepare a geometric model of this part that looks like Figure 10.28 so that MANUFACTURING PLANNING can design a holding fixture shown in Figure 10.29.

3. Prepare a geometric model for profile milling of the part as shown in Figure 10.30.

4. Next provide a tool path display for the rib milling required for this part as shown in Figure 10.31 .

5. Provide a production control display as shown in Figure 10.32.

6. Prepare manufacturing data base to produce the part shown in Figure 10.33.

7. Produce the display file structure shown in Figure 10.34.

8. Use this file structure to display a plane projection as shown in Figure 10.35.

9. Output the FEA model shown in Figure 10.36 that is contained in the ENGINEERING DESIGN.

10. Use the TANTO access routine to display Figure 10.37.

11. Use the DOT and LINES access routines to dispay Figure 10.38.

12. Use access routines to display Figure 10.3

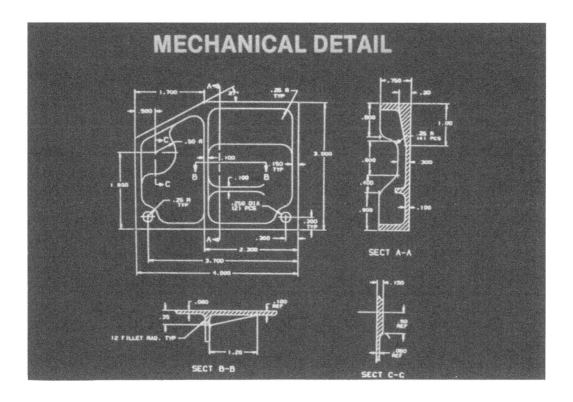

Figure 10.27   Problem 1 (Courtesy Gerber Systems Technology, Inc.)

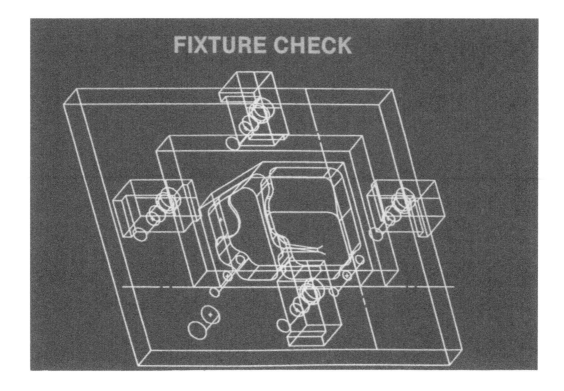

Figure 10.28   Problem 2 (Courtesy Gerber Systems Technology, Inc.)

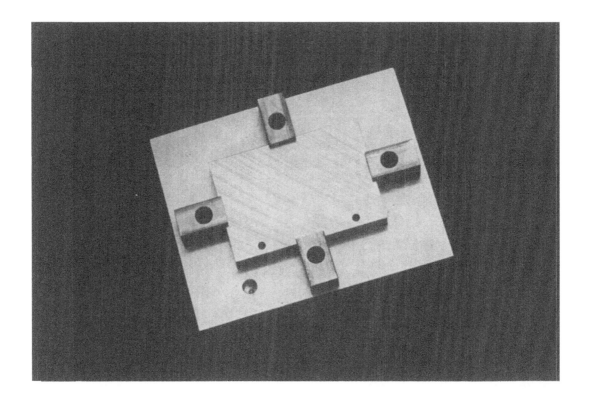

Figure 10.29 Problem 2 (Courtesy Gerber Systems Technology, Inc.)

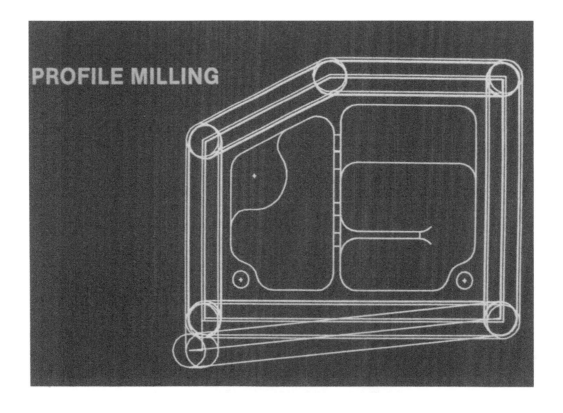

Figure 10.30 Problem 3 (Courtesy Gerber Systems Technology, Inc.)

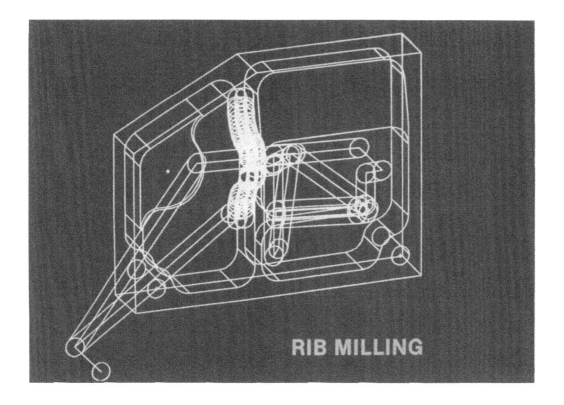

Figure 10.31   Problem 4 (Courtesy Gerber Systems Technology, Inc.)

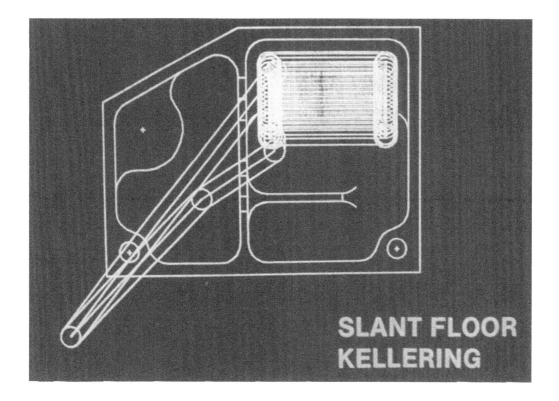

Figure 10.32   Problem 5 (Courtesy Gerber Systems Technology, Inc.)

Figure 10.33   Problem 6 (Courtesy Gerber Systems Technology, Inc.)

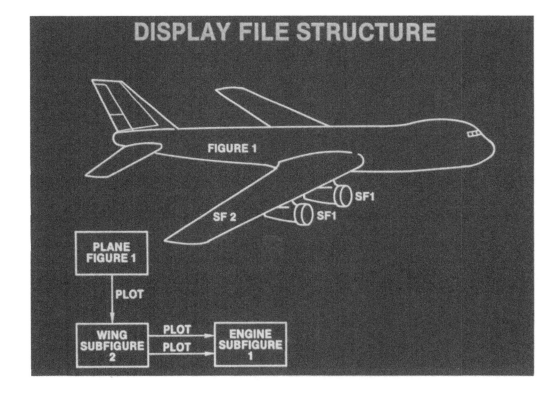

Figure 10.34   Problem 7 (Courtesy Gerber Systems Technology, Inc.)

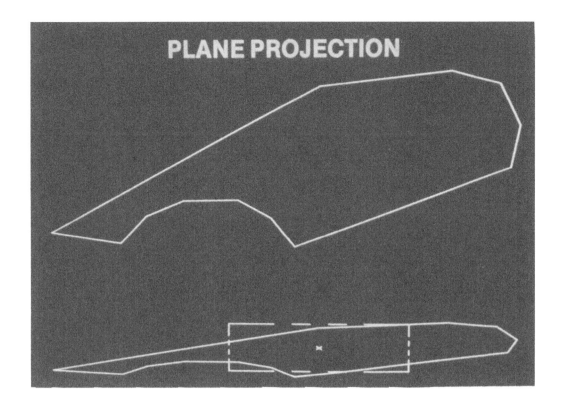

Figure 10.35   Problem 8 (Courtesy Gerber Systems Technology, Inc.)

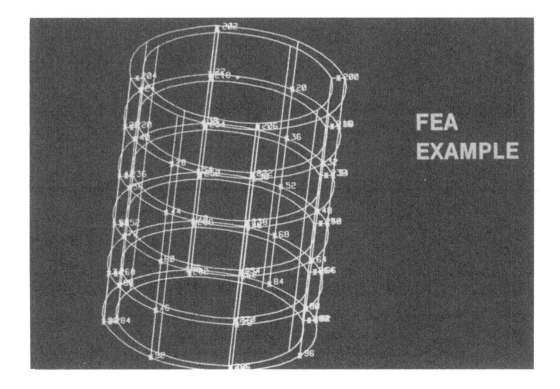

Figure 10.36   Problem 9 (Courtesy Gerber Systems Technology, Inc.)

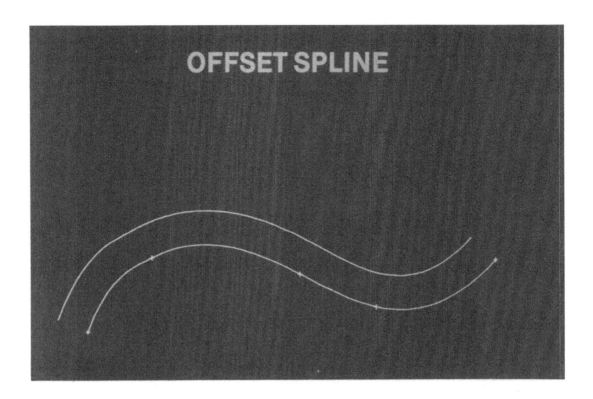

Figure 10.37   Problem 10 (Courtesy Gerber Systems Technology, Inc.)

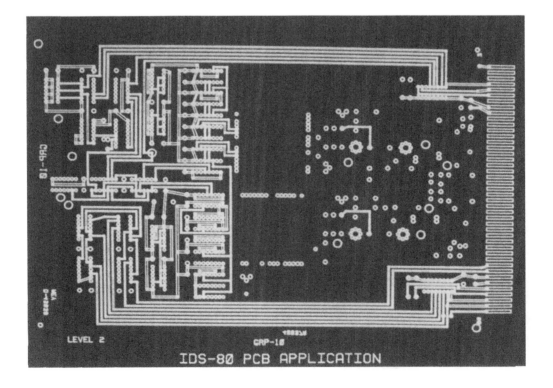

Figure 10.38   Problem 11 (Courtesy Gerber Systems Technology, Inc.)

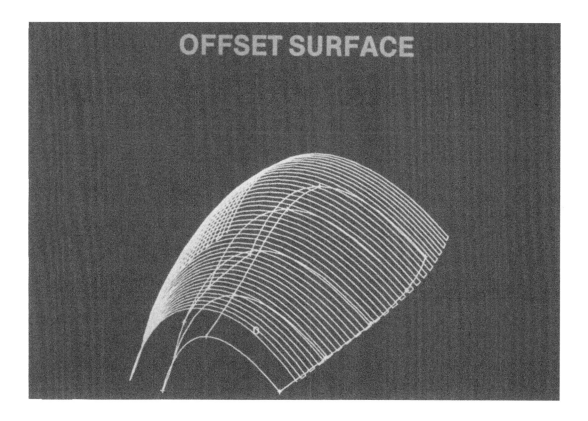

Figure 10.39   Problem 12 (Courtesy Gerber Systems Technology, Inc.)

10.25   CHAPTER SUMMARY

This chapter is presented so that you might see an application for the information presented in the first nine chapters. The chapter is entitled "advanced theory" and for some of you it will be advanced indeed. Read the chapter, review the problems at the end and study the answers to these problems in Figures 27 through 39. The application of any theory is always more enjoyable and easier to understand than the consideration of where the applications come from. Still you should aware that a design data base exists and that as a draftsperson you must enter and exit this data stream shown in Figure 10.1.

Automated drafting is a complex subject, it is not a new coat of paint for a few manual techniques used at the beginning of this century. By learning the proper theory, you will better understand your role in a computer age.

11

Pictorials

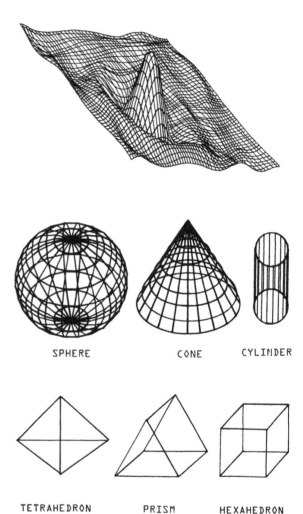

In the previous ten chapters the importance
of data bases to form plane surfaces was
discussed. The 3-dimensional data base was
used to present everything from working -
multiviews to planes used as section devices.
As demonstrated in some chapters, a plane
can be used to cut through any 3-dimensional
space problem and expose the interior detail
for dimensioning. The creation of solids
was not presented.

In Chapter 10, 3-dimensional concepts
were presented as transparent objects. Refer back to them again for further study;
they are the foundation concepts for pictorial representation of solid objects. In
this chapter we shall study the use of a
plane surface as a boundary for solids. A
surface forms a boundary completely enclosing
a portion of space in three dimensions and
becomes a solid. The term solid is used in
its mathematical sense. A mathematical
solid is considered to be hollow. It is
constructed of plane surfaces defined by
edges. These edges are lines defined by
datapts as described in Chapter 10. The
first step in solid construction is the

SPHERE   CONE   CYLINDER

TETRAHEDRON   PRISM   HEXAHEDRON

Figure 11.1 Common types of mathematical
solids; sphere, cone, and cylinder (finite
element models); tetrahedron, prism, and
hexahedron (wireform models). Engineering
models tend to be FEM or FEA (finite element), while drafting models are wireform.

information contained in the data base. Is it in the form of an engineering model, detail drafting model or some other form? In Figure 11.1 the common types of mathematical solids are shown. Notice that engineering models contain a large number of lines, and the drafting models tend to be much simplier in their data base construction. Drafting type models were shown in Chapter 6, pages 100 through 108 you might want to review how the data base is formed.

11.1 TYPES OF PICTORIAL DRAWINGS

Pictorial drawings are prepared for both the engineering and the drafting functions shown in Chapter 10. The engineering pictorials are mainly geometric models for finite element experimentation, the drafting pictorials fall into three main groups: axonometric, oblique, and perspective.

Common display techniques exist for all three of these types, and before study-each one separately, we will cover these common techniques. Referring again to Chapter 10, we note that a plane surface was defined as at least three points connected in space. This is true for a triangular plane. The addition of a fourth point not inside the triangular plane creates three additional plane surfaces or a four-sided polyhedron (tetrahedron shown in Figure 11. 1). A polyhedron is a composite of points, lines, and planes.

The addition of a sixth point in space creates a six-sided polyhedron (hexahedron shown in Figure 11.1). The most common hexahedron is the cube. Figure 11.2 represents a cube made up of eight points, twelve lines, or six planes. The purpose of the program shown in Figure 11.2 is to display the hexahedron on a digital plotter. The display may be anywhere on the plotter surface,

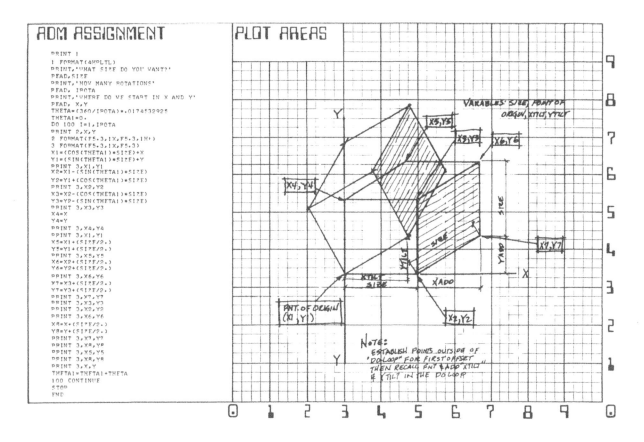

Figure 11.2 Preparing a pictorial display. (Courtesy DLR Associates, Clemson, South Carolina)

drawn any size, and rotated at any desired
angle for viewing. It should be emphasized
that any pictorial representation of a mathematical solid is merely an array of points,
lines, or planes. To display objects in a
pictorial projection system requires only
a simple extension of the 3-dimensional
orthographics developed for projecting lines
and planes discussed in Chapter 6. In review problem number 8, Chapter 6, a subprogram CUPID1 (commonly used pictorial
image display) was used to produce a type
of pictorial view. This view was plotted
along with the orthographic views for reference as shown in Figures 6.15, 6.25, 6.
38, 6.42, and 6.44. Depending upon the
type of pictorial desired; axonometric,
oblique or perspective, the construction
can be quite simple.

11.2   THEORY OF PICTORIAL CONSTRUCTION

A 3-dimensional subprogram like CUPID1
expands a plotting system's capability to
provide more meaningful information in the
drafting function. These types (CUPID1,
CUPID2, and CUPID3) of subprograms are not
fully developed for pictorial image processing (PIP). But they are an excellent base
from which a few programming instructions
can be added to produce the type of pictorial desired.

Oblique. This is the simpliest of the
PIP routines and is shown in Figure 11.3.
It should be added to the CUPID1 routines
used in Chapter 6 before it is placed in a
drafting software package to be used by all
drafting personnel.

The PIP routines can be compared directly to the CUPID routines FVIEW, HVIEW,
and PVIEW. For example, suppose that a
draftsperson needed just two views instead
of the normal three or six orthographic
views? Then the program statements:

```
        CALL FVIEW(NDATA,X,Y,Z,IPEN)
or
        CALL HVIEW(NDATA,X,Y,Z,IPEN)
```

could be used to position a single view,

```
C       SUBROUTINE FOR DISPLAYING OBLIQUE
C       PROJECTIONS INSIDE CUPID SOFTWARE
C
        SUBROUTINE PIP1(NDATA,X,Y,A,IPEN,ANG)
        DIMENSION X(100),Y(100),Z(100),IPEN
       +(100), XPLOT(100), YPLOT(100)
        DO 100 I=1,NDATA
        XPLOT(I)= X(I)+Z(I)*COS(ANG/57.3)
        YPLOT(I)= Y(I)+Z(I)*SIN(ANG/57.3)
    100 CALL PLOT XPLOT(I),YPLOT(I),IPEN(I)
        RETURN
        END
```

Figure 11.3 PIP routine for oblique pictorials. (Courtesy DLR Associates)

```
C       SUBROUTINE FOR DISPLAYING AXONOMETRIC
C       PROJECTIONS INSIDE CUPID SOFTWARE
C
        SUBROUTINE PIP2(NDATA,X,Y,Z,IPEN,
       +ANG1,ANG2)
        DIMENSION X(100),Y(100),Z(100),IPEN
       +(100),XPLOT(100),YPLOT(100), PIP(100)
        DO 101 J=1,NDATA
        XPLOT(J)= X(J)*COS(ANG1/57.3)+Z(J)*
       +SIN(ANG1/57.3)
        PIP(J)= X(J)*SIN(ANG1/57.3)*SIN(ANG2
       +/57.3)+Y(J)*COS(ANG2/57.3)
        YPLOT(J)= PIP(J)-Z(J)*COS(ANG1/57.3)
       +*SIN(ANG2/57.3)
    101 CALL PLOT(XPLOT(J),YPLOT(J),IPEN(J))
        RETURN
        END
```

Figure 11.4 PIP routine for axonometric
pictorials. (Courtesy DLR Associates)

say the top or front in an automated drawing display. The view is positioned by:

```
        CALL PLOT(XTRANS,YTRANS, -3)
```

where XTRANS is the X location on the dis-
play surface and YTRANS is the Y location.
Figure 11.5 is the routine for FVIEW and
Figure 11.6 is the routine for HVIEW.

Axonometric. This is the next PIP routine shown in Figure 11.3. It is very similar to PIP1 which is the logic for the oblique projection. You will notice that there is only one angle of projection used for oblique as shown in Figure 11.8, whereas two angles of projection are used in all axonometric projection.

An axonometric pictorial, also shown in Figure 11.8, is produced when both the horizontal angle (ANG1) and profile angle (ANG2) are greater than zero. If ANG1=0. and ANG2 is greater than zero, an oblique projection results. An axonometric always contains two angles of axis rotation, this produces lines of sight which are perpendicular to the plane of projection but the principal faces of the object are inclined to the plane of projection. The principal faces or axes may be displayed at any angle other than 0 or 90 degrees. Since these angles represent orthographic projections. Because these principal surfaces and edges of an axonometric pictorial object are inclined to the plane of projection, the general proportion of the object will vary, depending upon the placement of the object as shown in Chapter 6. Correction in proportion is done before the final display of the pictorial by foreshortening the lines that are inclined to the plane of projection. Therefore, the larger ANG1 and ANG2, the greater the degree of foreshortening.

Foreshortening is handled by adding an additional variable in the subroutine PIP1 or PIP2. It is called SF in PIP1 and reduces the Z data list so that different types of oblique projections such as cabinet or cavalier projection. In PIP2, SF is used to present isometric, dimetric and trimetric projections.

Perspective. A perspective drawing is one which more nearly presents an object as it appears to the eye. A third angle (ANG3) is required to be read into a display routine (PIP3). By this time you have guessed that PIP stands for pictorial image processor and 1 =ANG1, 2=ANG2, and 3=ANG3. In addition to ANG3, perspective pictorials are based on the fact that all lines (edges) which ex-

```
C     SUBROUTINE FOR DISPLAYING THE FRONT
C     ORTHOGRAPHIC VIEW INSIDE CUPID SOFT-
C     WARE SUBPROGRAM KNOWN AS CUPID1
C
      SUBROUTINE FVIEW(NDATA,X,Y,Z,IPEN)
      DIMENSION X(1000),Y(1000), Z(1000),
     +IPEN(1000)
      DO 90 K=1,NDATA
   90 CALL PLOT(X(K),Y(K),IPEN(K))
      RETURN
      END
```

Figure 11.5 FVIEW routine for multiviews.

```
C     SUBROUTINE FOR DISPLAYING THE TOP
C     ORTHOGRAPHIC VIEW INSIDE CUPID SOFT-
C     WARE SUBPROGRAM KNOWN AS CUPID1
C
      SUBROUTINE HVIEW(NDATA,X,Y,Z,IPEN)
      DIMENSION X(1000),Y(1000),Z(1000),
     +IPEN(1000)
      DO 95 L=1,NDATA
   95 CALL PLOT(X(L),Z(L),IPEN(L))
      RETURN
      END
```

Figure 11.6 HVIEW routine for multiviews.

```
C     SUBROUTINE FOR DISPLAYING THE PROFILE
C     ORTHOGRAPHIC VIEW INSIDE CUPID SOFT-
C     WARE SUBPROGRAM KNOWN AS CUPID1
C
      SUBROUTINE PVIEW NDATA,X,Y,Z,IPEN
      DIMENSION X 1000 ,Y 1000 ,Z 1000 ,
     +IPEN 1000
      DO 90 M=1,NDATA
   90 CALL PLOT Z M ,Y M ,IPEN M
      RETURN
      END
```

Figure 11.7 PVIEW routine for multiviews. See Figure 11.8 for a demonstration of Figures 11.3 through 11.7.

tend from the observer appear to converge or come together at some distance point.

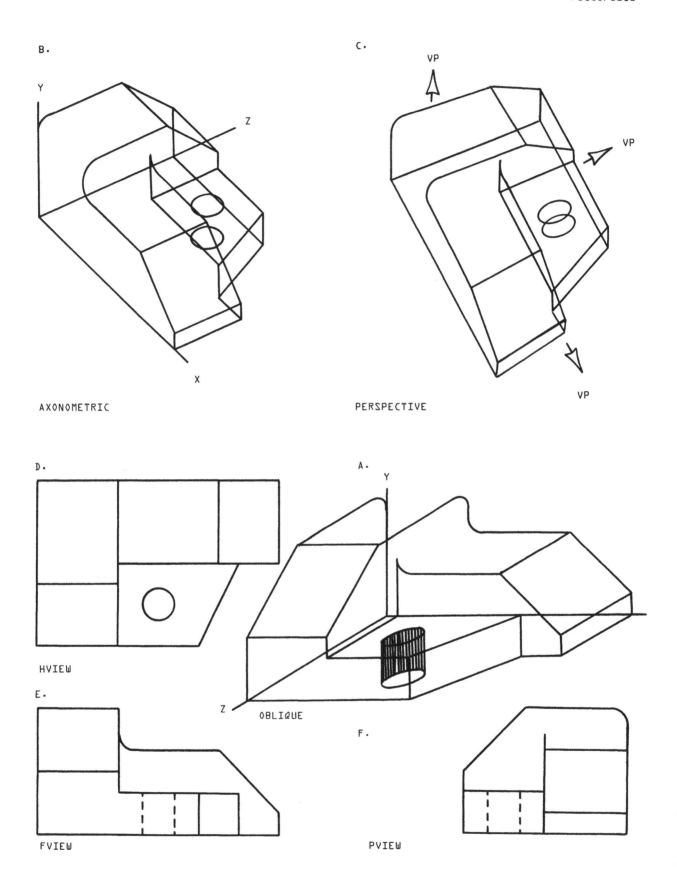

B.

AXONOMETRIC

C.

PERSPECTIVE

D.

HVIEW

A.

OBLIQUE

E.

FVIEW

F.

PVIEW

Figure 11.8 Display of A(PIP1), B(PIP2), C(PIP3), D(HVIEW), E(FVIEW), and F(PVIEW).

A stylized pictorial system as defined by the last six subprograms:

    PIP1, 2 and 3

    CUPID1, 2 and 3

provides 3-dimensional plots of any data which can be expressed as:

    1. PIP = X,Y,Z,IPEN

    2. CUPID = X,Y,Z,ICON

    3. Function of two variables to produce a third.

Using this system, a designer or draftsperson can quickly take data base from Chapter 10 and produce pictorial drawings of almost any part. Because the data base is stylized, many pictorials can be created from many different angles in a shorter time than would be required to manually prepare one. Consider Figure 11.9, it was produced and output to an electrostatic plotter in six seconds.

11.3 METHODS OF PICTORIAL DISPLAY

The pictorial shown in Figure 11.1,11.2, 11. 11.8 and 11.9 is known as a wireform because they are separate wires representing connections in the data base. There are other methods used to present pictorials and these are more common in drafting applications.

For example, a user wanting to streamline the production of pictorial images of reasonable simple, 'unscultured' parts and assembilies will find that another method can be used merely as a powerful drafting system. Another application can be the creation of complet pictorials so that a tool designer can verify that the part machined wil have the proper dimensions and shape. Other methods can also be used to 3-dimensionally illustrate a geometric formula, to plot distribution of heat and/or pressure over a surface, or to plot geometrically complete mathematical models of machine parts as shown in Figure 11.10. This improvement of visualization can instantly provide insights and correlations which might require long scrutiny during the manufacturing process planning, assembly planning, and functional design. A

user's imagination can generate almost an unlimited number of applications for pictorials. Pictorials can be an outstanding educational or presentation tool because

1. They can automatically exclude lines (wires), at user option, so that a reference axis may be used to show orientation. Plane view projections in the horizontal, frontal, and profile can be created without grid lines for realistic surface definition (see Figure 11.8).

2. Pictorials can be overlapped to show any irregularly spaced DATAPT's and to test the line elimination technique (see Figure 11.11).

3. Pictorial image generators can produce several thousand separate pictorial drawings of a 3-dimensional object, each with a separate viewing angle and viewing distance, including line elimination. This makes possible the creation of inexpensive animated scripts for product simulation or design modification (see Figure 11.12).

4. Additionally, a data base can be plotted from a completely flexible choice of angle to show orthographic and pictorials side-by-side (see Figure 11.8).

5. Pictorial routines are written in simple coding statements as shown in Figure 11.3 and 11.4, so they will operate on the vast majority of computers.

11.4 DISPLAY SPECIFICATIONS

To advance beyond the simple point plotting of wireform models, the user replaces the usual CALL PLOTS or CALL INITT with CALL CUPID. All subroutines used in CUPID are designed for X,Y, and Z data input. The program symbology outlined in Chapter 10 is used throughout the logic of CUPID software routines. There are two basic definitional concepts embodied into three layers or versions of CUPID. The first concept is that of a wireform already discussed as an 'ideal' part- a hypothetical solid having perfect form and shape. Nominal part models are also possible (wireforms with lines removed) and are defined via projective geometry. That is, each in-

Figure 11.9 Multiple wireform projection from common data base.

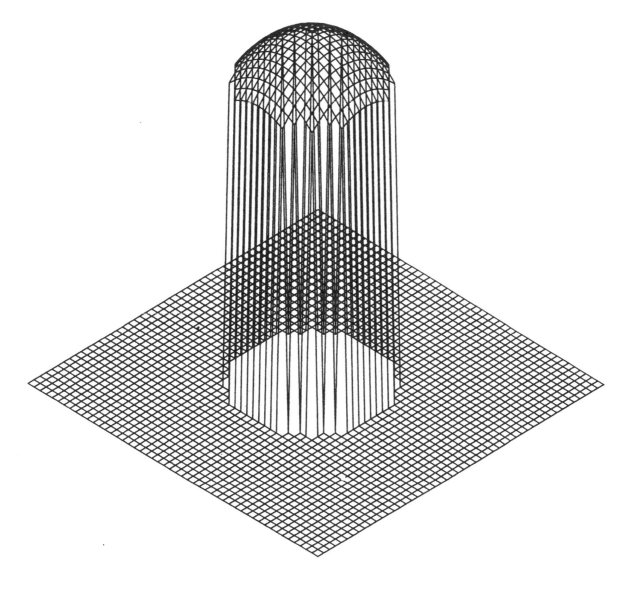

Figure 11.10   CUPID software can produce pictorials which represent geometric formulas, the distribution of heat and/ or pressure over a surface, or geometrically complete mathematical models of machine parts or tools.

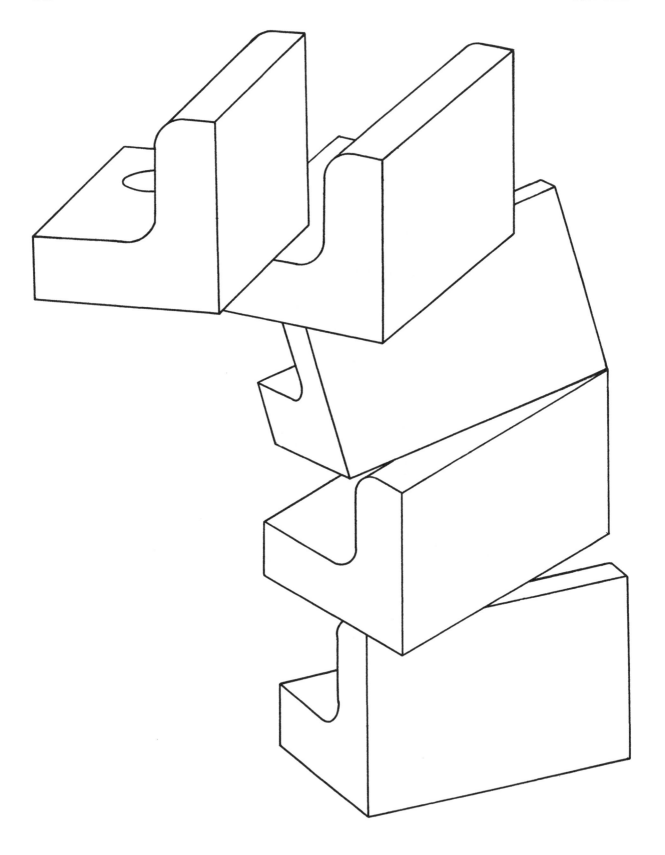

Figure 11.11   Testing the line elimination technique in CUPID software.

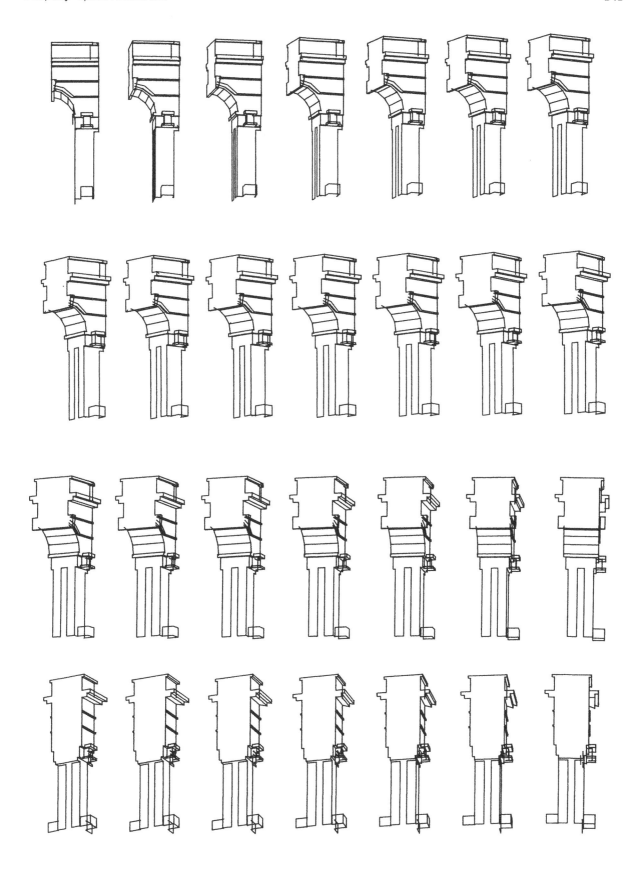

Figure 11.12   3-D objects for animation produced from a pictorial image generator.

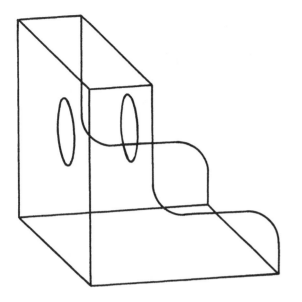

Figure 11.13 Wireform from XPLOT,YPLOT data.

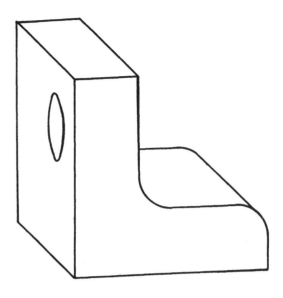

Figure 11.14 Solidform from ELINAT.

tersection of a wire is defined in units of
X, Y, and Z.

The second basic concept is that the
order of and method of point connection is
left entirely up to the user. As the part
model is placed on the worksheet shown in
Figure 11.2, each point location is given
a number and recorded as a data base. This
flexiblity produces several attributes

1. Models may be shown transparent
as in Figure 11.13 or as solid objects as
in Figure 11.14.

2. Modifications to part geometry can
be made quickly, notice a line was added in
Figure 11.14 to show rounded edge.

3. Line elimination is completed on
3-dimensional data base that has been transformed into plotter or screen coordinates;
XPLOT, YPLOT. This makes line elimination
an easier problem and can be done by the
following subroutine:

```
      SUBROUTINE ELINAT(XPLOT,YPLOT,IPEN,MODE)
C......................................
C   ON INITAL CALL SET MODE = 1 ..........
C   IPEN = DEVIATION ACCEPTED FROM DATABASE
C   XPLOT = PLOTTER COORDINATE FROM X,Y,Z
C   YPLOT = PLOTTER COORDINATE FROM X,Y,Z
C......................................
      GOTO(1,2,3)MODE
    1 MODE=1
      X1=XPLOT
      Y1=YPLOT
      PERCEN=FLOAT(IPEN)*.01
      CALL TPLOT(XPLOT,YPLOT,0,0)
      RETURN
    2 GOTO(4,5)MODE1
    4 MODE1=2
    7 X2=XPLOT
      Y2=YPLOT
      IPEN2=IPEN
      S2=(Y2-Y1)/(X2-X1)
      PS2=S2*PERCEN
      SU=S2+PS2
      SL=S2-PS2
      ISS2=(X2-X1)/ABS(X2-X1)
      RETURN
    5 IF(IPEN.EQ.IPEN2)GOTO 6
C.....PEN MODE IS DIFFERENT FROM PREVIOUS.
      CALL TPLOT(X2,Y2,IPEN,0)
      X1=X2
      Y1=Y2
      GOTO 7
C.....PEN MODE IS SAME AS LAST CALL ......
    6 S=(Y-Y1)/(X-X1)
      ISS=(X-X1)/ABS(X-X1)
      IF(S.GT.OR.S.LT.SL.OR.ISS.NE.ISS2)GO
     +TO 8
      X2=X
      Y2=Y
      RETURN
C.....MODE CHANGE, PLOT OLD VECTOR LOCATION
    8 CALL TPLOT(X2,Y2,IPEN2,0)
      X1=X2
      Y1=Y2
      GOTO 7
C.....LAST CALL ..........................
    3 CALL TPLOT(X2,Y2,IPEN,0)
      CALL TPLOT(XPLOT,YPLOT,IPEN,0)
      RETURN
      END
C
C.....CONTROL RETURNED TO CUPID SOFTWARE..
```

Subroutines like ELINATE are necessary if plotters or direct view storage tube displays are used. If CRT's like those shown in Chapter 9 are used, then the use of the subroutine ELINAT is not needed due to the selective erase function. The selective erase function should be used because it is faster for the drafter to point with a light pen than for a subroutine to logically check for visibility in each pictorial in a set of 100 or 1000.

Other methods than subroutine ELINAT and a light pen are also available. One such method is the use of a wireform and surface shading between wires. The draftsperson may call for shading as many times as necessary to describe the part as a solid. Figure 11.15 demonstrates a wireform model and the same data presented with shading between wires. The surface shading on visible sides covers the wires that appear in the rear of the pictorial. The drafter 'build up' the desired solid part by joining as many shaded surfaces as necessary. Because this approach produces geometrically complete models, the shading subroutines can generate any graphic representation of a part that can be defined in a mathematical data base. This data base is a mapping from 3- to 2- dimensional Euclidean space. The pictorials produced in this manner are still hollow, and if rotated from their base, a designer can go inside the pictorial and look out. Depending upon what the pictorial represents, this can be an advantage.

11.5 DATA GATHERING TECHNIQUES

Several methods of obtaining 3-dimensional data points for pictorial display are used. A worksheet method is comparable to the manual X,Y, and Z points obtained in Chapter 2. This is a tedious job at best but is used when other methods are not available. A comparison of 2-dimensional data gathering and 3-dimensional techniques is shown in Table 11.1.

Any of the methods described in Table 11.1 can be used to generate pictorial data. A combination of methods is oftentimes used

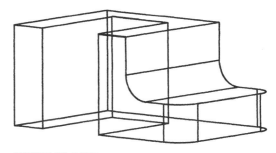

WIREFORM DATA

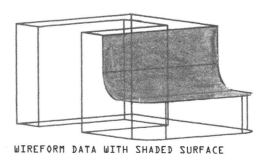

WIREFORM DATA WITH SHADED SURFACE

Figure 11.15   Soilds by surface shading.

Table 11.1

Data Gathering Techniques Used in Pictorials

Two-dimensional methods	Three-dimensional
2-d worksheet	3-d worksheet
Coordinatograph	Cordax machine
Graphics Tablet	3-d sonic digitizer
On-the-fly digitizer	Laser digitizer
Scanner	Model measurer

as shown in Figure 11.16. In Figure 11.16 the drafter is using a digitizer which will record the pictorial data from a movie film. Any similar combination of methods like in Figure 11.16 is a mechanical technique for recording data from an existing source; a drawing, model, machine part, movie or animation sequence, or the like. The most desired techniques were explained in Chapter 6, multiview projection. Here the data is gathered in three basic steps. First, the shape of the object is described in straight line segments, usually three or four lines will form a flat surface. This flat sur-

Figure 11.16 Tektronix researcher digitizes an object projected on a screen, building a pictorial data base. (Courtesy Tektronix Inc.)

face is then coded for the amount of shade required. Shades of gray can be produced as shown in Figure 11.17 and 11.15. These shading codes can be assigned in any order desired to each surface of a pictorial as illustrated in Figure 11.18.

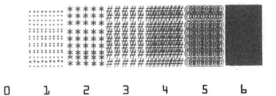

Figure 11.17 Shading code compares to new-pen values for line images.

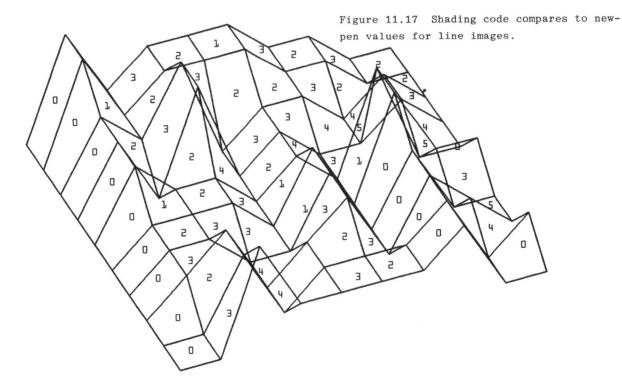

Figure 11.18 Shading codes assigned to pictorial surfaces.

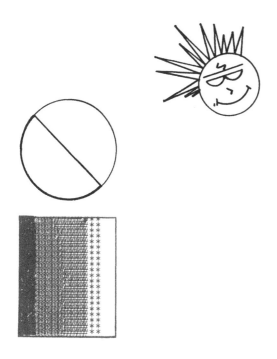

Figure 11.19 Shades of gray to indicate
light and dark areas for visualization of
a pictorial.

ADDED SURFACES IN TRANSPARENT WIREFORM.

Second, the coded areas are checked
for a logic pattern. Often, shades of gray
are used to show light (sun angles) and
dark areas of an object as shown in Figure
11.19. In this example, the object is a
simple cylinder, but the shade appears as
represented in Figure 11.17.

Third, the shaded area is presented as
a pictorial object. If the object needs to
be refined (object looks like Figure 11.18),
the number of flat surfaces can be increased
to improve the surface shape of the object,
as shown in Figure 11.20.

11.6 DISPLAY OF PICTORIAL DATA BASES

The data base which describes the pictorial object now contains shading data. A
data base can include several thousand separate number groups at this point. Always
display the smallest size data base that
will communicate the correct amount of pictorial information to the reader or viewer.
A series of examples will illustrate this

SELECTIVE ERASE FOR SOLID MODEL SHADING.

Figure 11.20 Examples of added surface definition prior to shading pictorial object.
(Courtesy DLR Associates, Clemson, SC)

point for you. Begin will Figure 11.20, as
the number of surfaces increase, so do the

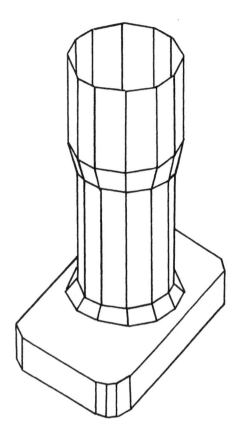

Figure 11.21 Surfaces ready for shading.

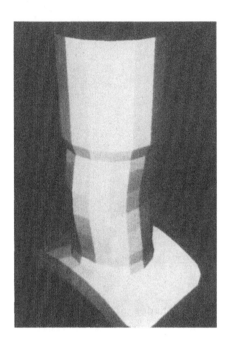

Figure 11.22 Surfaces displayed on CRT, no line elimination used. (Courtesy Evans and Sutherland)

data elements required to define them. A shaded area requires an additional amount of data to be processed, therefore, the shade produced along the bottom of the pictorial sphere in Figure 11.20 adds to the total data base for the pictorial sphere. One way to reduce the amount of data required to display a shaded pictorial is to reduce the number of surface planes. Notice that in Figure 11.21, the number of surface planes has begin reduced to a minimum by increasing the length along the Y axis of the object. This produces the shaded pictorial shown in Figure 11.22. The size of the data base is now smaller than Figure 11.20 before shading. However, a part of the pictorial still needs additional surface definition, namely where the sizes of the cylinders change diameter and where the smaller cylinder joins the base. The drafter adds the surface planes required and the object is displayed in Figure 11.23.

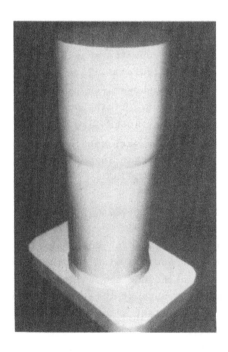

Figure 11.23 Added surfaces. (Courtesy E&S)

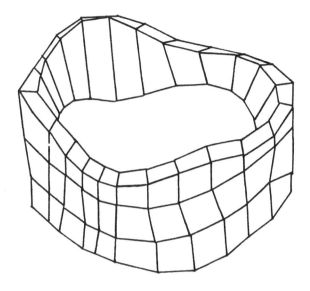

Figure 11.24 Pictorial data base.

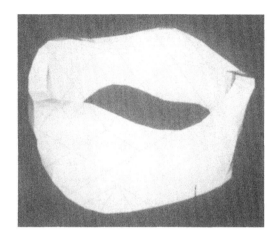

Figure 11.25 Shaded pictorial data base.
(Courtesy Evans and Sutherland Company)

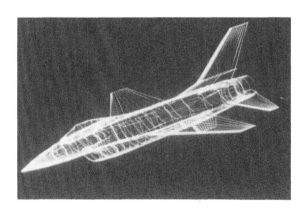

Figure 11.26 Pictorial data base.

(Courtesy Evans and Sutherland)

Figure 11.27 Shaded testing model.

(Courtesy Evans and Sutherland)

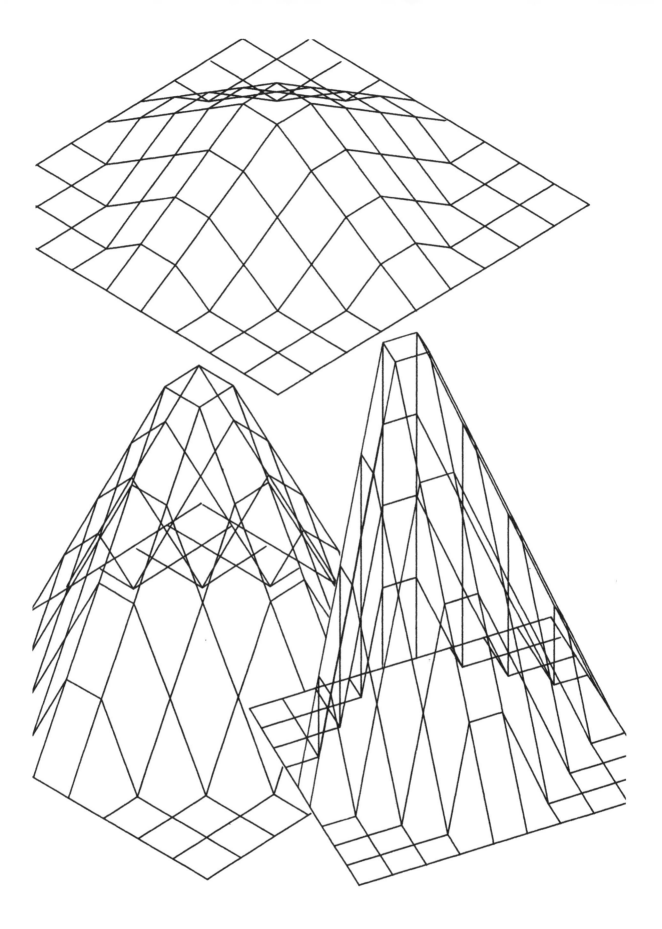

Figure 11.28   Use of SF along the Z axis before surface shading.

Other examples appear in Figures 11.24 and 11.25, where pictorial data base is reduced and then shaded before display. Depending upon the pictorial, this can be an excellent approach. In Figure 11.26, the pictorial is very complex and a reduction of the surfaces to be displayed is not as affective, as shown in Figure 11.27. Often the change of surfaces to be shaded and the SF in one axis will produce dramatic affects as shown in Figure 11.28. Here the same X and Y data base was used and Z was scaled before plotting.

11.7 REVIEW PROBLEMS

To understand automated drafting output of common pictorials, a series of problems should be worked. The following are designed for you to test your understanding of all of the chapters presented thus far in the text.

1. Input the following example program for producing pictorial views.

```
CALL PLOTS
CALL PLOT(0.,3.,-3)
CALL NEWPEN(4)
DIMENSION X(24),Y(24),Z(24),IPEN(24)
DIMENSION PNTS(100,3),ICON(100),VPNT(100,3),VP(3)
DATA X/0.,2.,2.,3*0.,1.,1.,4*0.,4*1.,1.2,1.8,1.2,1.8,4*2./
DATA Y/3*0.,0.,0.,6*2.,0.,2.,1.2,2.,1.2,4*1.,.8,0.,.8,0./
DATA Z/0.,0.,1.,1.,3*0.,1.,1.,0.,1.,1.,0.,0.,4*1.,4*0.,1.,1./
DATA IPEN/3,9*2,3,2,3,2,3,2,3,2,3,2,3,2/
ANG = 30.
NV=4
DO 1 I=1,NV
1   READ(1,*)VPNT(I,1),VPNT(I,2),VPNT(I,3)
CALL SYMBOL(0.,5.,.2,'DRAWING USING CUPID1',0.0,20)
CALL CUPID1(24,X,Y,Z,IPEN,ANG)
NDATA=24
CALL PLOT(4.,0.,-3)
CALL SYMBOL(4.,5.,.2,'ANIMATION USING CUPID2',0.0,22)
CALL CONVER(X,Y,Z,IPEN,NDATA,PNTS,ICON,NDATA2)
DO 25 I=1,NV
DO 3 J=1,3
3   VP(J)=VPNT(I,J)
CALL PLOT(3.,0.,-3)
CALL ZARC(X(23)-0.2,Y(23),-Z(23),.2,0.,90.,VP)
CALL ZARC(X(21)-0.2,Y(21),-Z(21),.2,0.,90.,VP)
CALL ZARC(X(19),Y(19)+0.2,-Z(19),.2,180.,270.,VP)
CALL ZARC(X(17),Y(17)+0.2,-Z(17),.2,180.,270.,VP)
CALL CUPID2(NDATA2,NDATA,PNTS,VP,ICON)
25  CONTINUE
CALL SYMBOL(5.,5.7,.2,'ANIMATION USING CUPID3',0.0,22)
CALL FACTOR(.75)
CALL PLOT(4.,2.,-3)
CALL CUPID3
CALL PLOT(0.,0.,999)
STOP
END
```

2. Store the following subprogram.

```
SUBROUTINE CUPID1(NDATA,X,Y,Z,IPEN,ANG)
DIMENSION X(100),Y(100),Z(100),IPEN(100),XPLOT(100),YPLOT(100)
COSA = COS(ANG/57.3)
SINA = SIN(ANG/57.3)
DO 10 I=1,NDATA
XPLOT(I) = X(I)+Z(I)*COSA
YPLOT(I) = Y(I)+Z(I)*SINA
10  CALL PLOT(XPLOT(I),YPLOT(I),IPEN(I))
CALL CIRCL(XPLOT(21),YPLOT(21),0.,90.,.2,.2,0.0)
CALL CIRCL(XPLOT(23),YPLOT(23),0.,90.,.2,.2,0.0)
CALL CIRCL(XPLOT(14),YPLOT(14),180.,270.,.2,.2,0.0)
CALL CIRCL(XPLOT(16),YPLOT(16),180.,270.,.2,.2,0.0)
RETURN
END
```

3. Store the following subprogram.

```
SUBROUTINE CONVER(X,Y,Z,IPEN,NDATA,PNTS,ICON,NDATA2)
DIMENSION ICON(100),X(100),Y(100),Z(100),IPEN(100),PNTS(100,3)
DO 7 I=1,NDATA
    PNTS(I,1)=X(I)
    PNTS(I,2)=Y(I)
    PNTS(I,3)=Z(I)
    IF(IPEN(I).EQ.2) ICON(I)=I
7   IF(IPEN(I).EQ.3) ICON(I)=-I
NDATA2=NDATA
RETURN
END
```

4. Store the following subprogram.

```
SUBROUTINE CUPID2(NP,NC,P,VP,IC)
DIMENSION P(100,3),IC(100),VP(3),PP(100,3)
A=ARTAN(VP(1),VP(3))
SA=SIN(A)
CA=COS(A)
DO 6 J=1,NP
    PP(J,3)=P(J,3)*CA+P(J,1)*SA
    PP(J,1)=P(J,1)*CA-P(J,3)*SA
6   CONTINUE
VPP=VP(3)*CA+VP(1)*SA
A=ARTAN(VP(2),VPP)
SA=SIN(A)
CA=COS(A)
CALL NEWPEN(4)
DO 7 J=1,NP
    PP(J,2)=P(J,2)*CA-PP(J,3)*SA
7   CONTINUE
DO 8 J=1,NC
    IF(IC(J).LT.0) GO TO 9
    CALL PLOT(PP(IC(J),1),PP(IC(J),2),2)
    GO TO 8
9       K=-IC(J)
        CALL PLOT(PP(K,1),PP(K,2),3)
8   CONTINUE
RETURN
END

FUNCTION ARTAN(Y,X)
DATA EPS/0.001/
AX=ABS(X)
AY=ABS(Y)
IF(AX.GT.EPS.AND.AY.GT.EPS) GO TO 1
IF(AX.LT.EPS.AND.AY.LT.EPS) GO TO 3
IF(AX.LT.EPS) GO TO 2
3   ARTAN=0.0
RETURN
2   ARTAN=(3.14159*AY)/(Y*2.0)
RETURN
1   ARTAN=ATAN2(Y,X)
RETURN
END
```

5. Store the following subprogram.

```
SUBROUTINE ZARC(XP,YP,ZP,R,SANG,EANG,VP)
DIMENSION P(100,3),IC(100),VP(3)
PI=3.14159265
N=IFIX(EANG-SANG)
THETA=(EANG-SANG)/N
THETAR=THETA*PI/180.
SRANG=SANG*PI/180.
DO 1 I=1,N
DX=R*COS(SRANG)
DY=R*SIN(SRANG)
P(I,1)=XP+DX
P(I,2)=YP+DY
P(I,3)=-ZP
SRANG=SRANG+THETAR
1   CONTINUE
NP=N
NC=N
IC(1)=-1
DO 999 I=2,NP
999 IC(I)=I
CALL CUPID2(NP,NC,P,VP,IC)
RETURN
END
```

6. Stop at this point and test part of the program shown in problem 1. Output only the portion that calls CUPID1 as shown on page 200.

DRAWING USING CUPID1

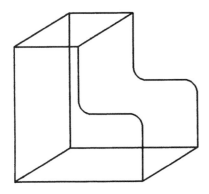

7.   Input the additional subprogram.

```
      SUBROUTINE CUPID3
      DIMENSION X(300),Y(300),Z(300)
      ICNT=0
      DO 200 J=1,10
      DO 300 K=1,10
      ICNT=ICNT+1
      X(ICNT)=J
      Y(ICNT)=K
      READ(1,*)Z(ICNT)
300   CONTINUE
200   CONTINUE
      S1=.707
      C1=.707
      DO 1 I=1,4
      CALL SURF(X,Y,Z,ICNT,S1,C1)
      CALL PLOT(13.,0.,-3)
1     S1=S1+.5
      RETURN
      END
```

10.   Input the last subroutine.

```
      SUBROUTINE SURF(A,B,C,N,S1,C1)
      DIMENSION A(300),B(300),C(300),BUFR(300),P(20,20)
      DATA AXLN/3.0/
      AXLN=5.
      AXLN1=AXLN+0.001
      XH=0.0
      YH=0.0
      ZH=0.0
      XL=100.
      YL=100.
      ZL=100.
      NN=10
      DO 1 I=1,N
      IF(XH.LE.A(I))XH=A(I)
      IF(YH.LE.B(I))YH=B(I)
      IF(ZH.LE.C(I))ZH=C(I)
      IF(XL.GE.A(I))XL=A(I)
      IF(YL.GE.B(I))YL=B(I)
      IF(ZL.GE.C(I))ZL=C(I)
1     CONTINUE
      DO 2 I=1,N
      A(I)=(A(I)-XL)/(XH-XL)*AXLN
      B(I)=(B(I)-YL)/(YH-YL)*AXLN
      C(I)=(C(I)-ZL)/(ZH-ZL)*AXLN
2     CONTINUE
      DO 3 I=1,NN
      DO 4 J=1,NN
      P(I,J)=0.
4     CONTINUE
3     CONTINUE
      DO 5 I=1,N
      IX=A(I)/AXLN1*NN+1.
      IY=B(I)/AXLN1*NN+1.
      IF(C(I).GT.P(IX,IY))P(IX,IY)=C(I)
5     CONTINUE
 C
```

```
      DO 6 I=1,NN
      X=I*AXLN/NN
      SX=X
      IP=3
      DO 7 J=1,NN
      X=SX
      Y=J*AXLN/NN
      Z=P(I,J)
      CALL ROTATE(X,Y,Z,S1,C1)
      CALL PLOT(X,Y,IP)
      IP=2
7     CONTINUE
6     CONTINUE
      DO 8 I=1,NN
      Y=I*AXLN/NN
      SY=Y
      IP=3
      DO 9 J=1,NN
      Y=SY
      X=J*AXLN/NN
      Z=P(J,I)
      CALL ROTATE(X,Y,Z,S1,C1)
      CALL PLOT(X,Y,IP)
      IP=2
9     CONTINUE
8     CONTINUE
      RETURN
      END
      SUBROUTINE ROTATE(X,Y,Z,S1,C1)
      DATA  C2/.404/,S2/-.587/
      SY=Y
      Y=Z*C2+X*C1*S2 +Y*S1*S2+5.
      X=SY*C1-X*S1+10.
      RETURN
      END
```

9.   Output the second part of the program

ANIMATION USING CUPID2

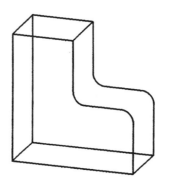 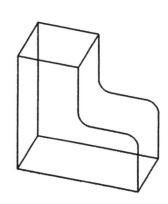

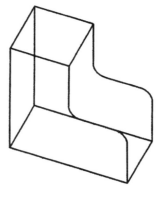 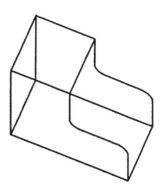

10. Test the last part of the program.

ANIMATION USING CUPID3

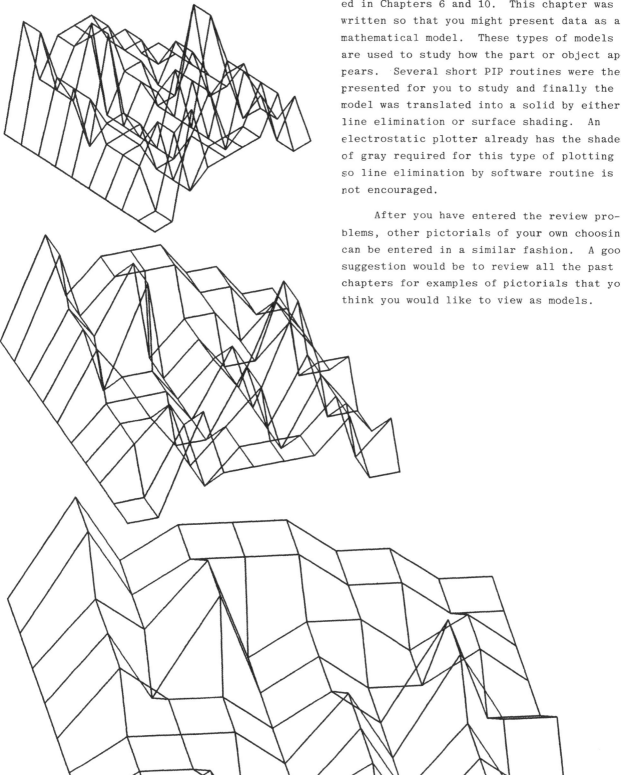

11.8 CHAPTER SUMMARY

Pictorial representations can be displayed from the computer data base discribed in Chapters 6 and 10. This chapter was written so that you might present data as a mathematical model. These types of models are used to study how the part or object appears. Several short PIP routines were then presented for you to study and finally the model was translated into a solid by either line elimination or surface shading. An electrostatic plotter already has the shades of gray required for this type of plotting so line elimination by software routine is not encouraged.

After you have entered the review problems, other pictorials of your own choosing can be entered in a similar fashion. A good suggestion would be to review all the past chapters for examples of pictorials that you think you would like to view as models.

12

Fasteners

Fasteners are displayed in automated drawings in an almost infinite number of types,
sizes, and characteristics to meet a wide
variety of design requirements. This wide
diversity is paralleled by the specialized
terminology associated with these displays,
their specification, and their manufacture.
This chapter along with its illustrations
is presented as an aid to proper display of
fasteners or fastener details.

Displays for fasteners are divided into
several groups:

1. Permanent fasteners (rivets, nails,
screws, welding and others)

2. Assembly fasteners (bolts, machine
screws, cap screws, set screws, keys, pins
and others)

3. Application of fasteners in automated drawing details.

12.1 FASTENER TERMINOLOGY

Fastener terminology is best learned
by a series of illustrations. These illustrations are used to show the various types
of fasteners, how to display them for an

Figure 12.1 Types of fasteners commonly
used. (Courtesy Russell, Burdsall & Ward
bolt and nut company, Port Chester, NY)

automated drawing and how fasteners are selected for various applications. Commonly
used fasteners are shown in Figure 12.1,
these and others will now be presented for
further study and illustration.

Bolt. Shown in Figure 12.2, this is
an externally threaded fastener designed to
go through holes in assembled parts, and
normally tightened or released by torquing
a nut.

Screw. Shown in Figure 12.3, this is
also an externally threaded fastener capable of being inserted into holes in assembled parts, of mating with a preformed internal thread or forming its own thread,
and of being tightened or released by the
torquing of the head as illustrated in Figure 12.2.

Stud. Also shown in Figure 12.3, this

Figure 12.2 Bolt and nut. (Courtesy Chandler Products Corp., Cleveland, Ohio)

is a cylindrical rod of moderate length, threaded on either ends, usually it replaces the bolt and screw. The bolt requires a nut as shown in Figure 12.2, a screw requires a threaded section with which to mate. The stud requires a nut, two nuts, or a nut and threaded section to mate. It therefore, is more widely used in certain applications.

Nut. Shown in Figure 12.2 and Figure 12.4, a nut is usually, but not always, a block or sleeve having an internal thread designed to assemble with the external parts of a bolt, screw, stud, or other fasteners.

Rivet. Shown in Figure 12.5, this a headed fastener of malleable material used to join parts by inserting the shank of the rivet through aligned holes in each piece and formed into a head by upsetting.

Tapping Screw. Shown in Figure 12.6, this is a hardened screw designed to cut its own thread in a plain drilled hole. Tapping screws are used on sheetmetal, thin materials of all types, and in other permanent joining applications.

Drive Screw. Similar to the tapping screw, except it is designed with an extremely fast lead thread which permits it to be put

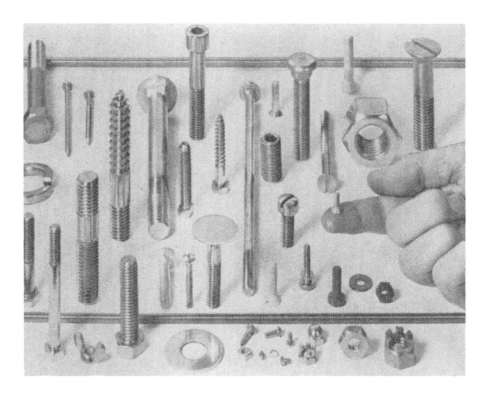

Figure 12.3   Screws, studs and bolts.   (Courtesy Harper Co., Morton Grove, Illinois)

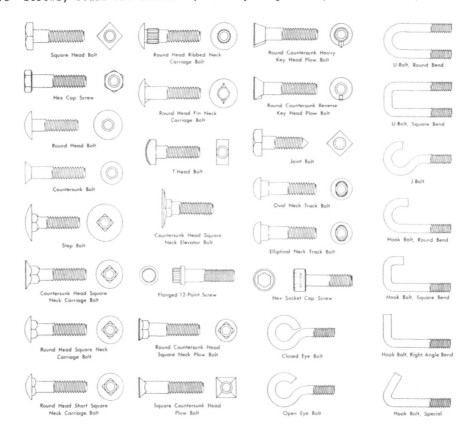

Figure 12.4   Nuts and bolts.   (Courtesy United Shoe Machinery Corp., Shelton, Conn.)

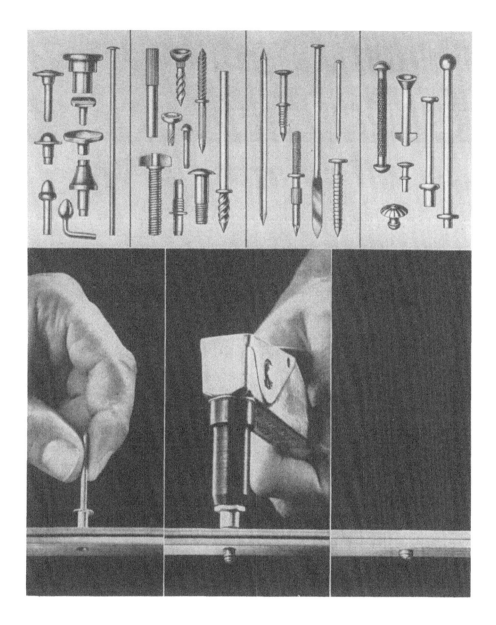

Figure 12.5 Rivets and installation. (Courtesy Industrial Fasteners Institute)

into place without the necessity of torquing.

   Blind fastener. A fastener, usually a rivet or bolt, which can be inserted from one side of the material. The head is formed on the far side by manipulation of the fastener from the front as illustrated in Figure 12.5.

   Locking Screw. Shown in Figure 12.7, this is an externally threaded fastener which has a special means within itself for (or within a mating part as in Figure 12.8)

gripping an internal thread so that rotation of the treaded parts, relative to each other is resisted as illustrated in the eight steps of Figure 12.7. Often, a locknut is used as demonstrated in Figure 12.7.

12.2 FASTENER PARTS

   A fastener, as shown in the Figures and illustrations of this chapter, is a mechanical device designed specifically to hold, join,

Teks®: eliminate costly punching, drilling and aligning operations, for they drill their own holes. Form a mating thread and make a complete fastening in a single operation.

Torx®: hexlobular head design makes driving easier. Completely eliminates camout. And endloading. Available as a recess or raised head design. Manufactured to tight tolerances for rigid fastener-driver alignment.

Orlo®: A new prevailing torque design. A rib on the non-pressure flank of the thread applies a uniform, spring-like torque action. Requires no special nuts, washers, inserts or deforming. Holds fast under severe vibration.

Swageform®: displaces material to form its own thread in drilled, cored or punched holes. Swaging action on thread produces a zero clearance fit and guarantees high resistance to backout.

Taptite®: here's another thread-forming design. It's a high-strength fastener with a trilobed thread that can be used successfully in ductile metals, die castings and plastics.

Powerlok®: an exceptional thread locking screw for pretapped holes. Creates areas of metal compression at the crest of each lobe. Result: no galling. Up to 400% better retained torque over IFI standards.

Hi-Lo®: an economical design for solving plastic fastening problems. Has a double lead of high and low thread. Low radial pressures minimize boss cracking. High pull-out strength.

Plastite® screws: fasten plastic parts without tapping. Start easy. Trilobed design first compresses material, then recovers to fill in behind lobes to provide powerful locking strength.

Pozidriv®: recess makes assembly operations easier. Reduces camout problems. Provides greater area of positive engagement. There's less operator fatigue, less bit wear.

Figure 12.6   Selecting the right tapping screw.   (Courtesy Elco Industries, Inc., Rockford, Ill.

or otherwise fasten together two or more parts of an object. The fasteners presented thus far have been for assembly purposes only, many others will be shown as we progress. For now, we must stop a moment and consider the parts of a fastener that must be included in an automated drafting presentation. They are:

Fastener type. The type of fastener must be noted on an automated presentation. The various types are usually cataloged in computer memory and recalled when needed. A list of types that should be included in a catalog are:

1. Insert, an internally threaded bushing or thread reinforcement.

2. External torque, shown in Figure 12.2.

3. Internal torque, which is assembled using an internal wrench in a socket or recess in the fastener (shown in Figure 12.8).

4. Aircraft, closely toleranced fastener used in high vibration applications.

5. Semifinished, looser tolerance than an aircraft fastener with only the threads finished to fit closely.

Nominal Size. The automated drawing designation used for the purpose of general identification, either metric or inch. For threaded sections it is usually the major diameter of the thread. For unthreaded fasteners it is usually the basic body diameter.

Length. The length of an automated display of a fastener is always measured parallel to the axis of the fastener. The length of a headless (Stud) fastener is the distance from one end to the other. The length of any headed fastener is from the base of the head to the end of the fastener.

Shank. The portion of an automated display of a headed fastener which lies between the base of the head and the end of the fastener (length) or the part of the fastener which will contain the threads, head location, and body is called the shank.

Body. The portion of a display that does not contain threads is called the body of the fastener. Bodies of fasteners may be reduced diameter, externally relieved, internally relieved or contain shoulders.

Shoulder. This is the enlarged portion of the display (body), usually next to

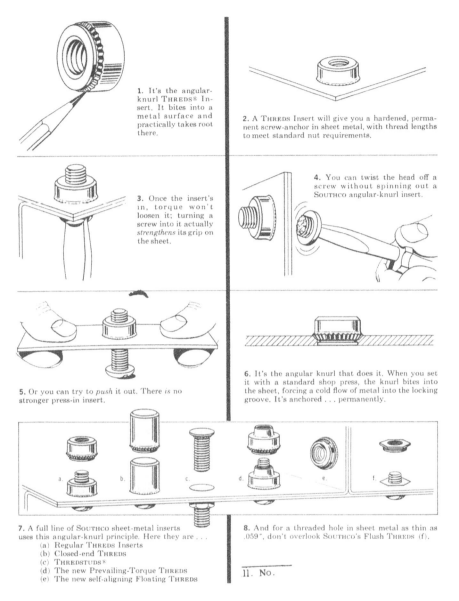

1. It's the angular-knurl THREDS® Insert. It bites into a metal surface and practically takes root there.

2. A THREDS Insert will give you a hardened, permanent screw-anchor in sheet metal, with thread lengths to meet standard nut requirements.

3. Once the insert's in, torque won't loosen it; turning a screw into it actually *strengthens* its grip on the sheet.

4. You can twist the head off a screw without spinning out a SOUTHCO angular-knurl insert.

5. Or you can try to *push* it out. There *is* no stronger press-in insert.

6. It's the angular knurl that does it. When you set it with a standard shop press, the knurl bites into the sheet, forcing a cold flow of metal into the locking groove. It's anchored . . . permanently.

7. A full line of SOUTHCO sheet-metal inserts uses this angular-knurl principle. Here they are . . .
    (a) Regular THREDS Inserts
    (b) Closed-end THREDS
    (c) THREDSTUDS*
    (d) The new Prevailing-Torque THREDS
    (e) The new self-aligning Floating THREDS

8. And for a threaded hole in sheet metal as thin as .059″, don't overlook SOUTHCO's Flush THREDS (f).

11. No.

Figure 12.7   Locking screws and nuts.   (Courtesy Southco, Inc., Lester, PA)

the head.   A shoulder can be a bearing face displayed on an automated drawing.

    Washer Face.   This is displayed as a circular boss on the bearing surface of a bolt or nut.

    Driving Recess.   Shown in Figure 12.8 and discussed under internal wrenching of a fastener, it is the indented portion of the bolt or screw head which is shaped to accept a driving tool.

    Grip.   The thickness of material or parts which the fastener is designed to secure when assembled should be labeled grip.

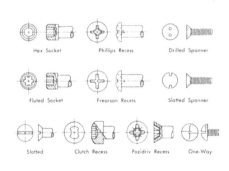

Figure 12.8   Driving recesses used.

12.3 DISPLAYING FASTENER PARTS

The fastener parts just described can be displayed on a DVST screen for the drafts-person to preview, change, or add information. The basic equipment used throughout the book and shown in Figures 4.1, 5.3, and 6.6 should be adequate for simple fastener parts. For example, Figure 12.9 can be displayed from software (programming) already in place from Chapter 11 pictorials. In Figure 12.10 the display is constructed as a combination of Chapter 6 Multiview Projection and Chapter 7 Dimensioning Practices. The DVST display shown in Figure 12.11, however, requires additional software descriptions for the fasteners used.

This additional programming is selected by the drafter from a menu list. For the display of Figure 12.11, a drafter selects the fasteners from a menu shown in Figure 12.12. A menu may be prepared on an overlay and placed directly on the graphics tablet. The stylus is used to point to the item desired. A more advanced system is the use of a touch tablet shown in Figure 12.13. In this example the menu shown in Figure 12.12 can be placed along the right or left side, allowing another menu to inset also. The drafting function becomes a point and place operation if all the desired fasteners described in section 12.1 are included on the menu. If special purpose fasteners, such as quick acting are desired then a general menu is

Figure 12.9 Fastener display.

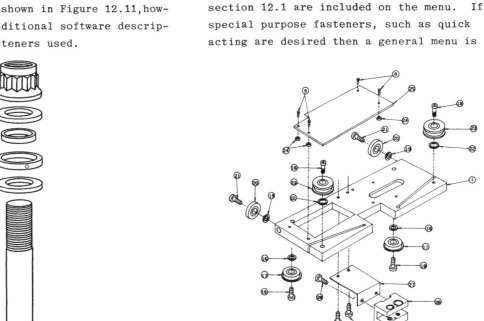

Figure 12.11 Multiple fastener display. (Courtesy Auto-trol Technology Corporation)

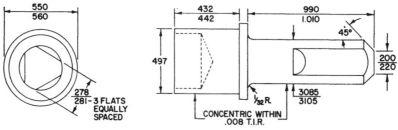

Figure 12.10 Dimensioned fastener display.

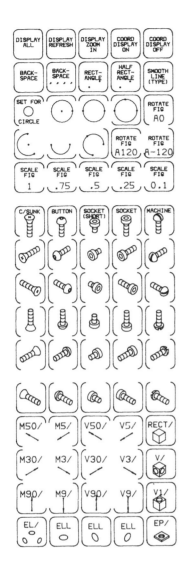

Figure 12.12 Fastener menu. (Courtesy Auto-trol)

Figure 12.13 Touch tablet. (Courtesy Auto-trol)

LAYERING			PEN SELECT	ROUNDOFF		BROKEN LINES	LINE WIDTH	DOUBLE LINES		DISPLAY CONTROL	SPECIAL FEATURES	MIRROR	CROSS-FEAT. HATCHING	3-D OPERATIONS			CIRCLES, ELLIPSES AND POLYGONS				
WORK LAYER	DISP LAYER	EDIT LAYER		DIGI-TIZER	CRT									90 DEGREE ROTATION			AUTOMATIC CENTER CROSS				
LAYER 1	LAYER 1	LAYER 1	PEN 0	NONE	NONE	.2-.1	WIDTH OFF	.025	.05	DISPLAY *ALL*	BACKSPACE (LAST ITEM)	MIRROR ABOUT ORIGIN				DEFINE POLYGON (TYPE SIDES)	NONE	.05	.1	MANUAL (TYPE SIZE)
LAYER 2	LAYER 2	LAYER 2	PEN 1	.05	.05	.12-.06	WIDTH ON	.1	.15	REFRESH DISPLAY	RECTANGLE (IND.2 DIAG. PT.S FOR EACH)	ABOUT -X- AXIS		RESTORE	RESTORE	RESTORE	MINOR TO MAJOR AXIS RATIO				
																	1:1 (CIRCLE)	.866	.707	.5	.259	MANUAL (TYPE RATIO)
LAYER 3	LAYER 3	LAYER 3	PEN 2	.0625	.0625	.1-.05	WIDTH .05	.25	MANUAL (TYPE HALF-WIDTH)	ZOOM-IN (INDICATE 2 DIAG.CORNERS)	auto-record	ABOUT -Y- AXIS		MANUAL ROTATION (TYPE IN ANGLE DESIRED)			ELLIPSE PROJECTION ANGLE				
																	90 (CIRCLE)	60	45	30	15	MANUAL (TYPE ANGLE)
LAYER 4	LAYER 4	LAYER 4	PEN 3	.1	.1	.1-.1	WIDTH .1			DISPLAY AT ACT. SIZE (IND.CENTER)	SMOOTH LINE	ABOUT A DIAG. LINE (INDICATE 2 POINTS)		ISO-METRIC VIEW	DI-METRIC VIEW	TRI-METRIC VIEW	ELLIPSE INCLINATION ANGLE				
																	0 (HORIZ)	30	45	60	90	MANUAL (TYPE ANGLE)
LAYER 5	LAYER 5	LAYER 5	PEN 4	.125	.125	.05-.1	MANUAL WIDTH (TYPE WIDTH)			MOVE WINDOW (IND.CENTER)	"FILLET" $\frac{1}{16}$ R .1 R	ABOUT A HORIZ. LINE (INDICATE 1 POINT)		RESTORE	RESTORE	RESTORE					
LAYER 6	LAYER 6	LAYER 6	PEN 5	.2	.2	CENTER LINE	DISPLAY MODE SINGLE			RAPID ACCESS IND. WINDOW	$\frac{1}{8}$ R .2 R	ABOUT A VERT. LINE (INDICATE 1 POINT)		perspective (TYPE IN DISTANCE)							
ALL LAYERS	ALL LAYERS	ALL LAYERS	PEN 6	.25	.25	PHANTOM	DISPLAY DOUBLE NORMAL OPEN			DISPLAY SORTED WINDOW (TYPE WINDOW NUMBER)	$\frac{1}{4}$ R $\frac{3}{8}$ R	ABOUT ANY POINT		*AUTOMATIC ARROWS*							
														START END BOTH							
MANUAL TYPE LAYER NUMBERS	MANUAL TYPE LAYER NUMBERS	MANUAL TYPE LAYER NUMBERS	PEN 7	MANUAL (TYPE DIST.)	MANUAL (TYPE DIST.)	MANUAL TYPE LENGTHS	DISPLAY DOUBLE NORMAL CLOSED			*CLEAR* WORKSPACE AREA	$\frac{1}{2}$ R	MANUAL (TYPE RADIUS)									

Figure 12.14 General drafting menu. (Courtesy Auto-trol Technology Corporation)

used as shown in Figure 12.14. The various parts of the fastener can be displayed like other images. The item that sets fasteners apart, is the display of threads.

12.4 DISPLAYING THREADED SECTIONS

The display of screw threads for an automated drawing is represented by a ridge of uniform section in the form of a helix on the external or internal surface of a cylinder. Figure 12.15 represents this concept. However, threaded sections also involve a helix in the form of a conical spiral on the external or internal surface of a cone. Figure 12.16 represents this concept.

An automated display may be programmed to represent any of the following thread types: sharp, unified, whitworth, square, acme, worm, buttress, knuckle, dardelet, or ISO (International Standardization Organization). But for fasteners only two types are worth mentioning, the Unified and the ISO. The Unified and the ISO have essentially the same profile and viewers can not tell one from the other on a display screen. Therefore, only one type of thread is programmed for display, that being the ISO.

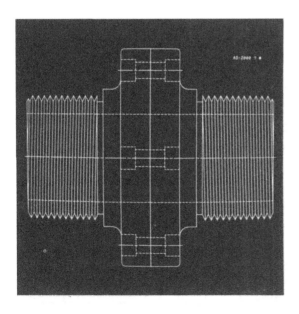

Figure 12.15 DVST display. (Courtesy Autotrol)

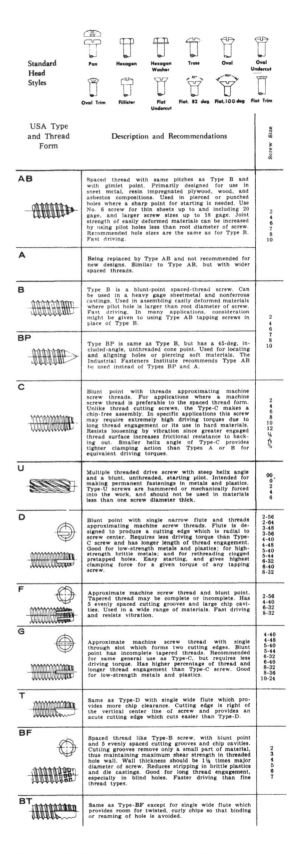

Figure 12.16 Helix and spiral screw thread. (Courtesy Industrial Fasteners Institute)

Both the Unified and ISO display threads
are considered out-of-date for modern fastener design. A new optimum metric fastener
system (OMFS) should be used for all newly
created fasteners. However, many excellent
fastener designs exist under the IFI-500
standard (Industrial Fastener Institute)
and these type fasteners can be displayed
on automated drawings with little effort on
the part of the drafter. Only the notes
are changed from one thread display to the
next. Regardless of the source of the display thread, some design considerations do
not change. These are:

Type of thread. Several types of displays exist for threads, they are; single,
multiple, external, internal, right-hand,
left-hand, complete, incomplete, effective
and total thread. Displays of threads must
consider the technical aspects of threads
and these do not change from Unified to ISO
to OMFS to IFI formats. A single or single
start thread has a lead equal to the pitch.
Single thread fasteners are shown in Figure
12.17. Multiple thread fasteners are shown
in Figure 12.18. External and internal display of threads is shown in Figure 12.19.

Right-hand threads are displayed in a
clockwise and receding direction from the
starting end, when viewed in a right profile
while left-hand winds in a counterclockwise
direction. A complete thread is displayed
having full form at both the crest and root
while incomplete threads are not fully formed
at either crest or root. Examples of complete, effective, total threads are shown
in Figure 12.20.

12.5 DISPLAY FORM OF A THREAD

The theoretical profile of a thread is
displayed according to basic forms. These
are:

Crest. Mentioned earlier, this is the
outermost tip of a male thread as seen in a
thread profile (Figure 12.15) or the inner-
most tip of a female thread (Figure 12.19).

Root. The bottom of the thread, identical with or immediately adjacent to the
cylinder or cone from which the thread pro-

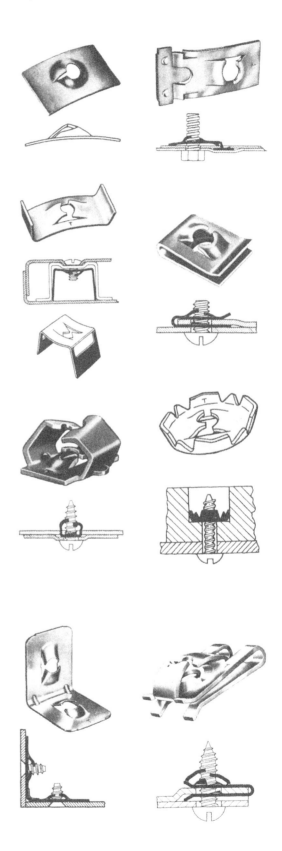

Figure 12.17 Single thread fasteners.
(Courtesy IFI)

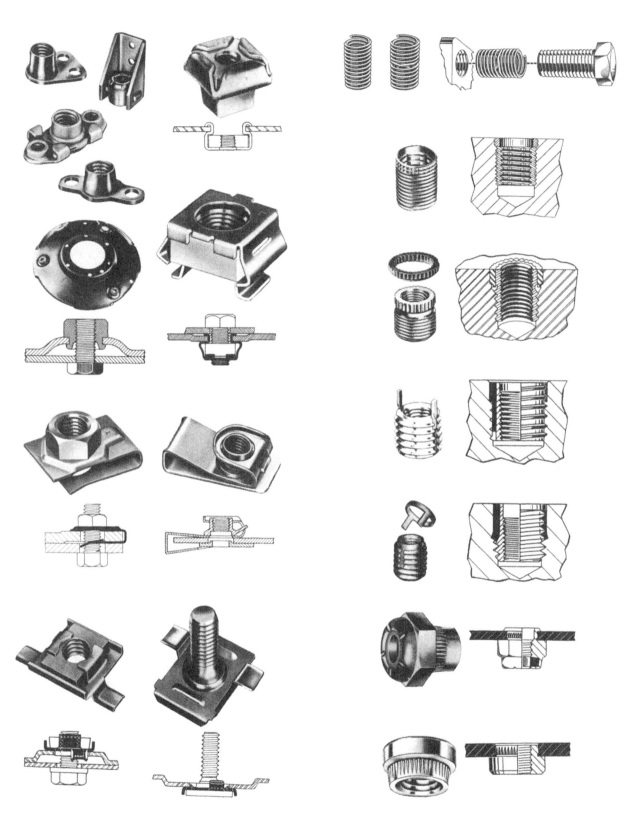

Figure 12.18 Multiple threaded fasteners.
(Courtesy Industrial Fastners Institute)

Figure 12.19 External/internal threaded
Fasteners. (Courtesy IFI)

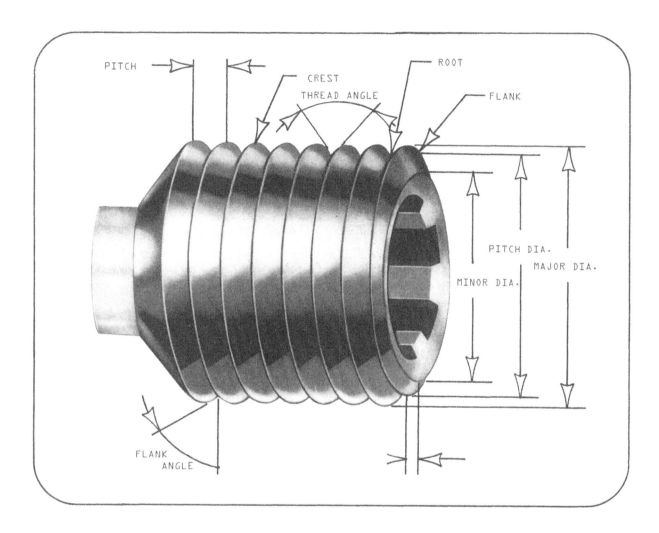

Figure 12.20 Complete display form of a thread. (Courtesy DLR Associates, Clemson, SC)

jects. See Figure 12.20.

Flank. The thread surface shown in Figure 12.20 which connects the crest with the root.

Flank angle. The angle between the flank surface and the perpendicular to the axis of the thread shown in Figure 12.20. A flank angle of a symmetrical thread is one-half the thread angle shown in Figure 12.20.

Pitch. The distance, shown in Figure 12.20 which is measured parallel to the thread axis, between corresponding points on adjacent thread forms on the same side of the thread.

Lead. The distance shown in Figure 12.20 that a thread moves axially in one complete revolution.

Lead angle. The angle made by the helix of a thread at the pitch line with the perpendicular to the axis.

Thread angle. The lead angle plus the helix angle which is the angle made by the helix of the thread at the pitch line. See Figure 12.20.

Major diameter. Shown in Figure 12.20 it is the diameter of the imaginary cylinder bounding the crest. In the case of internal threads, it is the imaginary cylinder bounding the root of the thread.

Minor diameter. Shown in Figure 12.20 it is the diameter of the imaginary cylinder bounding the root. In the case of internal threads it is the crest of the threads.

Effective thread. Dimensioned thread which includes the complete thread as shown in Figure 12.20 and that portion of the incomplete thread having fully formed roots. The threads in Figure 12.20 would be dimensioned with the letter M followed by the major diameter in millimeters and then the pitch in millimeters as:

M30 x 1.5

Pitch diameter. Shown in Figure 12.20 this is the diameter of the imaginary cylinder expressed in millimeters, whose surface passes through the thread profiles in such a way to make the widths of the thread and the thread groove equal. The pitch tolerance appears in the thread note after pitch as:

M30 x 1.5 - 5g

Allowance. The allowance is the interference between the dimensional limits of mating threads. For example:
     lowercase e = large allowance(male)
     uppercase E = large allowance (female)
     lowercase g = small allowance (male)
     uppercase G = small allowance (female)
     lowercase h = no allowance (male)
     uppercase H = no allowance (female)
The allowance is placed after the tolerance grade. Grades available are:
     4 = low tolerance
     6 = normal
     8 = high tolerance
A 4 millimeter tolerance is considered a closer fit than the old ANSI Unified 2A or 2B fits, while 8 should be used for large, course size metric threads.

12.6 SPECIFICATION OF FASTENER MATERIAL

Because modern industrial fasteners are available in almost any material, the specification for automated drafting notation is practically unlimited. The key to material specification for fasteners is in knowing where the fastener will be used.

Usually the use can be determined from answering the following questions:
     1. Will the fastener be subjected to corrosives?
     2. Will it experience high or low temperature ranges?
     3. Is weight critical?
     4. Should it be nonmagnetic?
     5. How much should it cost?
     6. Is it permanent, assembly or temporary life span?
The problem of corrosion can be met by using a protective coating or finish. If a fastener will experience a wide range of temperatures, a manufacturer can meet your specifications by cold working or heat treating one of the common materials on hand. A weight reduction is possible with nonferrous metals (aluminum). It can be nonmagnetic in this manner or the fastener may be made from nonmetallic materials. In an electric motor, a fastener of a magnetic material would not be used next to the coil windings.

When specifying a fastener, always be as general as possible, never over-specify chemical properties, and the like. These types of specifications add to the cost and availability of the fastener. The last item to consider is the life span. If the fastener is temporary (staple in a carton) it is thrown away after use. If the same use is designed for assembly, an adhesive could be used to fasten the carton. If a permanent carton fastener is to be designed, than quick acting fasteners like snaps or pins can be used over and over again to close the carton.

12.7 COMMON FASTENER MATERIALS

The greatest number of fasteners are made of steel (SAE 1010). Machine screws, carriage bolts, and other similar fasteners might use SAE 1018, 1020, or 1021. High strength bolts, studs, nuts and cap screws use SAE 1038.

Aluminum is ideal for applications where high strength-to-weight ratio and corrosion resistance is required. Cold formed bolts, screws, rivets, and nuts use 2024-T4 alloy. Milled fasteners use 2011-T3 alloys.

KIND OF MATERIAL	THREAD-FORMING								THREAD-CUTTING			SELF DRILLING	
	Type A	Type B	Type AB	HEX HEAD B	SWAGE FORM*	SWAGE FORM* B	Type U	Type 21	Type F*	Type L	Type B-F*	DRIL-KWICK	TAPITS*
SHEET METAL .015" to .050" thick (Steel, Brass, Aluminum, Monel, etc.)	✓	✓	✓	✓	✓	✓		✓				✓	✓
SHEET STAINLESS STEEL .015" to .050" thick	✓	✓	✓	✓	✓	✓		✓	✓			✓	✓
SHEET METAL .050" to .200" thick (Steel, Brass, Aluminum, etc.)		✓	✓	✓		✓	✓	✓	✓			✓	
STRUCTURAL STEEL .200" to ½" thick				✓	✓				✓				
CASTINGS (Aluminum, Magnesium, Zinc, Brass, Bronze, etc.)		✓	✓	✓	✓	✓	✓		✓				
CASTINGS (Grey Iron, Malleable Iron, Steel, etc.)					✓				✓				
FORGINGS (Steel, Brass, Bronze, etc.)					✓				✓				
PLYWOOD, Resin Impregnated: Compreg, Pregwood, etc. **NATURAL WOODS**	✓	✓	✓	✓				✓	✓	✓		✓	✓
ASBESTOS and other compositions: Ebony, Asbestos, Transite, Fiberglas, Insurok, etc.	✓	✓	✓	✓		✓						✓	✓
PHENOL FORMALDEHYDE: **Molded:** Bakelite, Durez, etc. **Cast:** Catalin, Marblette, etc. **Laminated:** Formica, Textolite, etc.		✓	✓			✓	✓		✓		✓		
UREA FORMALDEHYDE: **Molded:** Plaskon, Beetle, etc. **MELAMINE FORMALDEHYDE:** Melantite, Melamac						✓					✓		
CELLULOSE ACETATES and NITRATES: Tenite, Lumarith, Plastacele Pyralin, Celanese, etc. **ACRYLATE & STYRENE RESINS:** Lucite, Plexiglas, Styron, etc.		✓	✓		✓	✓	✓				✓		
NYLON PLASTICS: Nylon, Zytel						✓	✓	✓		✓			

Figure 12.21 Tapping screws for various materials. (Courtesy Parker Kalon Division, Campbellsville, Kentucky)

Brass, copper, nickel and titanium are the other common fastener materials. A titanium fastener has great strength but should not be used next to magnesium because of galvanic corrosion. The choice of a fastener material is often dependent on the material to be joined.

12.8 TAPPING SCREWS

The material to be joined must be considered before selecting the proper fastener. Often a self-tapping screw can be used in materials. These types of fasteners are thread-forming, cutting or self-drilling. Use Figure 12.20 as a guide when selecting a tapping screw in different types of materials. Notice that a wide range of fasteners are available for sheet metal, while castings and forgings are limited. Thread-forming and thread-cutting screws require predrilled holes in the material to be joined. The supplier will list the size of tap hole to be provided for each of the materials shown in Figure 12.21. As a general rule, the softer the material, the smaller the tap hole required, but just choosing a diameter between the major and minor diameters is not enough. Tap hole sizes are critical for most materials and have been established by the fastener manufacturer through laboratory tests.

12.9 SET SCREWS

Set screws are displayed on an automated drawing as semipermanent fasteners to hold a collar, sheave, or mechanical element on a shaft as shown in Figure 12.22. In comparison with other fasteners, a set screw is essentially a compression device. The forces shown in Figure 12.22 produce a strong clamping action that resists relative motion between assembled parts. The basic problem in set screw display is to program the best combinations of screw selection based on form, size, and point style to provide the holding power.

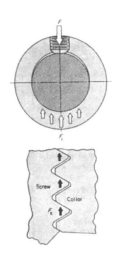

Figure 12.22 Set screw. (Courtesy IFI)

Shaft Diameter (in.)	No. 0	No. 1	No. 2	No. 3	No. 4	No. 5	No. 6	No. 8	No. 10	¼ in.	5/16 in.	3/8 in.	7/16 in.	½ in.	9/16 in.	5/8 in.	¾ in.	7/8 in.	1 in.
Seating Torque (lb-in.)	0.5	1.5	1.5	5.0	5.0	9.0	9.0	20	33	87	165	290	430	620	620	1225	2125	5000	7000
Axial Holding Power (lb)	50	65	85	120	160	200	250	385	540	1000	1500	2000	2500	3000	3500	4000	5000	6000	7000
1/16	1.5	2.0	2.6	3.2															
3/32	2.3	3.0	4.0	5.6	7.5														
1/8	3.1	4.0	5.3	7.5	10.0	12.5													
5/32	3.9	5.0	6.6	9.3	12.5	15.6	19	30											
3/16	4.7	6.1	8.0	11.3	15.0	18.7	23	36	51										
7/32	5.4	7.1	9.3	13.0	17.5	21.8	27	42	59										
¼	6.2	8.1	10.6	15.0	20.0	25.0	31	48	68	125									
5/16		10.0	13.2	18.7	25.0	31.2	39	60	84	156	234								
3/8			16.0	22.5	30.0	37.5	47	72	101	187	280	375							
7/16				26.3	35.0	43.7	55	84	118	218	327	437	545						
½					40.0	50.0	62	96	135	250	375	500	625	750					
9/16						56.2	70	108	152	261	421	562	702	843	985				
5/8						62	78	120	169	312	468	625	780	937	1090	1250			
¾							94	144	202	375	562	750	937	1125	1310	1500	1875		
7/8								109	168	236	437	656	875	1095	1310	1530	1750	2190	2620
1								192	270	500	750	1000	1250	1500	1750	2000	2500	3000	3500
1¼									338	625	937	1250	1560	1875	2190	2500	3125	3750	4375
1½										750	1125	1500	1875	2250	2620	3000	3750	4500	5250
1¾											1310	1750	2210	2620	3030	3500	4375	5250	6120
2											1500	2000	2500	3000	3500	4000	5000	6000	7000
2½													3125	3750	4370	5000	6250	7500	8750
3														4500	5250	6000	7500	9000	10500
3½															6120	7000	8750	10500	12250
4																8000	10000	12000	14000

Note: Experimental data were obtained by seating an alloy-steel, cup-point screw against a steel shaft with a hardness of Rockwell C 15. Screw threads were Class 3A, tapped holes were Class 2B. Holding power was defined as the minimum load necessary to produce 0.01 in. of relative movement between the shaft and the collar. Rustproofing oil on screws only.

Figure 12.23 Torsional holding power (lb-in) for cup point set screws. (Courtesy IFI)

STANDARD HEAD FORMS

Hexagon Socket

Standard size range: No. 0 to 1 in., threaded entire length of screw in ⅛-in. increments from ¼ to ⅝ in., ⅛-in. increments from ⅝ to 1 in., ¼-in. increments from 1 to 4 in., and ½-in. increments from 4 to 6 in. Coarse or fine-thread series, Class 3A.

Fluted Socket

Same as hexagon socket. Nos. 0 and 1 have four flutes. All others have six flutes.

Slotted Headless

Standard size range: No. 5 to ¾ in., threaded entire length of screw. Coarse or fine-thread series, Class 2A.

Square Head

Standard size range: No. 10 to 1½ in. Entire length of body is threaded. Coarse, fine, or 8-threaded series, Class 2A. Sizes ¼ in. and larger are normally available in coarse threads only.

STANDARD POINTS

Cup

By far the most widely used. For quick, permanent location of gears, collars, and pulleys on shafts, when cutting-in action of point is not objectionable. Heat-treated screws of Rockwell C 45 hardness or greater can be used on shafts with surface hardness up to Rockwell C 35 without deforming the point.

Cone

Used where permanent location of parts is required. Because of penetration, it develops greatest axial and torsional holding power when it bears against material of Rockwell C 15 hardness or greater. Usually spotted in a hole to half its length, so that penetration is deep enough to develop ample shear strength across cone section.

Oval

Used when frequent adjustment is necessary without excessive deformation of part against which it bears. Also used for seating against angular surfaces. Circular U-grooves or axial V-grooves are sometimes provided in the shaft to allow rotational or longitudinal adjustment. In other applications, shaft is spotted to receive the point. However, has the lowest axial or torsional holding power.

Flat

Used when frequent resetting of one machine part in relation to another is required. Flat points cause little damage to the part against which the point bears, so are particularly suited for use against hardened steel shafts. Can also be used as adjusting screws for fine linear adjustments. Here, a flat is usually ground on the shaft for better point contact. Also preferred where walls are thin or threaded member is a soft metal.

Half Dog

The half dog has more threads per length of screw than full dog. More widely used. Normally applied where permanent location of one part in relation to another is desired, spotted in a hole drilled in the shaft. Drilled hole must match the point diameter to prevent side play; holding power is shear strength of point. Occasionally used in place of dowels, and where end of thread must be protected. Recommended for use with hardened members and on hollow tubing, provided some locking device holds screw in place.

Full Dog

Same as half dog except for a longer point.

Figure 12.24 Set screw types. (Courtesy Industrial Fasteners Institute, Cleveland, Ohio)

The holding power can be determined from the chart shown in Figure 12.23 for cup points. The various types of points that can be displayed are shown in Figure 12.24. Here the standard points are shown as cup, flat, cone, oval and dog. The standard head forms for set screws are; hexagon, fluted, slotted and square. Set screws are, therefore, displayed on an automated drawing by these two characteristics; head form and and point style desired.

12.10 PINS

Pins used as fasteners offer an inexpensive and effective approach to assembly where loading is primarily in shear. Pins are either semipermanent, like set screws, or quick release. Semipermanent pins are called machine pins and are illustrated in Figure 12.25. Machine pins may be displayed on an automated drawing as:

1. Hardened and ground dowel pins.

HARDENED-&-GROUND DOWEL PIN

Standardized in nominal diameters ranging from 1/8 to 7/8 in. Standard pins are 0.0002 in. oversize on the nominal diameter; oversize pins are 0.001 in. oversize.

Standardized pin lengths vary with nominal diameter, ranging from a minimum of 1/2 in. (1/8 in. size) to a maximum of 5 1/2 in. (7/8 in. size).

Standard tolerance on all pin diameters is ±0.0001 in. A typical manufacturer's specification for such pins is: Material, heat-treated alloy steel; surface hardness, Rockwell C 60-62; core hardness, Rockwell C 50-54; surface finish, 6 mu

in. max; and average single-shear strength, 150,000 psi.

A press or tap fit into reamed holes recommended. Secondary reaming operation is often omitted where production requirements do not justify the refinement.

Used for:

1. Holding laminated sections together with surfaces either drawn up tight or separated in some fixed relationship.
2. Fastening machine parts where accuracy of alignment is a primary consideration.
3. Locking components on shafts, in the form of a transverse pin key, where a secure, high-strength joint with minimum stress concentration under severe torque loads is desired.
4. Can be used for combination functions as a locking fastener and hinge, wrist pin, or small stub shaft in moving parts.

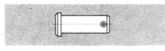

TAPER PIN

Standard pins have a taper of 1/4 in. per ft measured on the diameter. Basic dimension is the diameter of the large end. Diameter d of the small end is given by $d = D - 0.02083L$ where $D =$ diameter of large end, in., and $L =$ pin length, in.

A series of numbered pin sizes has been standardized, ranging from No. 7/0 (0.0625 in. large end) to No. 10 (0.7060 in. large end). Lengths for these pins, depending on diameter of large end, vary from 3/8 to 6 in. Full range of manufactured sizes includes No. 8/0 (0.047-in. large end) to No. 14 (1.523 in. large end) in standard and nonstandard lengths.

Two manufactured grades are recognized: Commercial type, without a concavity tolerance, in sizes 7/0 to 14; and Precision type, with concavity tolerance, in sizes 7/0 to 10. Available pin materials include mild, stainless, and alloy steel and brass.

Holes for taper pins are customarily sized by reaming. A through hole is formed by step drills and straight fluted reamers. Present trend is toward the use of helically fluted taper reamers which provide more accurate sizing and require only a pilot hole the size of the small end of the taper pin. Pin is usually driven into the hole until it is fully seated. Taper of the pin aids hole alignment in assembly.

Used for light-duty service in the attachment of wheels, levers and similar components to shafts. Torque capacity is determined on the basis of double shear, using the average diameter in the tapered section in the shaft for area calculations.

Modifications have been developed to meet nonstandard design situations. For joints where a slotted construction is employed for assembly, a taper pin milled flat along one surface provides an effective wedge fastener for locking the members securely. In locations where access to either end of the taper pin is blocked a threaded section may be incorporated in the accessible end to facilitate pin installation or removal. Other common modifications include special head shapes and split-end constructions.

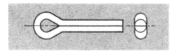

CLEVIS PIN

Standard nominal diameters for clevis pins range from 3/16 to 1 in. Corresponding shank lengths vary from 11/16 for 3/16-in. size to 2 3/4 in. for 1-in. size. Standard material is steel, either soft or cyanide-hardened to meet service conditions.

Basic function of the clevis pin is to connect mating yoke, or fork, and eye members in knuckle-joint assemblies. Held in place by a small cotter pin or other fastening means, it provides a mobile joint construction which is readily disconnected for adjustment or maintenance.

Several designs have been developed to meet special conditions. Common modifications include slotted and noncircular-head constructions as well as threaded ends for assembly with standard nuts or into internal thread sections tapped into one of the yoke arms.

COTTER PIN

Eighteen sizes have been standardized in nominal diameters ranging from 1/32 to 3/4 in. Available materials include mild steel, brass, bronze, stainless steel, and aluminum. Available in a number of point

styles, shown in *Standard Cotter-pin Point Styles*.

Locking device for other fasteners. Used with a castle or slotted nut on bolt, screws, or studs, it provides a convenient, low-cost locknut assembly.

Hold standard clevis pins in place. Can be used with or without a plain washer as an artificial shoulder to lock parts in position on unthreaded or threaded rods and shafts. Cotter-pin sizes used with various shafts are shown in *Recommended Assembly Practice for Standard Cotter Pins*.

Recommended Assembly Practice for Standard Cotter Pins

Nominal Thread Size (in.)	Nominal Cotter Pin Size (in.)	Cotter Pin Hole (in.)	End Clearance* (in.)
1/4	1/16	5/64	7/64
5/16	5/64	3/32	7/64
3/8	3/32	7/64	9/64
7/16	3/32	7/64	11/64
1/2	1/8	9/64	11/64
9/16	1/8	9/64	13/64
5/8	5/32	11/64	15/64
3/4	5/32	11/64	17/64
7/8	5/32	11/64	9/32
1	3/16	13/64	5/16
1 1/8	3/16	13/64	25/64
1 1/4	7/32	15/64	13/32
1 3/8	7/32	15/64	7/16
1 1/2	1/4	17/64	31/64
1 5/8	1/4	17/64	31/64
1 3/4	5/16	5/16	35/64
1 7/8	5/16	5/16	35/64
2	5/16	5/16	41/64
2 1/4	5/16	5/16	41/64
2 1/2	3/8	3/8	3/4
2 3/4	3/8	3/8	3/4
3	1/2	1/2	3/4

*Distance from extreme point of bolt or screw to center of cotter pin hole.

Standard Cotter Pin Point Styles

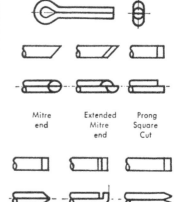

Mitre end   Extended Mitre end   Prong Square Cut

Bevel point   Hammer lock   Chisel point

Figure 12.25 Machine pin types and specifications. (Courtesy Industrial Fasteners Inst.)

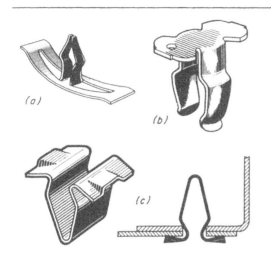
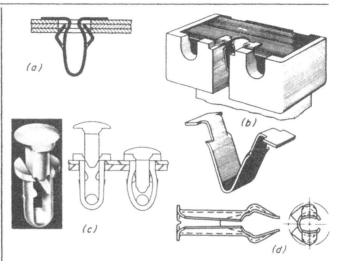

Holding power of dart fasteners can be kept within pre-determined limits by controlling stock thickness and width, and the configuration of the retaining elements. Clip *a* has moderate holding power and will not damage delicate components, nor distort finished parts. Clip *b* is designed for more strength. Heavy-duty removable clip *c* has raised portions in the head where a screwdriver may be inserted to pry it up.

Varieties of Dart Design

To accommodate wide variations in stock thickness, extra arms are lanced in the faces of dart *a*. These expand beyond the dart faces to clamp thin stacks, or can be pinched flush with the dart face to fit a thicker buildup. Hole variations can also be tolerated. Barbs in the edges of dart *b* bite into stock to hold the fastener where there is no "other side." One-piece plastic dart fastener is noncorrosive, insulating, reuseable, and easy to apply and remove, *c*. Semitubular dart clip, *d*, fits panel hole more closely than a square clip would, and restricts movement after assembly.

Figure 12.26 Spring clips, dart design. (Courtesy Industrial Fasteners Institute)

2. Taper pin.
3. Clevis pin.
4. Cotter pin.

12.11 SPRING CLIPS AND RINGS

Spring fasteners are particularly important to the mass producing industries. Clips and rings perform multiple functions, often eliminate the handling of several smaller parts and, therefore, reduce the assembly costs. Figure 12.26 is an example of one type of spring clip known as the dart. Automated drawings contain reference to dart clips, the display is related to the use of dart clips as shown in Figure 12.27a through d.

Another popular type of spring clip is shown in Figure 12.28, it is the cable and tube clip. Special spring clip designs are

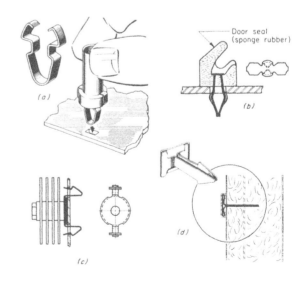

Figure 12.27 Use of dart clip. (Courtesy IFI)

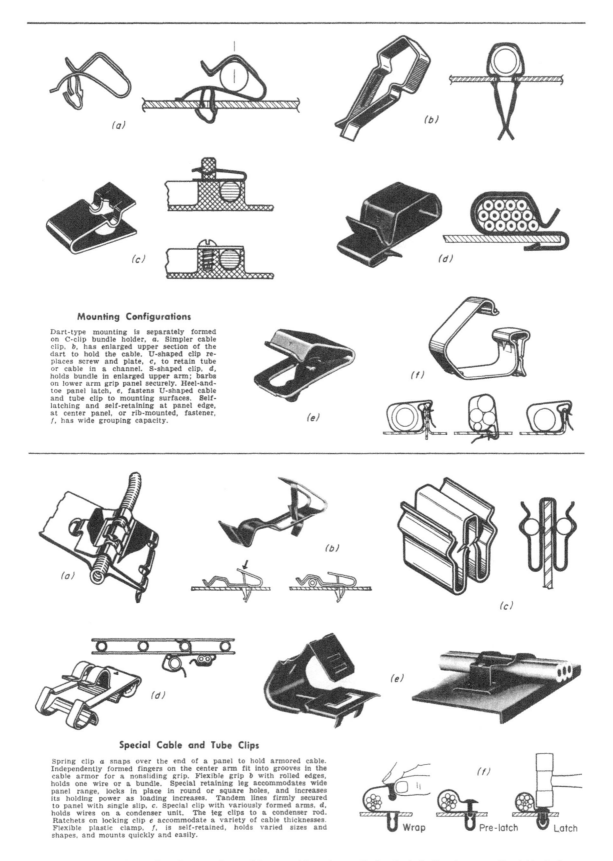

Mounting Configurations

Dart-type mounting is separately formed on C-clip bundle holder, *a*. Simpler cable clip, *b*, has enlarged upper section of the dart to hold the cable. U-shaped clip replaces screw and plate, *c*, to retain tube or cable in a channel. S-shaped clip, *d*, holds bundle in enlarged upper arm; barbs on lower arm grip panel securely. Heel-and-toe panel latch, *e*, fastens U-shaped cable and tube clip to mounting surfaces. Self-latching and self-retaining at panel edge, at center panel, or rib-mounted, fastener, *f*, has wide grouping capacity.

Special Cable and Tube Clips

Spring clip *a* snaps over the end of a panel to hold armored cable. Independently formed fingers on the center arm fit into grooves in the cable armor for a nonsliding grip. Flexible grip *b* with rolled edges, holds one wire or a bundle. Special retaining leg accommodates wide panel range, locks in place in round or square holes, and increases its holding power as loading increases. Tandem lines firmly secured to panel with single slip, *c*. Special clip with variously formed arms, *d*, holds wires on a condenser unit. The leg clips to a condenser rod. Ratchets on locking clip *e* accommodate a variety of cable thicknesses. Flexible plastic clamp, *f*, is self-retained, holds varied sizes and shapes, and mounts quickly and easily.

Figure 12.28   Cable and tube spring clips.   (Courtesy Industrial Fasteners Institute)

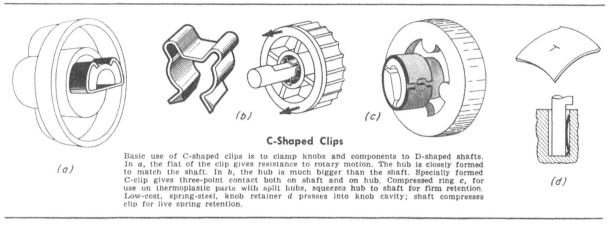

C-Shaped Clips

Basic use of C-shaped clips is to clamp knobs and components to D-shaped shafts. In *a*, the flat of the clip gives resistance to rotary motion. The hub is closely formed to match the shaft. In *b*, the hub is much bigger than the shaft. Specially formed C-clip gives three-point contact both on shaft and on hub. Compressed ring *c*, for use on thermoplastic parts with split hubs, squeezes hub to shaft for firm retention. Low-cost, spring-steel, knob retainer *d* presses into knob cavity; shaft compresses clip for live spring retention.

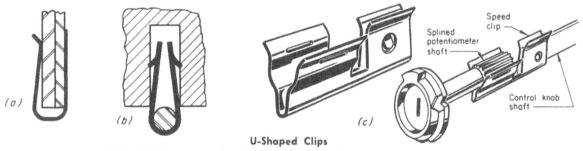

U-Shaped Clips

U-shaped clips exert a clamping action, *a*, to hold two pieces of stock together at the edge, or a spreading action, *b*, to make a tight fit in a slot or hole. This could be half of a lightweight hinge. A split U, *c*, couples a flat control knob shaft to a potentiometer shaft. Splines fit into slots in the arms of the U to drive the potentiometer, but spring out to let the control shaft overtravel when the potentiometer reaches its limit of motion. A barb snaps into a hole in the flat control shaft to hold it in place.

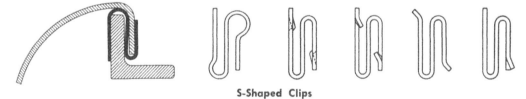

S-Shaped Clips

A basic S-shaped type spring clip and five variations. S-shaped clips can be developed in various widths and stock thicknesses to provide specific holding powers. The bend radiuses are controlled to accommodate thicknesses of panels and flanges. Variations of the S-shaped clips include turned up corners or ends for easier insertion over panels and flanges, built-in retaining barbs that bite into metal, retaining dimples to help hold the clip in place and widened radiuses.

Figure 12.29 Special CUS-shaped spring clips. (Courtesy Industrial Fastener Institute)

shown in Figure 12.29, while ring designs are illustrated in Figure 12.30.

12.12 NUTS AND WASHERS

Nuts and washers are used in automated drawings as locking fasteners which mate with other threaded fasteners. Figure 12.31 is an example of some of the types and styles of nuts available for display. A nut be-comes a locking device when the proper type of washer is selected for display. Figure 12.32 is an example of some of the washer types available. A plain washer provides little locking action. A spring washer design increases the locking potential of the washer surface. Not all washers are locking types. Many washers serve other functions such as; calibrated load distribution, and surface protection or insulation.

Axially Assembled Rings Used in Grooves

Internal   External

Basic Types

Designed for axial assembly. Internal ring is compressed for insertion into bore or housing, external ring expanded for assembly over shaft. Both rings seat in deep grooves and are secure against heavy thrust loads and high rotational speeds. External rings for shafts over 4 in. diam are supplied with special lugs to maintain ring balance during rotation.

Internal   External

Bowed Rings

For assemblies in which accumulated tolerances cause objectionable end play between ring and retained part. Bowed construction permits rings to provide resilient end-play take-up in axial direction while maintaining tight grip against groove bottom. Used widely to provide spring tension on adjustment devices and prevent rattle in machine linkages. Useful also for salvaging assemblies designed originally for flat rings where groove has been cut too wide.

Internal   External

Beveled Rings

Designed for rigid end-play take-up. Rings have a 15-deg bevel on groove-engaging edge and are installed in grooves having corresponding bevel on load-bearing wall. Ring acts as a wedge between retained part and groove wall, seating deeper in the groove to compensate automatically for tolerances or wear. Used widely in applications where a tight seal is critical.

Standard size   Miniature

Heavy-Duty Rings

Heavy duty external ring resistant to high thrust and impact loads. Much thicker than basic type and has greatly increased section. Suitable for extreme loading conditions, and for retaining bearings and other parts with large corner radii or chamfers. Eliminates need for spacer washers. Also available as miniature high-strength external ring. Forms tamperproof shoulder on small-diameter shafts subject to heavy thrust and impact loads.

Internal   External

Inverted Rings

Same tapered construction as basic types, with lugs inverted to abut bottom of groove. Section height increased to provide higher shoulder, uniformly concentric with housing or shaft. Rings provide better clearance, more attractive appearance than basic types. Can accommodate parts with curved abutting walls, large corner radii or chamfers. Because of lug design, rings have less contact surface with groove wall than basic types, somewhat lower thrust load capacity.

Free Ring   Ring Assembled

Permanent-Shoulder Ring

When compressed into V-shaped groove, notches deform into small triangles, closing gaps and simultaneously reducing ID and OD. Ring grips groove tightly and provides 360-deg shoulder against heavy thrust loads. Eliminates machined shoulders and permits reduction in shaft diameters. Recommended for small-diameter shafts, studs, and similar parts.

Radially Assembled Rings Used in Grooves

Crescent Ring

Has a tapered section similar to the basic axial types. Remains circular after installation on a shaft and provides a tight grip against the groove bottom. Because of narrow section height and uniformly circular shoulder, used widely in assemblies where clearance is limited.

Flat   Bowed

E-Rings

Provides a large bearing shoulder on small-diameter shafts and often is used as a spring retainer. Three heavy prongs, spaced approximately 120 deg apart, provide contact surface with groove bottom. Seated in deep groove for increased thrust capacity. Bowed E-rings are used for providing resilient end-play take-up in an assembly.

Reinforced E-Ring

Provides approximately 5 times greater gripping power and 50 per cent higher rotational speed limits than conventional E-rings. Heavy web section and tapered bending arms develop substantially greater spring pressure with no increase in permanent set.

Locking Prong Ring

Derives its name from two prongs which grip shaft and prevent ring from being forced from groove. To be removed, ring must be flattened so that prongs clear groove. Fastener has relatively high thrust load capacity and may be used as shoulder against rotating parts. Bowed construction provides resilient end-play take-up and permits ring to function as spring as well as shoulder.

Interlocking Ring

Balanced two-part ring designed to withstand high rotational speeds, heavy thrust loads. Identical semicircular halves are held together by interlocking prongs at free ends. Forms high circular shoulder uniformly concentric with shaft. Secure against relative rotation between retained parts. Attractive appearance makes fastener especially suitable for exposed applications.

Self-Locking Rings Which Do Not Require Grooves

Circular External Rings

Push-on type fasteners with inclined prongs which bend from their initial position to grip the shaft. Ring at left has arched rim for increased strength and thrust load capacity; extra-long prongs accommodate wide shaft tolerances. Ring at right has flat rim, shorter locking prongs, smaller OD. Recommended for assemblies where flat contact surface with retained part is required or clearance dimensions are critical.

Grip Ring

Exerts a frictional hold against axial displacement from either direction. Tapered section permits ring to remain circular after expansion and assure maximum contact surface with shaft. Ring is substantially thicker than basic external type and has larger section for increased spring pressure and shoulder height. Fastener is adjustable on shaft, does not mar surface and is reusable following disassembly.

Triangular Retainer

Provides larger shoulder than circular push-on types and has greater gripping strength. Dished body holds retained part under spring pressure, eliminating need for bowed washers and other tensioning devices. In some sizes, inclined prongs are ribbed for increased stiffness and holding power.

Circular Internal Ring

Designed for use in bores and housings. Functions in same manner as external types except that locking prongs are on outside of rim. Like other self-locking rings, useful for taking up end-play caused by accumulated tolerances in retained parts.

Figure 12.30   Ring fasteners. (Courtesy Industrial Fasteners Institute)

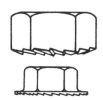

When the upper half of this nut is tightened, it presses the collar of the lower half against the bolt.

Grooved washer causes the threaded collar to radially lock on the bolt.

Two positions of a diaphragm-type locknut, before and after seating. Bending action causes upper threads to grip the bolt.

Deformed bearing surface. Teeth on the bearing surface "bite" into work to provide a ratchet locking action.

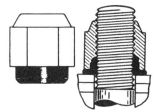

Nylon insert flows around the bolt rather than being cut by the bolt threads to provide locking action and an effective seal.

Nut with a captive toothed washer. When tightened, the captive washer provides the locking means with spring action between the nut and working surface.

Locking action is produced by deflection of the upper beams when the nut is tightened against the work.

Arched bottom of this square free-spinning type nut causes the top to pinch in and bind.

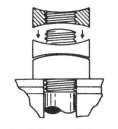

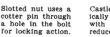
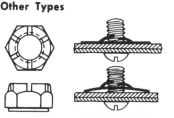

Other Types

Seating locknut applied over a regular nut. Locking force comes from thread distortion when firmly seated.

Jam nut, applied under a large regular nut, is elastically deformed against bolt threads when the large nut is tightened.

Slotted nut uses a cotter pin through a hole in the bolt for locking action.

Castle nut is basically a slotted nut with a crown of reduced diameter.

Single-thread locknut, which is speedily applied, locks by grip of arched prongs when bolt or screw is tightened.

Figure 12.31 Display of nuts. (Courtesy Industrial Fasteners Institute)

12.13 RESISTANCE-WELDED FASTENERS

A resistance welded fastener is a part designed to be fused permanently in place by standard production welding equipment. Two popular methods of welding are used; projection and spot welding. With either, the fusion of the threaded fastener to a metal part surface is the result of the natural resistance of metal to a controlled current under pressure. Figure 12.33 illustrates both of this methods.

Resistance welding can be used with an automated drawing fastener displayed earlier. For example, screws and nuts or pins can be resistance welded. Figure 12.34 shows the basic types of weld fasteners.

In projection welding, heat is localized through embossments or projections on the fastener. In spot welding, the current is directed through the entire area under the electrode tip. Welding is usually performed by a rocker-arm type spot welder. The length of the arms may range from 12 to 60 inches. The main advantage of spot welding is that spot welders cost less.

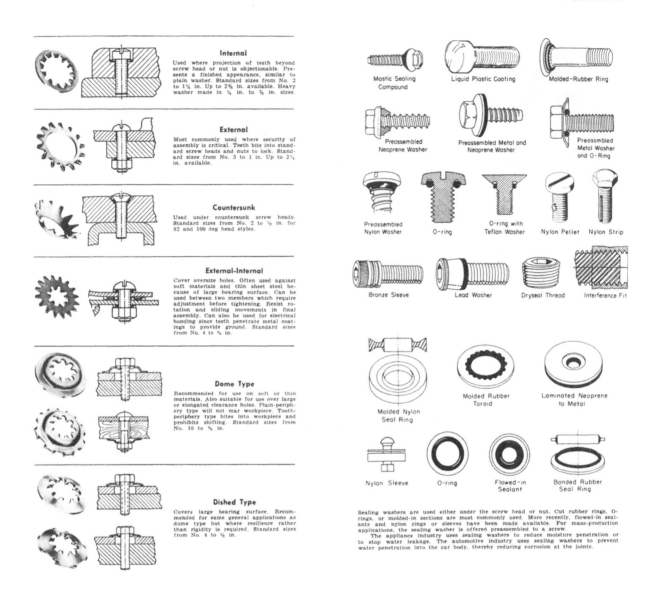

Figure 12.32   Washer types available.   (Courtesy Industrial Fasteners Institute)

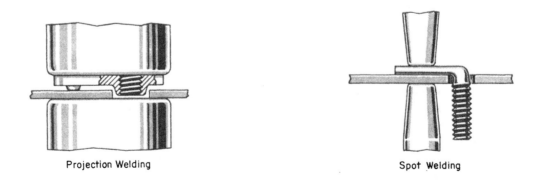

Figure 12.33   Two types of resistance welded fasteners.   (Courtesy IFI)

PROJECTION-WELD SCREWS

Through-Hole
Self-locating for through-hole applications. Provides high strength and flush surface for mating parts.

Watertight
Self-locating screw which provides complete hermetic seal. Ideal for welding to perforated sheets.

Button-Projection
Easily welded to curved surfaces. Recommended for use on heavy sheets, ¾ in. thickness and over.

Blind-Location
Provides unmarred exterior. Requires no holes in sheets and no electrode insulation.

Spade
Used where a threaded section is required on the edge of a sheet.

Right-Angle Spade
Used in confined locations or channels.

PROJECTION-WELD NUTS

Flanged
Projections on top or under flange. Provides extra-long thread engagement and also serves as bearing surface, mounting surface, or spacer. T-shape head used primarily in confined areas.

Watertight
Ring projection provides hermetic seal. Round body nut has blind threaded hole for completely watertight fastening. Widely used on air conditioning equipment and transformers.

Piloted, Dual-Line Projection
Self-locating pilot. Ideal in applications where tension is against the weld. Readily adaptable to automatic feeding.

Piloted, Single-Button Projection
Self-locating pilot. Ideal in confined areas and channels. Easily welded to curved surfaces and heavier gage sheets.

Pilotless
Ideal for blind locations where there is no hole in the sheet or on slightly curved surfaces.

Right-Angle (Bracket)
Used when it is necessary to attach a nut at a 90 deg angle.

SPOT-WELD NUTS

Piloted, Target
Self-locating pilot and recessed target electrode area. Ideal where other spot-welds are being made because the nut can be attached with the same equipment and settings. Can be welded with low KVA capacity equipment.

Piloted, Double-Tab
Self-locating pilot. Ideal for bridging corners or sections, or where extra strength is needed.

Piloted, Single-Tab
Self-locating pilot. Can be welded with same settings and electrodes used to join components.

Dual Piloted
Self-locating pilot for use where two tapped holes must be located close together.

SPOT-WELD SCREWS

Right-Angle Spade
Ideal where other spot-welds are being made and a threaded section can be attached at the same time with the same equipment and weld set-up.

Figure 12.34 Types of weld fasteners. (Courtesy Industrial Fasteners Institute)

12.14 REVIEW PROBLEMS

Begin your review of this chapter by providing a screen display of each of the fasteners shown below.

1.

2.

3.

4.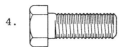

5.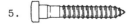

6.

7.

8.

9.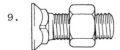

Continue your review by displaying the following from your display menu.

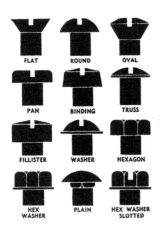

Use the general drafting menu or 2-D image processor to code the washer profiles shown below. Your instructor will select those you are required to code.

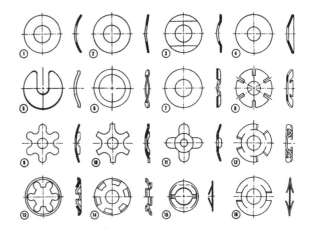

Practice the display technique for:

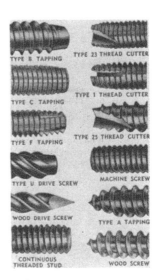

Design a screen window for rivets as:

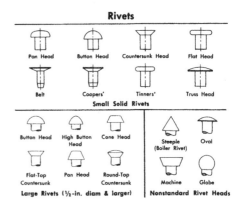

12.15 CHAPTER SUMMARY

This chapter presents a wide array of
industrial type fasteners and how to display
them inside an automated drawing. The Industrial Fasteners Institute has provided
an almost infinite number of types, sizes,
and characteristics for fastener display.
The author is indebted to the generousity
of the institute and hopes that the contents
of this chapter will represent a summary of
the 298 pages of material researched.

To fully understand fasteners, a student must become familiar with its terminology. Stop and review section 12.1 if you
still are not clear on fastener terms. The
excellent examples from the Institute and
the supporting industries should make the
understanding of fastener parts easier.

Once the terms and parts are known, we
can begin to apply the knowledge gained in
earlier chapter to display the fastener parts.

Review sections 12.3 and 4 for new information on display techniques. The display
form of a screw thread was then presented
in section 12.5, Figure 12.20. Study the
terms shown in this figure. Specifications,
notes for fasteners were then discussed
along with the material makeup for fasteners.

The chapter ends with a short presentation of common fasteners such as self-
tapping screws, set screws, pins, clips
and rings, nuts and washers, and welding
techniques. While this chapter is not a
complete study of fasteners, it does represent those items that will appear on an
automated drawing surface. Remember that
the programming for fasteners is contained
the the automated drafting system, that is
not the main issue here. We must be know-
ledgable about what fasteners are, how to
use them in drawings, and how to specify
them for an application.

13

DETAIL DRAWING

Detail drawing is a process whereby the
draftsperson describes in detail the geometry, construction, fabrication or assembly of an object. If this object is part
of an automated drafting procedure, it is
generally thought of as a working drawing.
A working drawing contains items from each
of the previous chapters, for example, a
working drawing contains orthographic views
(Chapter 6), section views (Chapter 8), an
auxiliary view or two (Chapter 9). The views
are dimensioned (Chapter 7), presented on
automated equipment (Chapters 1,2 and 4),
the working drawing contains lettering in
various amounts (Chapter 3), the details
are arranged according to a plan (Chapter
5 and 10), some details are pictorial in
nature (Chapter 11) and finally working
drawings contain a great deal of fasteners.

Each of the preceeding chapters can
come together to produce a new working arrangement as shown in Figure 13.1 or the
individual concepts can be used to modify
or expand existing drawing files. Let's
begin the working drawing concept by looking at existing drawing files. Figure 2.9

is a good example of this. Turn to page 24
and study Figure 2.9, it is a single detail,
but it is not a working drawing. Now look
at Figure 13.2, shown on page 230, it is a
working drawing. Can you see that Figure
2.9 is a section view taken from Figure 13.
2? Figure 2.9 represents a detail of Figure 13.2. In order to understand the working relationship of the parts from the multiviews shown in Figure 13.2, a number of separate details should be prepared. These
are shown in Figure 13.3. A final automated
drawing would be a combination of Figures
13.2 and 13.3 and it would require the pre-
parer to:

1. Use an automattic view generator.

2. Plot a section view called A-A.

3. Use an automatic dimensioning package and note generation.

4. Use special symbols such as welding
fasteners, structural shapes, and others.

5. Create subpictures such as washers,
bolts, screws, nuts and other items.

6. Provide a plate and sleeve assembly.

7. Plot profile views for detail information.

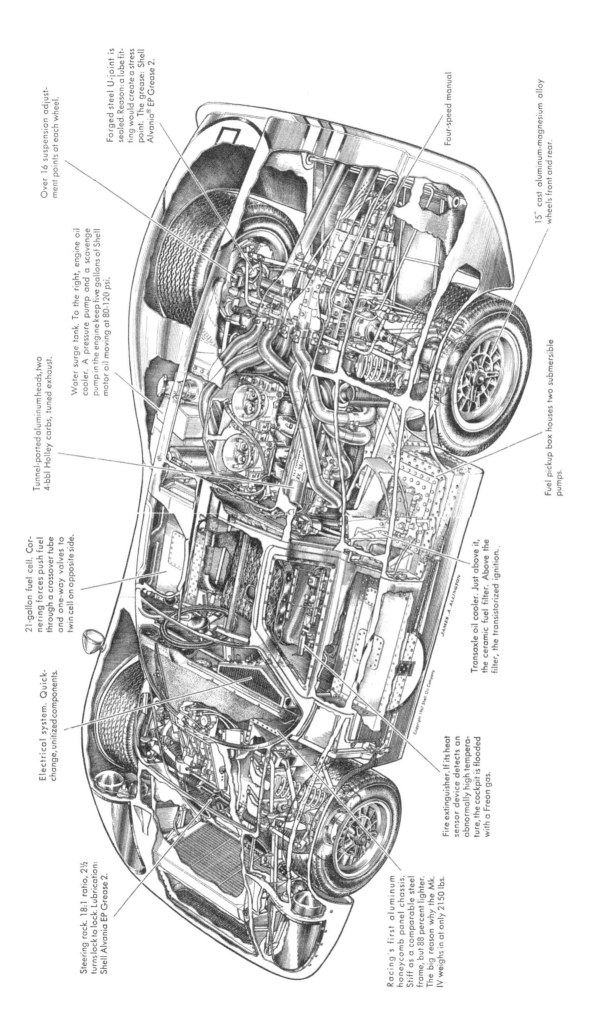

Over 16 suspension adjustment points at each wheel.

Forged steel U-joint is sealed. Reason: a lube fitting would create a stress point. The grease: Shell Alvania® EP Grease 2.

Four-speed manual

Water surge tank. To the right, engine oil cooler. A pressure pump and a scavenge pump in the engine keep five gallons of Shell motor oil moving at 80-120 psi.

Tunnel-ported aluminum heads, two 4-bbl Holley carbs, tuned exhaust.

15" cast aluminum-magnesium alloy wheels front and rear.

21-gallon fuel cell. Cornering forces push fuel through a crossover tube and one-way valves to twin cell on opposite side.

Fuel pickup box houses two submersible pumps.

Electrical system. Quick-change, unitized components.

Transaxle oil cooler. Just above it, the ceramic fuel filter. Above the filter, the transistorized ignition.

Steering rack. 18:1 ratio, 2½ turns lock to lock. Lubrication: Shell Alvania EP Grease 2.

Racing's first aluminum honeycomb panel chassis. Stiff as a comparable steel frame, but 88 percent lighter. The big reason why the Mk. IV weighs in at only 2150 lbs.

Fire extinguisher. If its heat sensor device detects an abnormally high temperature, the cockpit is flooded with a Freon gas.

Figure 13.1 Working drawing of a complete product. (Courtesy Shell Oil Company, Trenton, N.J.)

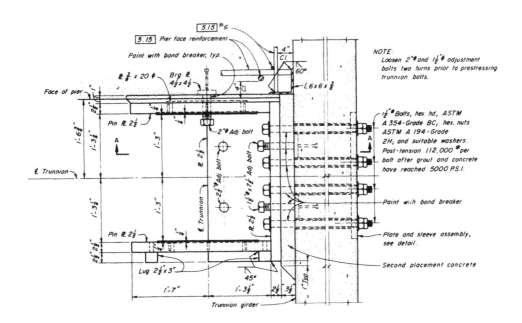

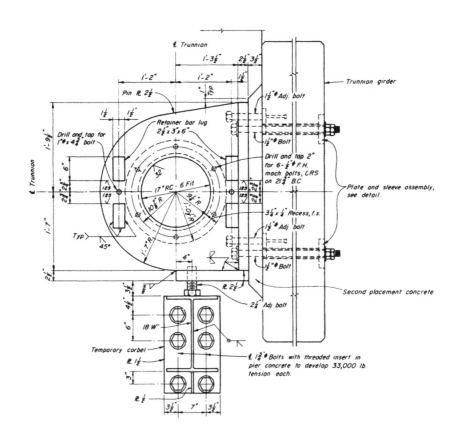

Figure 13.2 Working drawing of industrial fabrication of parts. (Courtesy DLR Associates)

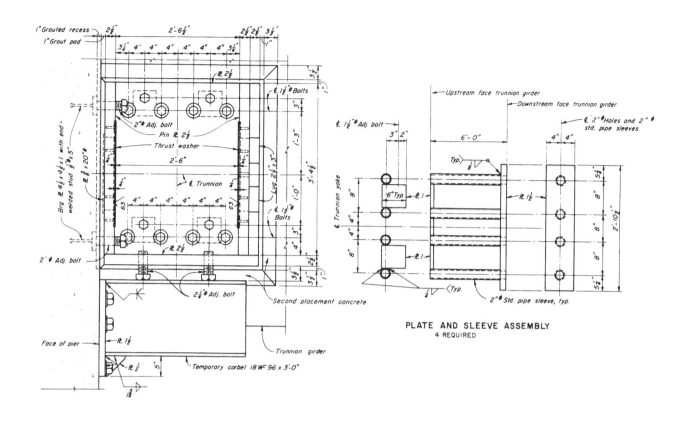

PLATE AND SLEEVE ASSEMBLY
4 REQUIRED

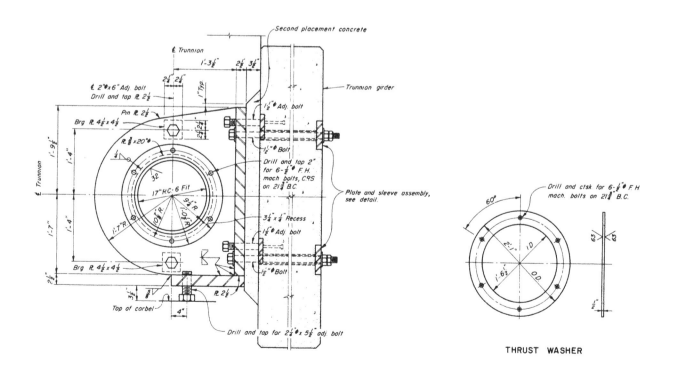

THRUST WASHER

Figure 13.3 Details for working drawing shown in Figure 13.2. (Courtesy DLR Associates)

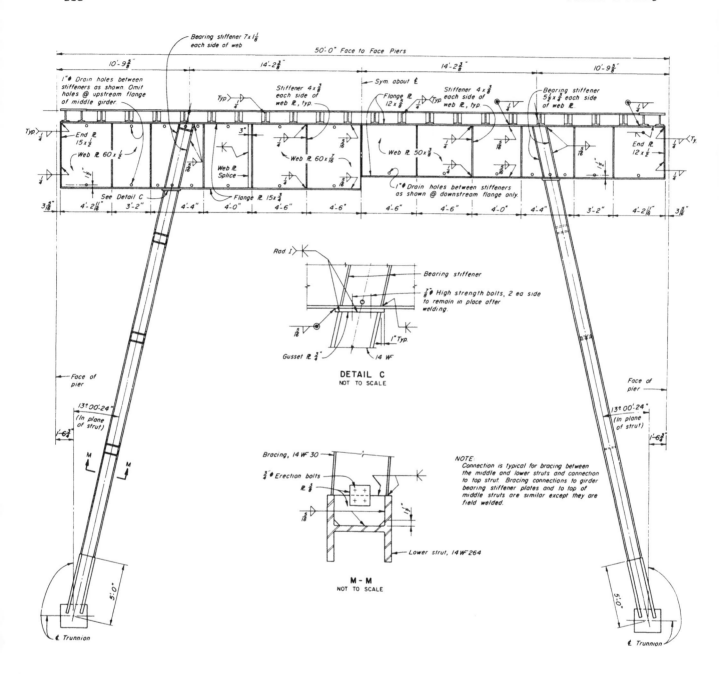

Figure 13.4 Selecting the working views. (Courtesy DLR Associates)

13.1 DETAIL DRAWING DISPLAY

A detail drawing supports a working drawing arrangement. In the case of the machine drawing for mechanical engineers it is the working drawing containing all the necessary shop information for the production of the individual pieces of the machine. A detail drawing display should give complete information for the construction of the design intent. In the case of a civil engineer, the detail drawing may be a wall section or stair detail. Details almost always involve sections to fully explain the assembly of an object. Figure 13. 4 is a working view of an object along with detail C and section M-M. Together these views would be called a detail drawing. But

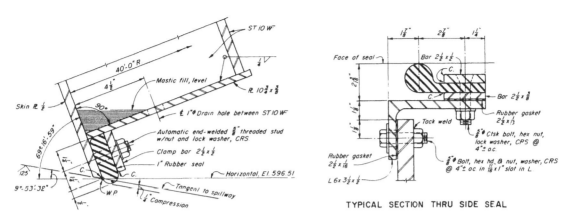

TYPICAL SECTION THRU BOTTOM SEAL

TYPICAL SECTION THRU SIDE SEAL

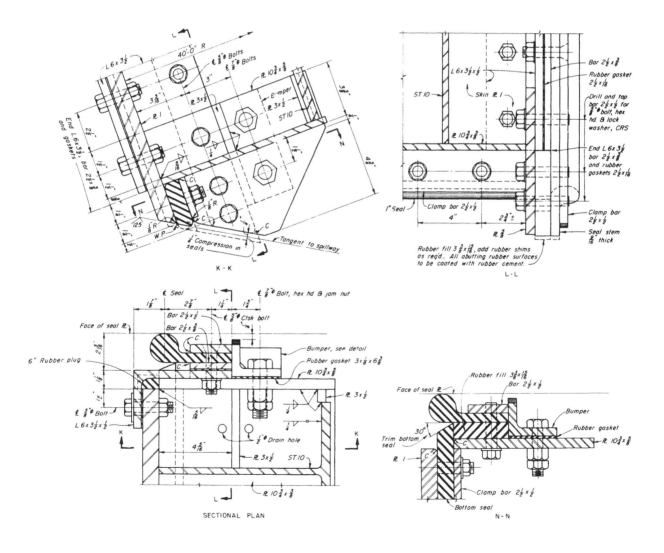

Figure 13.5 Details used to support Figure 13.4.   (Courtesy DLR Associates)

the number of details is lacking. Examine
Figure 13.5 and place these details with
the working view shown in Figure 13.4. Now
a more complete picture of the fabrication
process exists. Add the few remaining details shown in Figure 13.6 and the process
is complete.

13.2 PROCEDURE FOR DETAIL DRAWINGS

The working drawing concept is used
in all fields of engineering documentation
for detailing (describing) the design layout. The designer has input the overall
design intent (Figure 10.1, step 1), anal-
yzed the design parameters, and tested the
validity of the design. Figure 10.1 illustrates the output from this process and the
passing of the data base to step 2 (drafting). A set of detailed drawings can now
be made from the design data passed from
step 1, Figure 10.1. To display a detail
drawing, all the sections and other images
must be contained within the data stream
as a series of data points. These points
will be used to display the various items
contained in the working drawing layout.

Once this has been done, the procedure
for making a detailed drawing is as follows:

1. Select the views, remembering that
in Chapter 6 the front view is described
by plotting the X and Y array from the DATA-
PT list. The other views are plotted by
selecting the X and Z array from the DATA-
PT list for the horizontal view and the
Z and Y list for the profile. In addition
to the main orthographic views, auxiliary
views may also be constructed as shown in
Chapter 9.

2. Choose a drawing scale which will
allow an arrangement of the views and the
location of needed dimensions, notes, and
part labels.

3. Call out the location for centerlines of features such as holes, slots, or
machined sections and block in all fillets
and rounds as shown in 13.7.

4. Add dimensions by the DIMEN routines shown in Chapter 7, remembering that
this routine contains extension lines, dim-

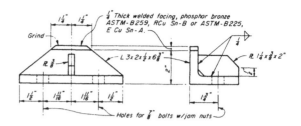

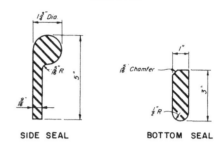

BUMPER
8 REQUIRED EACH GATE
SCALE: 6" = 1'-0"

SIDE SEAL   BOTTOM SEAL

Figure 13.6 Remaining items for detailing.

ension lines, arrowheads, and annotation for
placement and specification of dimensions.

5. Select the proper notes by the use
of CALL LABEL or CALL NOTE and fill in the
automated title block and related notes necessary to the drawing.

6. Check the output by previewing on
the DVST, make any corrections in the drawing file data base, and output on the digital
plotter.

The procedure for each of the six steps
is contained in six separate chapters studied
earlier, if you are unsure, review those
items now.

13.3 PLACING DETAILS ON DRAWING SHEETS

Figure 13.7 is an example of correct
detail placement. It is an expansion of
an earlier file (Figure 2.13). The details
necessary for construction have been added
to the drawing data base to represent this
complete working drawing sheet. Notice that
the drawing image has been shifted to the
left so that proper placement of the section
views can take place. Compare the sheet
drawing shown on page 235 (Figure 13.7) with
page 27 (Figure 2.13). Auxiliary views are
contained in Figure 2.13 and they will re-

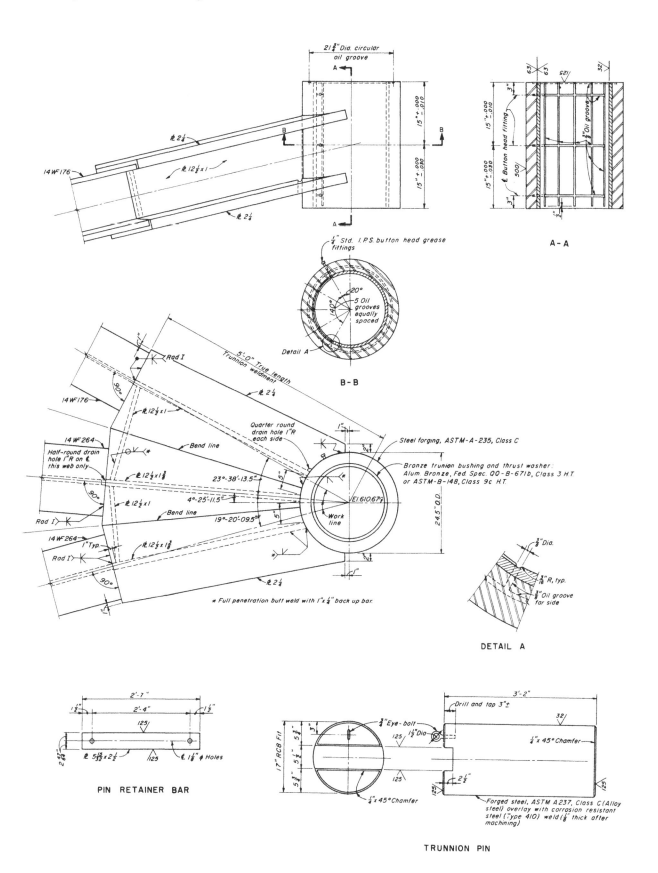

Figure 13.7 Complete working drawing showing merged details during layout phase.

appear when the two data files are merged.
A concept known as file merge is necessary
because of the display device called the
DVST. You will remember from earlier chapters that this is a direct display device
and, therefore, objects can not be moved
in real-time. They can be moved in memory,
the screen erased, and the drawing replotted
for the draftsman.

Another graphics concept called PAN
is used on display devices like CRT s. A
typical CRT response is in real-time and
the display object can be moved from side
to side or up and down by the press of a
bottom. If Figures 2.13 and 13.7 are the
same data file than by pressing a button,
the missing items contained to the left of
Figure 13.7 can be shifted (panned) onto
the screen as shown in Figure 13.8. Of course the screen does not get larger and information is lost as it disappears off the
right side of the display screen. In the
same fashion, information will be lost as
we pan up and down. Pan is useful to locate individual details, however. Once the
detail has been located, it can be made to
fill the screen by pressing a button. This
is called ZOOM. The zoom function is also
in real-time on a CRT screen and would be

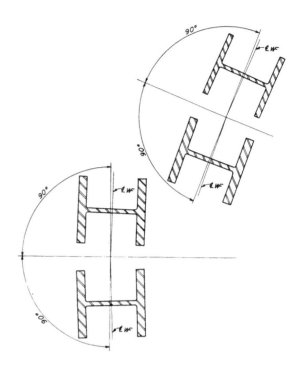

Figure 13.8 Information panned onto screen.

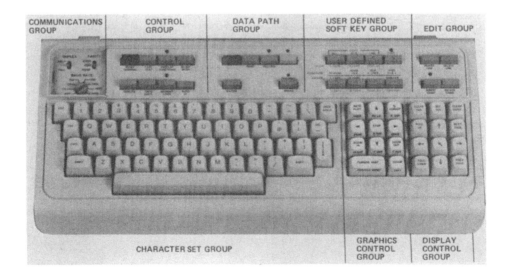

Figure 13.9 Typical CRT keyboard. (Courtesy Hewlett-Packard Company)

impossible to duplicate on a DVST.

13.4   DISPLAYING DETAILS ON CRTS

Placing the individual details on the final drawing sheet is the last step in the working drawing sequence, but it is shown early in this chapter because it is easy to understand and it introduces what a detail is. Once the detail is understood, the techniques for displaying it become easier. The CRT keyboard shown in Figure 13.9 is ideally arranged for working drawings and detail placement. The DVST keyboard is less useful and lacks some display techniques, but remember, if the DVST has served as the only display device to this point, it would be silly to abandon it for a CRT. However, some computer center environments provide both types of output devices and in these cases the use of the CRT keyboard can be very helpful. See Figure 13.9.

You will notice that the keyboard is divided into various key groups, these are:

1. Communications
2. Control
3. Data path
4. User defined (drafting functions)
5. Edit
6. Character set
7. Graphics control
8. Display control.

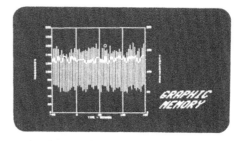

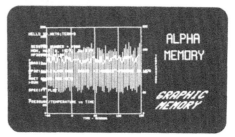

Figure 13.11   Features of a CRT display. (Courtesy of Hewlett-Packard Company)

Figure 13.10   Groups 7 & 8. (Courtesy Hewlett-Packard)

Figure 13.11 illustrates the display techniques useful for detail drawing and detail placement within a working drawing.

13.5 DISPLAYING DETAILS ON DVSTS

The preview mode, whereby a drafter
looks at a data file before it is sent to
a digital plotter is the most useful method
for displaying details on a DVST. The DVST
is a quick acting screen display of what is
stored. It is not a dynamic work surface
that will allow the panning (shifting) of
details or zoom function. A DVST as used
in Chapter 6 is a larger screen size than
a typical CRT and it can be divided into
separate window sections called viewports.
A separate viewport may contain a detail,
and in this manner, several details may be
previewed in any order that you wish. If
a change is made in one window, the entire
screen must be erased (this is a disadvant-
age - compared to the CRT).

CRTs are preferred by engineers for
use during the design phase where many types
of changes are possible. DVSTs are preferred by drafters, who typically have fewer
changes to make. Figure 13.12 illustrates
that the DVST approach is very useful in a
detail drawing situation. In this example
only a half plan is contained within a viewport. Figure 13.4's data file has been
placed in two separate viewports. The left
side is previewed in Figure 13.12. Notice
that detail C and section M-M are missing,
they can be added to this viewport or moved
to the right side of the working drawing.
By storing information in separate viewports
the drafter can move between them very
easily, or display both inside the same
window sized screen area. By decreasing
the size of each viewport and increasing
the number of viewports contained inside
the display window, a drafter can simulate
the CRT pan function. Of course there is
still the delay while the DVST flashes (removes unwanted viewports) and replots the
new viewport contents and locations. Figure 13.13 represents a pan to the right.
Only a CRT is dynamic, a textbook illustration is like a DVST screen. There is no
motion provided in a book illustration, so
Figure 13.13 is a pan to the right and Figure 13.12 is a pan to the left. Figure 13.

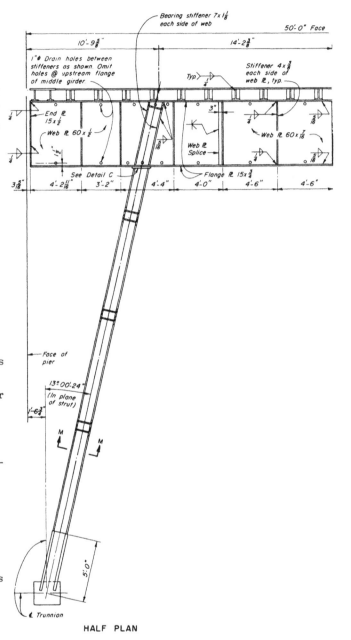

HALF PLAN

Figure 13.12 DVST viewport arrangement.

14. is a pan downward, Figure 13.15 is a
pan downward and to the right. Figure 13.
16 is a pan downward and to the left, while
Figure 13.17 is a pan upward and to the
right, Figure 13.18 is a pan upward and to
the left. The simulated pans are possible
because the number of viewports contained in
the working drawing is nine. In the CRT
diagram of Figure 13.10 you will notice a

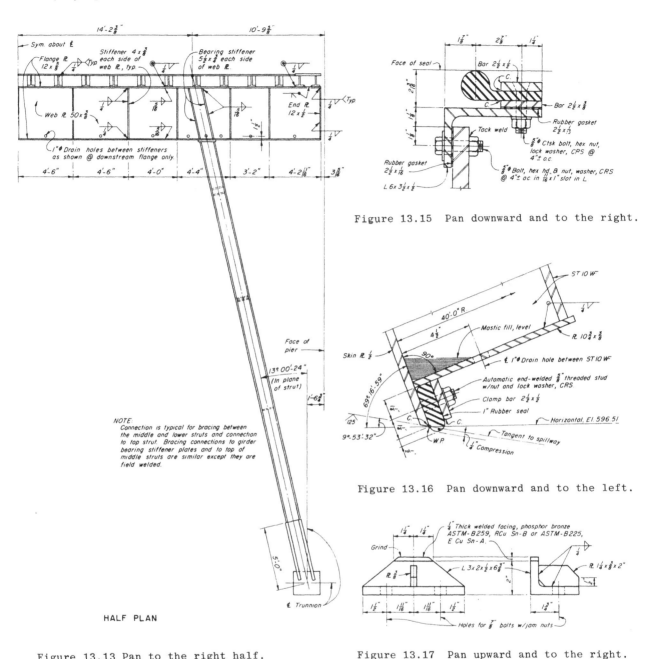

HALF PLAN

Figure 13.13 Pan to the right half.

Figure 13.15 Pan downward and to the right.

Figure 13.16 Pan downward and to the left.

Figure 13.17 Pan upward and to the right.

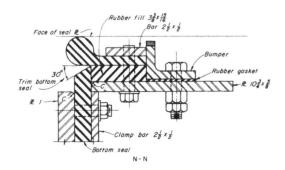

N – N

Figure 13.14 Pan downward.

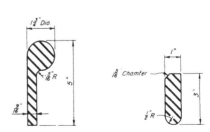

Figure 13.18 Pan upward and to the left.

notice that the key arrangement is based upon nine basic positions also. A cluster of
five keys is used to direct pan action as:

where a DVST pan action is shown in 13.19

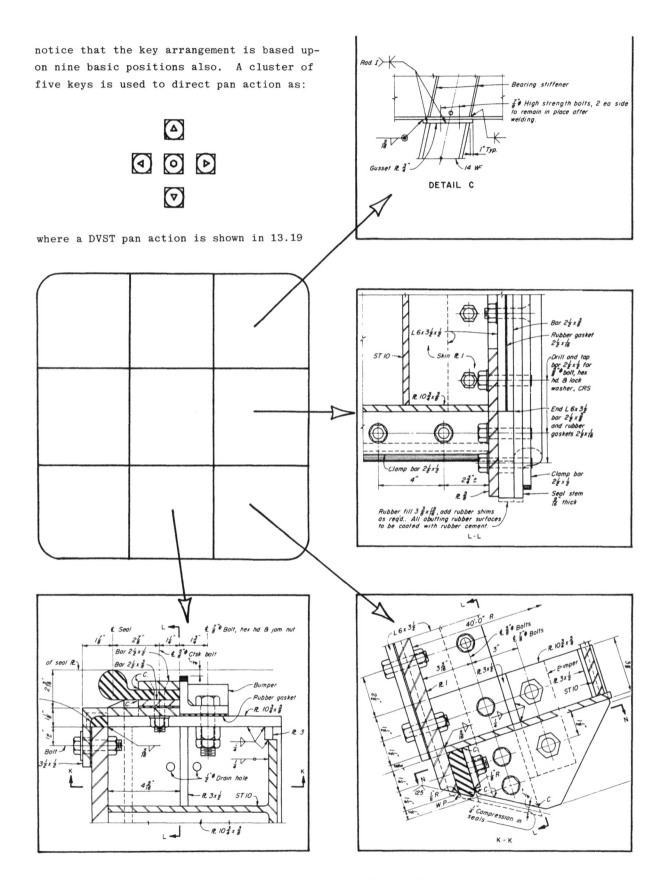

Figure 13.19 Simulated pan action on large screen DVST. (Courtesy DLR Associates)

13.6 PREVIEWING WORKING DRAWINGS

The advantage of previewing working
drawings on a DVST is that the mulitview
procedure (learned in Chapter 6) is already
available for use. A working drawing is a
combination of multiviews and working details that describe the assembly or construction of an object. This being the
case, the 3-D manipulation of data files
on DVST screens is a big advantage. Suppose that we needed another multiview of
Figure 13.4? This figure's output was from
the data files X and Z registers, it produced a plan or horizontal view. Let's use
that same file but display the X and Y data
to produce a frontal or elevation view. A
certain amount of details can now be displayed based upon this new orthograhic information. Using a preview method, the
DVST screen would appear as shown in Figure
13.20. Details A and B plus sections B-B,
C-C, D-D, F-F, G-G, H-H and J-J can be sent
to a digital plotter.

13.7 REVIEW PROBLEMS

Considerable data base construction has
been done in the review problems prior to
this chapter. Your instructor will assign
several working drawings based upon the
data files completed for Chapter 6 and 10.

1. Display the simple block data for
problem 1 in Chapter 6, modify this data
so that a keyed shaft will be located inside the block. Provide the necessary
details, dimensions, section views, and
fasteners as assigned by your instructor.

2. Repeat this sequence on problems
2 through 7 on page 114. Different types
of details will be assigned by your instructor. Use the preview DVST method or
CRT method for display of the details.

3. Choose any of the details shown
at the right of this page and enter each
in a separate viewport. Arrange those
details assigned by your instructor to
create a detailed drawing sheet from a
digital plotter. Preview each viewport or
use a CRT to check each detail.

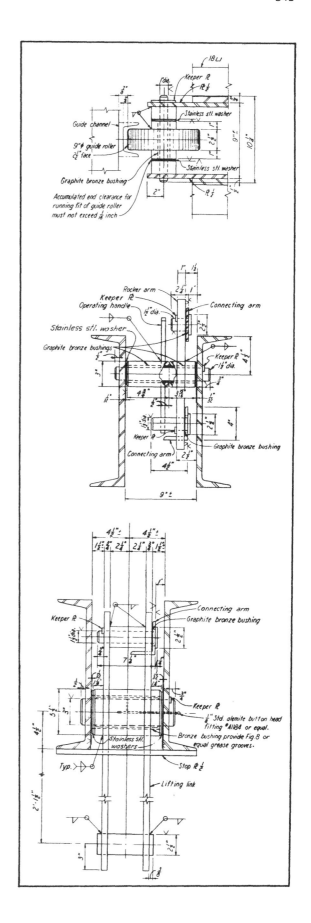

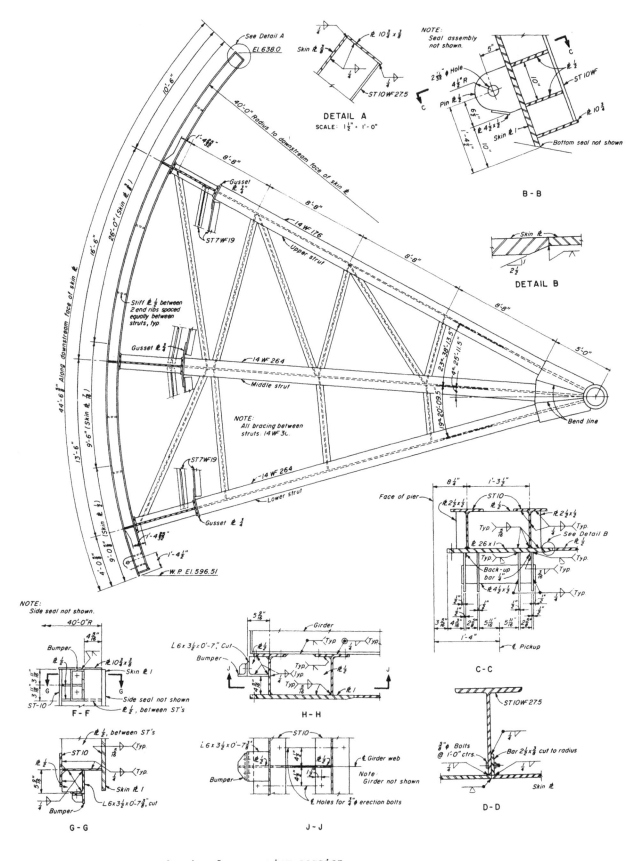

Figure 13.20   Working drawing from preview session.

4. Add the details from problem 3 to the working drawing shown below.

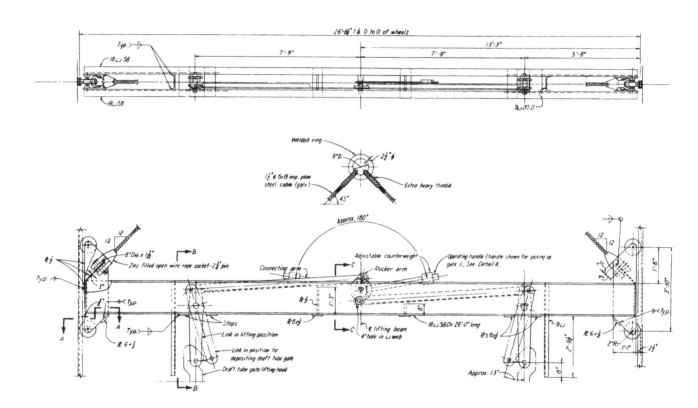

13.8 CHAPTER SUMMARY

Detail drawing is considered to be the "bits and pieces" of drafting. Some experts have said, "You can never have too many details of an object, because the more you know about the object to be manufactured the better off you are as a designer." As a student of automated drafting, the construction of a detail is quite simple because the data file exists for the overall object. This object can be displayed as a multiview drawing by simply pressing the right buttons on the right hardware. The formation of details requires experience on the part of the drafter. This person must understand both the construction or fabrication methods plus the techniques of automated display of computer data base.

Several steps to remember when preparing a working drawing are listed on page 228. If you follow these procedures and select a fairly simple object from Chapter 6 data base, the assembly of a working drawing with details should be an enjoyable experience.

Once you have mastered the combination of multiviews and simple details, you are ready to try more difficult details. A procedure for detail drawings was given on page 234. Follow these six steps whenever you prepare a full sheet of detailed parts. Next you may place the details on a finished plotter drawing. The details are often previewed on a DVST or created on a CRT. At this point the details are checked for errors and information is added as required. The final step is the completed drawing itself. While this sounds rather simple, it is time consuming until you master the skills.

14

Production Drawing

Production drawing, like detail drawing is a display process whereby the draftsperson describes the production plan for the part geometry, construction, fabrication, or the assembly of an object. The object may be as large as that pictured in Figure 14.1, or as small as that pictured in Figure 14.2. If this object is also part of an automated planning process, it is generally thought of as a production type drawing. Commonly displayed production drawings always involve parts fabrication. Parts fabrication includes, but is not limited to the following:

1. Casting
2. Forging
3. Extruding
4. Cold forming (extruding and heading)
5. Stamping
6. Deep drawing
7. Spinning
8. Roll forming
9. High-velocity forming
10. Machining

In addition to the fabrication processes listed above, production drawings contain items from each of the previous chapters,

for example, a production drawing contains orthographic views arranged in detail fashion (Chapters 14 and 6), section views, auxiliary views, dimensioning, lettering, pictorial views and other concepts presented on automated drafting equipment and programs.

Production processes must be understood by the person who prepares the production drawing. For example, the object shown in Figure 14.1 is forged from a 92-inch ingot of vacuum-degassed, nickel-chrome-molybdenum steel, weighing 174,000 pounds. It is a 35-ton, mine-hoist replacement shaft, shipped to the customer as 70,140 pounds, 33 feet in length with a 30 inch diameter and a 54 inch flange diameter. This example required several production drawings which described the forging, heat treatment, and machining processes. Normal delivery time for such a part is 50 to 100 working days. Compare this with the production process shown in Figure 14.2, here a much smaller piece part is positioned and drilled -- total production time less than one minute.

Figure 14.1 Large mine-hoist replacement shaft. (Courtesy Bethlehem Steel, Bethlehem, PA)

14.1 CASTING

Casting processes can be classified
either by the type of mold or pattern, or
by the pressure or force used to fill the
mold. Conventional sand, shell, and plaster molds utilize a permanent pattern, but
the mold is used only once. Permanent molds
and diecastings are machined in metal sections and are used for a large number of
mold making sessions. Investment casting
and full-mold processes involve both an
expendable mold and pattern.

Most castings are produced by filling
the mold cavity with molten metal by the
action of gravity. In diecasting, the
metal is forced into the die cavities under
pressure. Some sand, shell, and investment
molds are filled with molten metal by centrifugal force.

The most widely used casting process
uses a permanent pattern that shapes the
mold cavity when a loose molding material
is pressed over the pattern. The compact-
ing of this material is called jolt and
squeeze. A typical sand mold, with the
various provisions for pouring the molten
metal is shown in Figure 14.3. Sand molds
consist of two or more sections called drag,
cope and cheek (bottom, top and intermediate). The molten metal is poured into the
sprue, and connecting runners called gates
provide flow channels for the metal flow to
enter the mold cavity. Riser cavities are

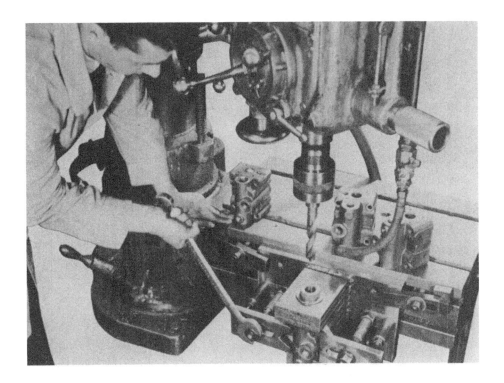

Figure 14.2 Typical manual operation for drilling a piece part in a short time frame.

located over the heavier sections of the casting as shown in Figure 14.3. These risers fill with molten metal and provide reservoirs to compensate for the contraction of the metal during solidification. The top of the mold is held down with weights as shown or held with clamps, the chaplets are metal supports that, along with core prints, hold a core in position. Chaplets fuse into and become part of the casting. The core is removed after cooling and forms an opening in the cast part. To aid in the cooling, chill blocks called chills are placed into the sand so that the solidification happens rapidly. Once the cast part is solid, it is removed from the sand mold, the core is removed along with the sprue, gates, runners and risers.

A sand type mold usually will require more than one type of sand. For example;

Green sand. A sand mixture which contains 2 to 8 % water is used directly under the weights. It forms a bounding to support the rest of the mold cavity.

Dry sand. The sand directly in con-

tact with the pattern which forms the mold cavity. Usually the pattern is split along a centerline and mounted on a metal plate. This metal plate is called the match plate. The match plate extends beyond the flask and is heated to remove some of the water con-

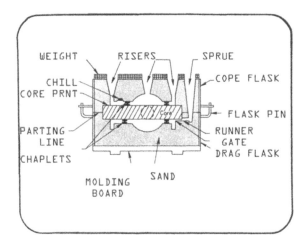

Figure 14.3 Typical display for sand casting.

tained in the sand mixture. The sand does
not fall apart because dry sand mixtures
also contain various organic binders, together with clay.

CO$_2$ sand. Cores are often molded with
special sand containing 2 to 5% liquid sod-
ium silicate. This binder becomes very
hard when gassed with carbon dioxide. This
process is used in place of another core
technique which requires baking in an oven.
A baked core is thrown away after it is
used, while the CO$_2$ sand is mulled (mixed)
and used again.

Shell sand. The refactory sand used
in shell molding is bonded by a thermo-
setting resin that forms a relatively thin
shell over the pattern as shown in Figure
14.4. Here again a match plate is used
to heat the sand. Green sand is then used
to back-up the shell covering.

Of course other substances besides
sand are used to form casting molds, some
but not all types are listed below:

1. Plaster, consisting primarily of
gypsum with various amounts of fillers.

2. Investment, consisting of wax,
plastic, or frozen mercury are used as
patterns in permanent metal molds.

3. Die, where molten metal is forced
in a permanent metal mold.

4. Centrifugal, where the mold is
turned rapidly and the molten metal clings
to the sides of the mold (pipe casting).

14.2 FORGING

Forging consists of deforming a heated metal ingot, bar or billet by squeezing
or hammering. Practically all ductile
metals can be forged if heated. In order
to better understand production drawings
and displays, the types of forgings are:

Closed-die. Forgings that are made
by hammering or pressing metal until it
conforms closely to the shape of the die.

Open-die. Forgings are made by hand
with simple tools like hammers and flat
dies. The metal is progressively worked
to shape by hand and depends upon the
skill of the forger.

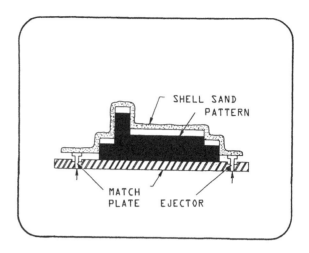

Figure 14.4 Typical production display.

Blocker-type. Forgings that are a
combination of closed and open dies are
called blocker. They require closed dies
but are worked by hand.

Close-tolerance. Also known as draft-
less, precision forging; this type of forging is used for fine quality tool parts.

Upset. These types of forgings have
a certain amount of material pushed back on
itself. This may occur as a flange at one
end or as collars or bosses along the body.

Forging displays are used to describe
the type of production that is to be used.
For example, Figure 14.5 illustrates a
parting line (sometimes called flash line).
This is where the forge dies meet and separate. The parting line, or lines, need
not be in the same plane as shown in Figure 14.6. Here a parting plane (also called forging plane or die plane) is shown.
It is a plane perpendicular to the direction
of pressure. This may or may not be in
the same plane as the parting line. An
upset forging, for example, has two parting planes. Figure 14.7 illustrates die
closure. This is the variation that occurs
when forging dies do not close completely.
It is always measured perpendicular to the
parting plane. Figure 14.8 is a display
showing the mismatch or die shift possible.
Mismatch is usually measured parallel to
the parting plane, although it may be in
any direction.

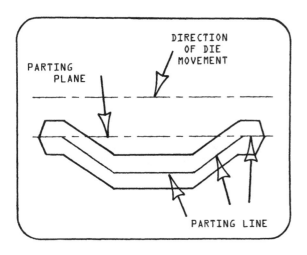

Figure 14.5   Display of forged part line.

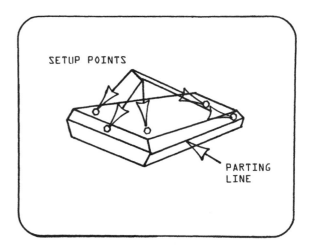

Figure 14.8   Mismatch or die shift.

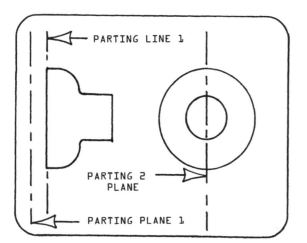

Figure 14.6   Parting plane example.

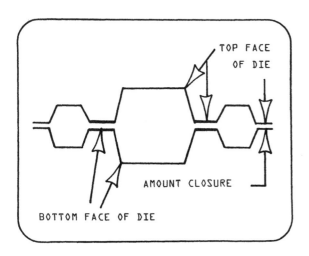

Figure 14.9   Forging ingot or starting blank.

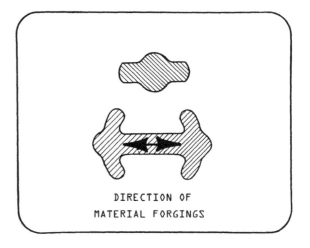

Figure 14.7   Die closure.

Figure 14.10   Typical cross-section parts.

Figure 14.9 is a display of the forging blank of material ready for shaping.
The setup points are used to aline the die
position relative to the parting line. And
in Figure 14.10 the typical cross-sections
of the forged parts are displayed. While
the displays shown in Figures 14.7 and 8
might indicate that forgings have sharp
edges and corners, this is not true. They
are typical of the forms shown in Figure 14.
10 and 14.6.

14.3 EXTRUDING

Extruding is a special form of forging
where hot metal is forced through a die to
form a shape having a cross-section matching
the opening in the die. Extrusion presses
are usually horizontal, hydraulically operated. This type of press consists of a
heated container or holder for a billet, a
ram that applies pressure to the billet, a
hardened die and a shear to cut the material
to length. A die can be visualized by looking at Figure 14.10 again and saying to yourself, this is the material being pushed
through the face of the die opening. Most
dies are round blanks the diameter of the
billet (raw material) and are inserted as
shown in the display of Figure 14.11.

Temperature, pressure and extrusion
speed depend on the metal and cross-section
shape of the die. The most popular metal
is aluminum, its heat range is 400 to 1600
degrees F; pressure 30,000 to 160,000 psi;
speed 1 to 1000 fpm.

14.4 COLD FORMING

Two main types of cold forming can be
used. They are extrusion and cold heading.
These processes are also called impact extruding, shown in Figure 14.12 and cold
forging. Cold extruding is a process in
which a metal slug is shaped in a die and
punch as illustrated in Figure 14.12. Most
parts formed by the process are character-
ized by fairly long tube-like cross-sections
as displayed in Figure 14.13.

Cold heading was developed primarily

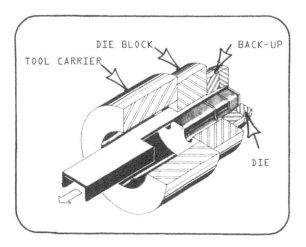

Figure 14.11 Extrusion process display.

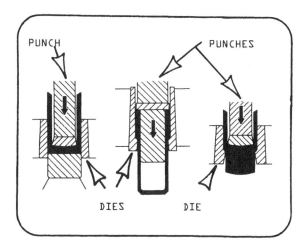

Figure 14.12 Backward, forward, combination.

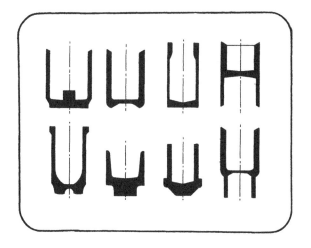

Figure 14.13 Cold extrusion shapes.

to make nuts, bolts, screws and other type
fasteners illustrated in Chapter 12 and reviewed in Figure 14.14. In cold heading, a
blank is fed into a die and worked by a
series of die motions so that it assumes
the shape of varying diameters. The process generates almost no scrap, and is therefore economical once tooling costs have been
recovered.

14.5 STAMPING

Stamping of material includes cutting,
bending, or forming done at room temperatures. Many types of presses are used for
performing stamping, ranging from small
bench presses to large automatic presses.
In order to create stamping displays, the
user must understand and able to describe:

Shearing. The production of cut sheets
from industrial presses containing knives.

Cut-off. Cutting strips to length in
a press.

Blanking. Producing accurate punched
out portions.

Notching. Separate operation for shaping edges of punched parts.

Lancing. Punching which leaves attach-
ments to the parent material (tabs).

Hemming. Bending material back on itself to form a smooth edge.

Flanging. Forming the edge of material
at the 90 degree bend.

These stamping techniques are illustrated in Figure 14.15.

14.6 DEEP DRAWING

Deep drawing and stamping are production processes often displayed interchange-
ably, yet the processes are basicly different. In deep drawing, material is held
between a pair of rings and then made to
flow over a punch of the desired shape.
Figure 14.16 indicates that parts can be
formed to greater depths than stamped parts
because movement of the material is controlled. The drawing operation can be done
either by a pressing or a pulling action.
Pulling actions require open dies.

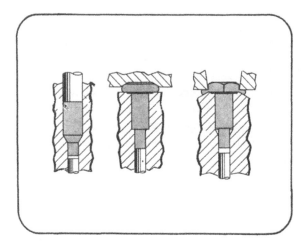

Figure 14.14 Cold heading a bolt.

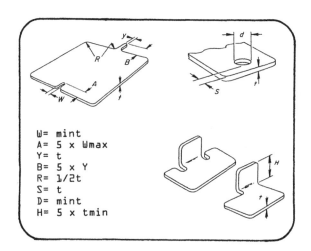

Figure 14.15 Stamping techniques.

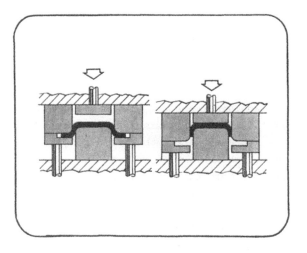

Figure 14.16 Deep drawing.

14.7 SPINNING

Spinning is one of the most simple production methods. It is a process where a flat disk of material (aluminum, copper, magnesium, brass, or stainless steel) is shaped over a mandel. The mandrel is located on a rotating spindle, much like a lathe, held by tailstock pressure. The material is placed between the mandrel and tail stock pressure plate and spun onto the face of the mandrel. The mandrel can be thought of as a mold which catches the spinning material. The material matches the shape of the mandrel surface as shown in the display of Figure 14.17.

14.8 ROLL FORMING

Roll forming is a continuous, high production process that shapes strips of material by means of progressive forming rolls as demonstrated in Figure 14.18. In this photograph a series of rollers produces shapes of uniform cross-section with close tolerances. If each of the stations shown in Figure 14.18 could be displayed separately as in isolated sections, the display shown as Figure 14.19 might result. It is more typical to view details of finished roll formed parts, however, and these are shown displayed in Figures 14.20, 21, and 22.

14.9 HIGH-VELOCITY FORMING

High-velocity forming (HVF) includes high-energy-rate forming (HERF), which is an older term used to describe these processes. HERF is awkward because it is the high deformation velocity- not energy, which does the forming. Also included in HVF is:
1. explosive forming
2. electrohydraulic forming
3. electromagnetic forming
4. pneumatic-mechanical forging
HVF production drawings have been used extensively in aerospace applications, but other industries have been using these types of automated drawings. Nonaerospace uses have been electro - hydraulic and magnetic.

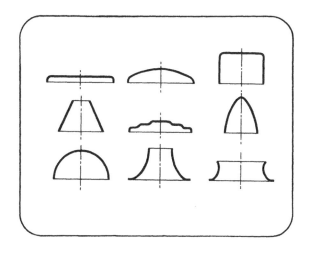

Figure 14.17 Typical spun cross-sections.

Figure 14.18 Roll forming machine. (Courtesy Industrial Fasteners Institute)

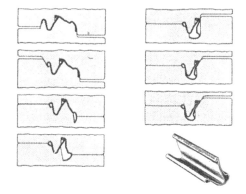

Figure 14.19 7 station location shapes.

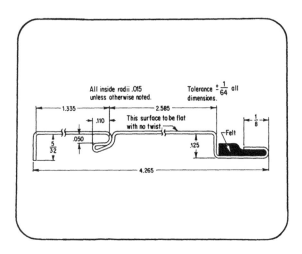

Figure 14.20   Rolled detail display.

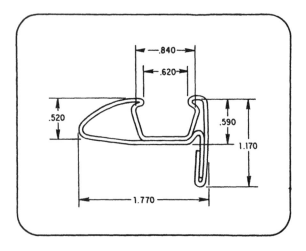

Figure 14.21   Cross-section detail.

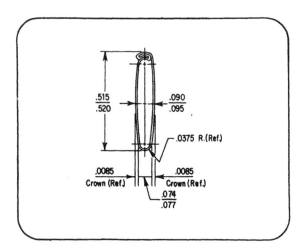

Figure 14.22   Tolerance detail.

14.10   MACHINING

Automated drawings for forming of parts by metal removal can produce shapes with smooth surfaces and dimensional accuracies not generally attainable by most forming methods. For this reason, most drafting books present only this production technique for manual drawing. Automated displays are often a combination of one or more of the ten listed production types. Automated drawings for machining of large-volume production parts is best accomplished by screw machines detail drawings. Lathes, drilling machines, milling machines, and grinders are also used to perform automated operations.

Screw machines can do turning, threading, forming, facing, boring, cutting off, burnishing, and specialized milling, slotting and cross drilling. Like a lathe, all machining is done with the axis of the part. Examples of machined parts were found throughout Chapters 10. 12 and 13, refer to them now.

14.11   REVIEW PROBLEMS

Indentify each of the following displays as casting, forging, extrusion, cold heading, stamping, deep drawing, spinning, roller, or high-velocity forming.

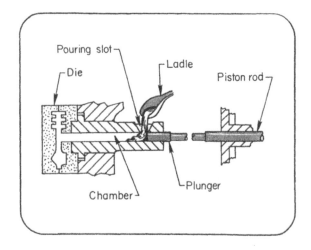

1. _____

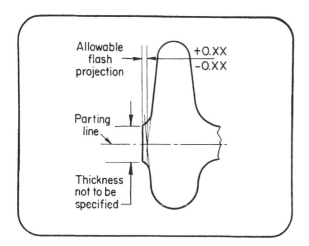

2. _____

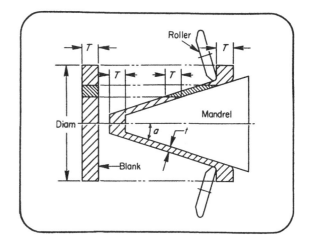

5. _____

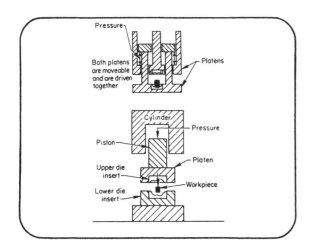

3. _____

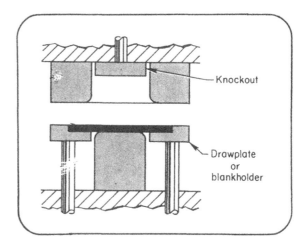

6. _____

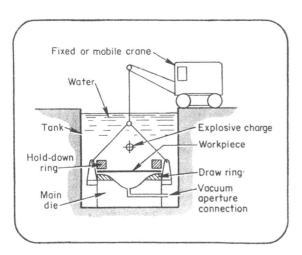

4. _____

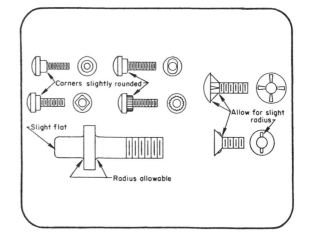

7. _____

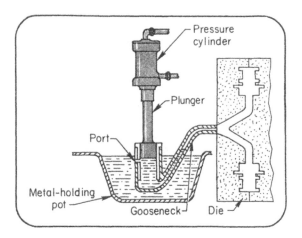

Pressure cylinder

Plunger

Port

Metal-holding pot

Gooseneck   Die

8.

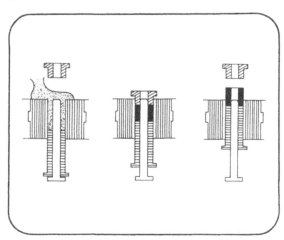

9.

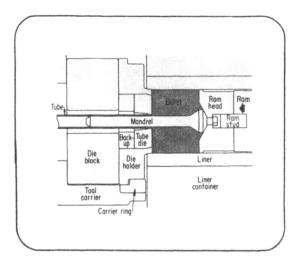

Tube

Billet   Ram head   Ram

Mandrel   Ram stud

Back-up Tube die

Die block   Die holder   Liner

Tool carrier   Liner container

Carrier ring

14.12 CHAPTER SUMMARY

Production drawing, like detail drawing was presented in this chapter as a display process whereby the draftsperson describes the production plan for the part geometry, construction, fabrication, or the assembly of an object. The objects shown in this chapter were as large as that in Figure 14.1 or as small as Figure 14.2. If these objects were also part of an automated planning process, they were production drawings. Commonly displayed production drawings always involve parts fabrication. Parts fabrication includes, but was not limited to the following:

1. Casting
2. Forging
3. Extruding
4. Cold forming
5. Stamping
6. Deep drawing
7. Spinning
8. Roll forming
9. High-velocity forming
10. Machining

While most traditional graphics books deal primarily with the last (machining); this chapter dealt with all ten production types. In addition to the ten types listed above, production drawing contains items from each of the previous chapters. For example, a production drawing will contain orthographic views arranged in detail fashion, section views, auxiliary views, dimensioning, lettering, pictorial views and other concepts presented on automated equipment and programs.

10.

15

Gears and Cams

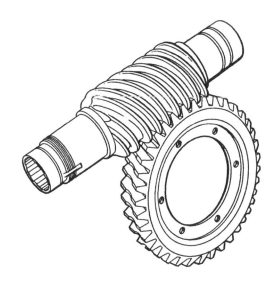

Figure 15.1 Gear display nonparallel shafts (Courtesy DLR Associates, Clemson, SC)

The automated display of gears and cams are often taken for granted because of their apparent simplicity. Gears or cams are placed on shafts, one shaft turns (driver) and causes the other to turn by means of two bodies in pure rolling contact. The shafts do not have to be parallel, but are often perpendicular or parallel. Special types of gears are displayed for shafts that are not parallel as shown in Figure 15.1. Here a worm gear (involute helicoid thread form) makes this nonparallel shaft location possible. Cams (nonthreaded, non -toothed gears) are often displayed on the basis of this principle. If the speed ratio must be exact or a rotary motion must be transferred as a rotation instead of a linear motion, the toothed wheels called gears are used in place of a cam.

A cam can be displayed as a plate, cylinder, or other solid with a surface of contact so designed as to translate rotary motion to linear motion. The cam is mounted to the driving shaft, which rotates about a fixed axis. By the cam rotation, a follower is moved in a definite path. This is

one of the applications for this chapter. The follower may be a point, a roller, or a flat surface. The follower may be attached to another part of the machine displayed by a linkage, or the follower may move radially (part of gear teeth).

15.1 TYPES OF GEAR DISPLAYS

Figure 15.1 indicates that rolling bodies may be used to connect axes that are intersecting. If the shaft axes are parallel than another type of gear is used. Different names are given to the display of gears according to the situation for which they are designed. Gear displays may be classified, as shown in Table 15.1. The term pinion used in Table 15.1 is often used to label the smaller of a pair of gears. The various other kinds of gear displays enumerated in Table 15.1 will be discussed

Table 15.1 Types of Gear Displays

Axes connection	Type	Classification
Parallel	External Internal Helical Herringbone Rack & Pinion	Spur
Intersecting	Miter Crown Spiral	Bevel
Different Plane	Hypoid Helical	Worm

in more detail after the terminology that applies to gearing in general has been considered.

15.2 GEAR TERMINOLOGY

The following display terms and their definitions will be easier to understand if Figure 15.2 is referred to occasionally.

Pitch diameter. The theoretical point of the contact between two gears as they mesh, also known as the pitch circle. It corresponds to the outside diameter of a cyclinder that would transmit force or motion through a friction contact with another surface.

Diametral pitch. The ratio of the number of teeth to the number of display units in the pitch diameter.

Outside diameter. The diameter of the gear teeth at their outer edges.

Pressure angle. The angle displayed by the gear tooth contour and line of action from a line parallel or perpendicular to the center of the gear.

Circular pitch. The distance from the center of one tooth to the center of the next tooth measured on the pitch diameter of the gear.

Chordal or tooth thickness. The thickness of the tooth displayed at the pitch diameter.

Addendum. The display distance from

the pitch diameter of the gear tooth to its outer edge.

Dedendum. The display distance from the pitch diameter of the gear tooth to the bottom of the tooth form.

Whole depth. The total display height of the tooth equal to the addendum plus the dedendum.

Working depth. The depth the gear tooth extends into the open space between teeth when mating.

Face width. The length of the tooth from one side to the other.

Face of tooth. The surface area on the face width of the tooth between the pitch diameter line and the top of the tooth

Flank of tooth. Similar to the display face except that it extends from the pitch diameter line to the bottom of the tooth. This includes the normal small fillet that is displayed at the bottom of the tooth.

Backlash. The space or clearance between the gear teeth as they are displayed at the pitch diameter line. This clearance is displayed on the back or nonworking face of the tooth. The amount of backlash will vary with the different gears shown in Table 15.1.

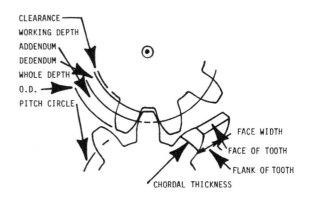

Figure 15.2 Gear terminology.

15.3 DISPLAY RELATIONSHIPS

Automated drawings require that all gear teeth be displayed with a slightly curved surface. This is known as an involute tooth form. This display form is necessary to assure ample clearance as the teeth mesh. If the teeth had straight sides, they would become damaged within a short time if a large amount of clearance were not provided. This display form is common to all types of gears, with the exception of worm gears.

The involute display contour, in addition to providing tooth clearance, also forms the tooth pressure angle. Gear teeth are usually displayed with two pressure angles: $14\frac{1}{2}$ and 20 degrees. A special 25 degree pressure angle is used for some of the gear types shown in Table 15.1. A pressure angle is displayed in all of the table types, including worm gears.

Another common display fundamental is that all gear teeth are spaced by diametral pitch. In a display program, DP is the storage location for the ratio of the number of display units in the pitch diameter of the gear, and would be entered as

DP=NT/PD

where NT is the number of teeth and PD is the pitch diameter. There is also a direct relationship between the DP of a gear tooth and its circular pitch, written in a program as

DP=3.1416/CP

where CP equals the circular pitch of the teeth. As a display example, the number of teeth on a gear, divided by the pitch diameter of the gear, would result in the display DP. Therefore, a modification using the outside gear diameter could also be used as

DP=(NT+2)/OD

where OD equals the outside diameter of the gear blank.

Study Figure 15.2 for display rela-

tionships and terminology.

15.4 GEAR ELEMENTS AND DISPLAY TECHNIQUES

Figure 15.2 shows a partial pair of external spur gears in mesh with each other. Since these are the simplest form of gears, the following discussion will be based on this type of gear. Spur gears were the first gear type developed by machine designers. In this automated drawing approach to gears, it is the next logical progression. Although there are currently many other displays for gears, spur gears are the most common. It must be kept in mind, however, that the basic gear elements and display techniques are general and apply to the other types of gears shown in Table 15.1 as well as to spur gears. The advantage of spur gears include ease of manufacture, reliability, minimum redesign, and ease of design assembly. Because of their design simplicity, spur gears are made of many types of metallic and nonmetallic materials.

Their straight or parallel tooth design enables the spur gear to be made in large diameters and broken down or split for assembly as segments in a field application without regard to special machinery. In addition, the straight tooth design eliminates all end thrust or side movement of the gears compared to gears having tapered tooth faces.

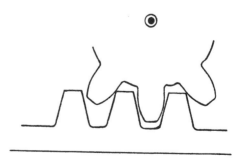

Figure 15.3 Spur gears.

15.5 SPEED RATIOS

The profiles of spur gear teeth must
be such that the speed ratio is constant.
A computer file must be generated for each
tooth displayed. The file contains the
display routines for straight tooth design.
These tooth designs are simple to display
on a DVST, but because of their simplicity,
they have some design disadvantages. Because of the minimum number of teeth in contact between the pinion and driven, they
are usually designed only for slow or med-
ium-speeds. As shown in Figure 15.3, spur
gears are displayed as regular gears and
also as straight gears called racks. The
are normally designed to transmit rotary
motion from the driving pinion into horizontal or linear motion of an object.

Now that the profile for each of the
common DPs can be generated by

CALL TOOTH(X,Y,DP,THETA)

where CALL TOOTH is the name given to the
computer routine for location in storage
and the list of arguments are:

X=horizontal screen location for tooth
Y=vertical location
DP=diametral pitch
THETA=rotation of tooth around gear blank

The designer may display working spur gears,
such a pair is shown in Figure 14.4. Here
the larger gear has 16 teeth and the smaller
12. Assume that the designer did not call

CALL GEAR(X,Y,NT,OD)

stored in memory and that shaft S is being
animated from computer memory. The 16 tooth
gear, labeled A, will turn with shaft S.
As the gear blank turns through (360/16)
degree segments, or 22.5 degree for each
theta; the display calls the routine TOOTH.
Each tooth of gear A is displayed as it
rotated on shaft S. If the center locations of gear blank A and B are properly
placed in the display program, the teeth

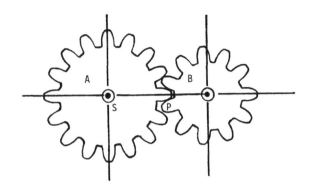

Figure 15.4 Display of gears.

of gear blank B may be displayed in the same
manner. As gear blank B is pushed by gear
tooth A through (360/12) degree segments;
the display program calls the routine TOOTH
to display each of the teeth on gear B. As
the program is executed, the viewer notes
that the teeth on A will push the teeth on
B, a tooth on A coming in contact with a
tooth on B and pushing that tooth along until the gears have turned so far around
that those two teeth swing out of reach of
each other or come out of contact. But before these two teeth come out of contact,
another pair of teeth must come in contact,
so that gear A will continue to drive gear
B. For B to make a complete revolution,
each of its 12 teeth must be pushed past
the centerline. Therefore, while B turns
once, gear A turns 12/16 of a turn, since
A has 16 teeth in all. It is evident from
viewing the display that in order for the
teeth to mesh, the distance from the center
of one tooth to the center of the next tooth
on both A and B must be alike.

The point at which gear A pushes gear
B is labeled P in Figure 15.4. Through this
point the programmer may display circles
about SA and SB as centers. These circle
diameters are stored as DA and DB. Then
the distance between gear centers is

CD = DA/2.+DB/2.

where the two gears are turning will have

the same speed ratio as two rolling cylinders of diameters DA and DB. The point P that divides the line of centers of the two gears is called the pitch point. The display circles are called pitch circles for gear A and B.

The distance from the center of one CALL TOOTH to the center of another CALL TOOTH, measured on the pitch circle, is called the circular pitch. This may be computed from the display program as

PC = 3.1416*DA/NT

15.6   DISPLAY OF GEAR ELEMENTS

The display routine CALL TOOTH is the software ordinarily used to designate the tooth size, location, and rotation. It is equal to the number of teeth divided by the diameter of the pitch circle. Often in desinating the size of a computer-generated gear the word pitch is used without the adjective diametral. For this reason, the diametral pitch is sometimes called the pitch number. The diametral pitch is expressed in the display program as

DP = NT/PCD

where DP is the diametral pitch or pitch number, NT the number of teeth, and PCD the pitch circle diameter. Two types of teeth routines may be used, as shown in Figure 15.5. With the routines shown here and Table 15.1, a complete gear representation can be displayed.

The display curves that form the profile of the gear teeth may, in theory, have any form whatever, provided that the profiles conform to the illustration of Figure 15.5. The teeth on the centerline position touch each other at P in Figure 15.4. That is, the curves of the CALL TOOTH are tangent to each other at this point. If a line is displayed tangent to the two curves at P, and the angle measured between this line and the centerline, the viewer will see the angle of the path of contact.

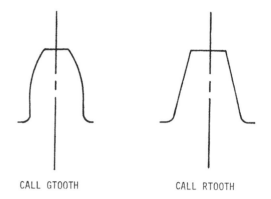

CALL GTOOTH   CALL RTOOTH

Figure 15.5   CALL TOOTH displays.

The actual path of contact is a line displayed through all the points at which the teeth of a gear touch each other during the display. This path may be straight or curved depending upon the nature of the curves that form the routine TOOTH. For every different position the CALL TOOTH occupies during the display of one pair of teeth, the curves have a different point of contact. However, in all properly constructed gear displays, the pitch point P is one point on the path of contact.

The angle of the path of contact is not the pressure angle. The pressure angle is the angle between a line displayed through the pitch point P perpendicular to the line of centers and the line displayed from the pitch point P to a point where a pair of teeth are displayed on contact. In some forms of gear teeth display, this angle remains constant; in others, it varies. This is because the direction of the force that the driving tooth exerts on the driven tooth is always along the line displayed from the pitch point to the point where a pair of teeth are displayed in contact. The smaller the pressure angle, the greater will be the component of the force in the direction to cause the driven gear to turn, and the less will be the tendency to force the shafts apart. In other words, a large pressure angle tends to produce a large pressure on the shafts holding the gears.

15.7   DISPLAY OF GEAR TEETH

When the tooth routine for teeth outlines have been displayed along with the
circular pitch, backlash, addendum, and
clearance it should appear as Figure 15.6.
To produce a display like Figure 15.6, the
draftsperson completes a number of tasks:

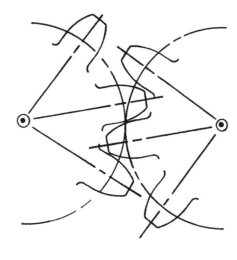

Figure 15.6   Displaying gear teeth.

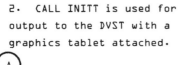

1.   The proper JCL is selected for the DVST.

2.   CALL INITT is used for
output to the DVST with a
graphics tablet attached.

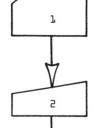

3.   Display the addendum
circle of each gear with a
radius of pitch circle plus
addendum.

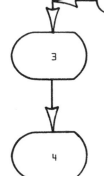

4.   Display the addendum
circle of each with a radius
equal to PC minus an amount
equal to the addendum of the
mating gear plus clearance.

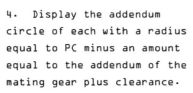

5.   Space off the circular
pitch on either side of P on
each PC. Use tablet. Start
with line tangent to PCs at
P and locate each tooth.

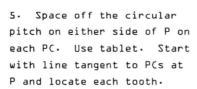

6.   Display each starting
location and enter each from
keyboard for storage matrix.

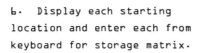

7.   From the tablet, enter
lines to represent pressure
angle, line of centers and
PCs.

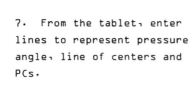

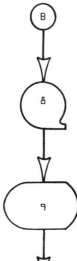

8.   From the program storage
matrix, supply the argument
lists for each tooth to be
displayed on the DVST.

9.   Display the gear teeth
as selected in program step 5.

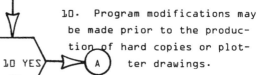

10.   Program modifications may
be made prior to the production of hard copies or plotter drawings.

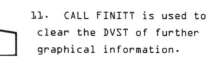

11.   CALL FINITT is used to
clear the DVST of further
graphical information.

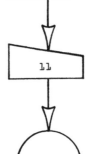

12.   The program stops.

Using the process described, the draftsperson checks for such things as tooth curves that have flanks that extend into the
root line. This will result in a weak tooth
and the draftperson avoids this by placing
a small fillet at this location. The size
of this fillet is limited by the arc of a
circle connecting the root line with the
flank and lying outside the actual path of
the end of the face of the other gear tooth
which mates. This path is called the **true
clearing curve.** This curve is the path
traced by the outermost corner of one tooth
on the plane of the other gear. This is
always known from information given in programming segments 5 and 6. The path can be
displayed in segment 10 for easy reference.

15.8 GEAR STUDIES AND ANALYSIS

The form of the gear display most commonly given to computer analysis is that
known as the involute of a circle. Gear
teeth constructed with this curve will conform to the fundamental concepts of automated display discussed earlier. This display curve and methods of producing it will,
therefore, be studied before the method of
applying it to gear teeth is discussed.

A typical curve may be displayed by
placing a jar lid on the graphics tablet
used in programming segments 5 and 7. Attached to the edge of the jar lid is a very
fine wire which is wrapped around the tablet
stylus. If the operator keeps the wire
taut and begins to wrap the wire around the
jar lid, causing the wire to become shorter
as the stylus traces a curve on the graphics
tablet; an involute of the circle represented by the jar lid is displayed. The
same result is obtained from a computer
routine stored in memory. Using either
method, all involutes from the same circle
are alike, but involutes displayed from
circles of different diameters are different. The greater the diameter of the circle, the flatter will be its displayed involute.

In considering the involute of a cir-

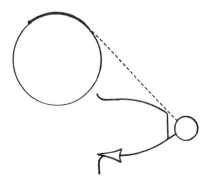

Figure 15.7 Display of an involute.

cle to be used in automated drafting, it is,
of course, impractical to stop and wrap a
wire around a jar lid. Instead, Figure 15.
7 shows how the computer routine uses this
concept to produce the involute for the
routine GTOOTH. Several points will need
to be calculated along the involute curve
line. These points are calculated based
on the size of the base circle shown as the
jar lid in Figure 15.7. Once the points
have been located they are connected through
a CALL SMOOT routine discussed in Chapter 5
and illustrated in Figure 5.22. The compu-
tations have been reduced to the number and
size of gear teeth required to automatically
display diametral pitches 2 through 16.

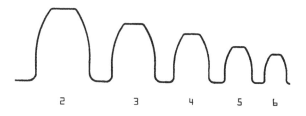

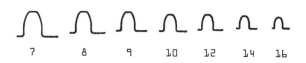

Figure 15.8 Involute shapes 2 through 16.

15.9 TYPES OF CAM DISPLAYS

A cam and its follower form an application of the principle of transmitting the motion of one object to another by direct sliding contact. As in the case of gears, various situations arise for the use of cams, and special names are given to the display of cams according to the situation for which they are designed. Like gears, the maximum pressure angle strongly affects the size of a cam. The display method presented in this portion of Chapter 15 can be used to determine cam size within the rescritions of a maximum value for the pressure angle.

Four roller-follower cam geometries are common; sliding and swinging axial followers driven by cylindrical cams, sliding and swinging driven by plate cams. The display routines are approximate, but define cam size accurately enough for automated drafting purposes. The display routines are written for:

1. Cylindrical cam - axial follower
2. Cylindrical cam - swinging follower
3. Plate cam - sliding follower
4. Plate cam - swinging follower

If the following information is known:

A = follower arm length
B = cam follower face width
C = distance from center of cam to
    center of follower arm
R = cam radius to follower centerline
X = axial position of follower
α = angular position of follower
Δ = cam displacement
ϕ = cam pressure angle
Θ = cam position during rotation

   Θ_i initial value
   Θ_1 limits
   Θ_m mean
   Θ_0 zero or lowest position

Table 15.2 lists the routine logic used in the display of the four types of cams, while Figures 15.9 through 15.12 demonstrate the types of cam displays. The review problems at the end of this chapter provide additional exposure to the types of cam displays.

Table 15.2 Types of Cam Displays

Cam system used	Routine logic used
Cylindrical, sliding follower	$R = \dfrac{\frac{\Delta X}{\Delta \Theta \ max}}{TAN(\phi_1)}$
Cylindrical, swinging follower	$\dfrac{R}{A} = \dfrac{\frac{\Delta \alpha}{\Delta \Theta \ max}}{TAN(\phi_1)}$
Plate, sliding follow	$R_m = \dfrac{\frac{\Delta R}{\Delta \Theta \ max}}{TAN(\Theta_1)}$
Plate, swinging	$\dfrac{R_m}{A} = \dfrac{\frac{\Delta \alpha}{\Delta \Theta \ max}}{TAN(\Theta_1)}$

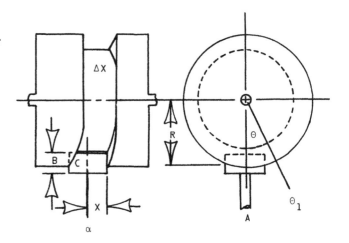

Figure 15.9 Cylindrical cam display information.

15.10 DISPLAY OF CAM PROFILES

A cam imparts motion to a follower guided so that is is constrained to move in a plane that is perpendicular to the axis about which the cam rotates. This type of cam is best defined as a flat plate. Cams may also occupy a plane coincident with or

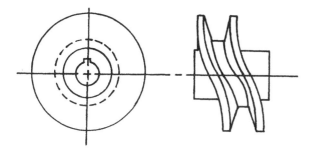

Figure 15.10 Cylindrical cam profile produced from Table 15.2 display routines.

parallel to a plane in which the cam rotates.
This type of cam is best displayed as cylindrical. The displacement transmitted to the
follower depends upon the shape of the cam.

The graphic relationship between the
sucessive positions of the follower is shown
in Figure 15.12. Displacement diagrams are
part of the automated drafting display of
the follower. The follower may move contin-
uously or intermittelntly; it may be displayed with uniform speed or variable speed;
or it may have uniform speed part of the
time. A knowledge of the various types of
cams, and an idea of the manner of attacking
the automated display for any specific purpose can best be obtained by studying a display method for a number of examples.

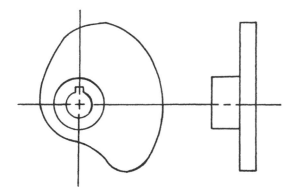

Figure 15.11   Plate cam profile produced
from Table 15.2 display routines.

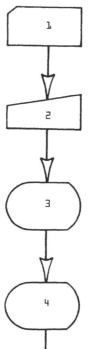

1. The proper JCL is selected for the DVST.

2. CALL INITT is used for
output to the DVST with a
graphics tablet attachment.

3. Display the base circle
for locating the positions
of the camshaft axis and follower. The BC is with radius
R or C whichever is equal to
the pitch circle.

4. Display the pitch profile
during one rotation of the
cam. Use X and ∆X reference
points.

5. Display the opposite view
of the cam or add a section
view.

Continued on page 264

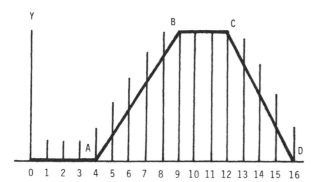

Figure 15.12 Displacement diagram.

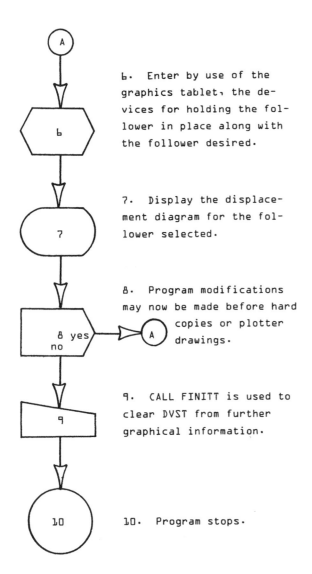

6. Enter by use of the graphics tablet, the devices for holding the follower in place along with the follower desired.

7. Display the displacement diagram for the follower selected.

8. Program modifications may now be made before hard copies or plotter drawings.

9. CALL FINITT is used to clear DVST from further graphical information.

10. Program stops.

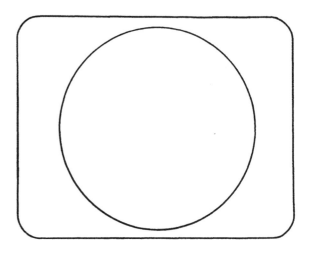

Figure 15.13   Display of outside diamater.

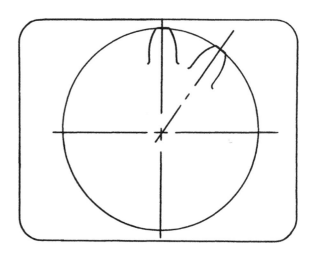

Figure 15.14   Display of OD and DP=4 teeth.

15.11   REVIEW PROBLEMS

Begin the review of gears and cams by selecting a DVST workstation and complete the steps shown below.

1. Use the image processor to display an outside circle for a spur gear blank as shown in Figure 15.13.

2. Use an outside diameter for step 1 above so that ten teeth with diametral pitch 4 can be displayed as shown in Figure 15.14.

3. Complete the gear blank for ten teeth (N), OD = 3.0, DP = 4., root circle, and pitch circle display as shown in Figure 15.15.

4. Dimension the gear in steps 1 -3;

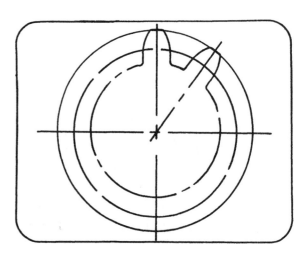

Figure 15.15   OD, PC, DP, and RC displayed.

so that the following items are known:

DP = Number of teeth/ pitch diameter

PD = Number of teeth/ diamteral pitch

OD = Number of teeth+ 2/ diametral P

N = (OD*DP)-2

Addendum = 1/DP

Dedendum = A-WD

Whole depth = A+D

Working depth = 2*A

Working clearance = WD-2*A

   5. Display the gear and rack shown in Figure 15.3. or 15.16.

   6. Display the gears shown in Figure 15.6. or 15.17.

   7. Display the cam shown in Figure 15.9..

   8. Display the cam shown in Figure 15.10.

   9. Display the cam shown in Figure 15.11.

   10. Display the cam follower displacement diagram shown in Figure 15.12.

15.12 CHAPTER SUMMARY

   Gears and cams can be displayed from techniques learned in earlier chapters if the correct terminology and procedures are learned. This chapter has presented the minimum level of display techniques so that a draftsperson might provide elementary gear and cam displays.

   The types of gears that can be displayed are limited to the types of display routines that are available to the draftsperson. In this chapter only the spur gear tooth profile routine was used. Other types are available and the new automated draftsperson should become familiar with these once the workstation is comfortable and the simplier spur gears are mastered.

   A cam profile, likewise, can be displayed as a plate or cylinder. Two types of cam followers were presented, but here again other types are used. The number of different types was kept to a minimum so that the chapter could focus on how to display two major types of machine parts, gears and cams.

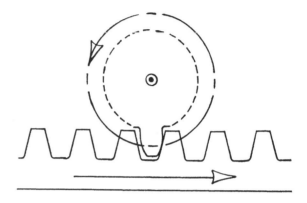

Figure 15.16 Rack and pinion display.

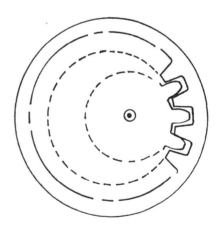

Figure 15.17 Spur gears.

16

Flexible Connectors

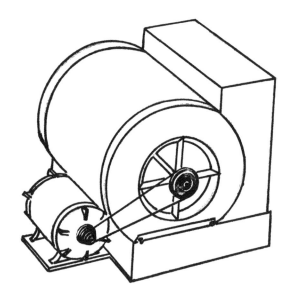

Figure 16.1 Most common flexible connector used in transmitting motion. Pulley and belt.

When the distance between the driving shaft in an automated drawing and the driven shaft is too great to be connected by gears or by cams, a flexible connector is displayed as shown in Figure 16.1. As in this CRT display, the most common replacement for a cam or gear is a pulley and belt. If a particular function, say in Figure 16.1, can not be performed by direct drive, surface to surface, as in the case of gear trains; another method must be found. A decision is made as to the type of parts that will transmit the motion. Design engineers usually establish a scale (list) of desirable types of movemnents and believe that of all the types available, flexing is perhaps the most trouble-free and gives the longest life.

The proof of this is shown in Figure 16.2 where a gear train and a pulley system provide the same movements. Gear train power-transmissions system provides rotary motion to a number of shafts, but require very accurate shaft center locations, lubricants, and high-cost components. Modern flexible belts running on smooth pulleys, solve many of these problems. Tolerances

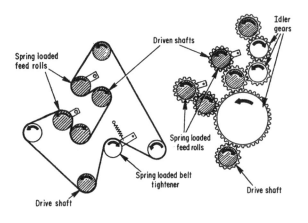

Figure 16.2 Comparison of gears and pulleys.

of all kinds can be absorbed in one spring-
loaded belt-tightening idler. The life of
a properly designed system of this type is
exceptionally long. The reason for this is
that the belt is thin and pliable and readily wraps around the pulleys. It is not in
rubbing contact with any element. A drive
system of this type is not synchrous. One
answer is to make the entire machine asynchronous or use a timing belt.

16.1 PULLEY AND BELTS

A pair of pulleys can be displayed in
an automated drawing as shown in Figure 16.1
and a felxible connector (belt) can be displayed between them. The drafter may select flexible connectors for pulleys from
several classes.

1. V-belt drives (Figure 16.1), where
power is tranmitted by the amount of friction
present. V-belts include double angle, frac-
tional horsepower, link, standard, and high
capacity.

2. Flat shown in Figure 16.3, makes
sheaves more efficient. The size of the
flat belt is determined by its cross-sectional area. Belts like those shown in Figure 16.3 are used for higher horse power
applications.

3. Timing belts, shown in Figure 16.
4. are special belts whose applications include drives that require specific timing
of related moving parts, while others require a positive transfer of power. To indicate this, timing belts are presented on
a tooth-grip principle in much the same manner as the teeth on a gear. The molded teeth
of the belt are designed to make positive
engagement with the grooves in a smooth,
rolling manner. Unlike most other types
of belts used in an automated drawing, timing
connectors do not derive their strength from
their driving force or thickness.

16.2 DISPLAY ITEMS FOR PULLEYS AND BELTS

Whether a belt drive is displayed as
V, flat or timing type, a few common names
and definitions still apply to all of them.

These include the following:

Driver Sheave. The driver sheave is
displayed on the motor as shown in Figure
16.1 and 16.3. It is usually the smallest
sheave with the highest power rating.

Driven Sheave. The sheave displayed
on the machine being driven, blower in Figure 16.1. Usually, this is the largest
sheave and has the lowest power rating.

Idler Sheave. An idler sheave is displayed to increase the drive tension, de-
tour the belt strand, reduce belt vibration,
and provide takeup as shown in Figure 16.2.

Belt Pitch Length. A V-belt's pitch
length is the length of the belt at the neutral axis of the display. This neutral axis
is located approximately two-thirds of the
distance from the bottom (narrow) to the
top (wide) portion of the belt.

Sheave Pitch Diameter. This is the
sheave diameter at a point where the neutral
axis of the belt contacts the sheave. Also,
this is where the belt and sheave speeds are
the same.

Arc of Contact. The arc of contact on
an automated drawing is the number of degrees
of wrap or contact of the belt around the
sheave. Reduction in the arc of contact of
the belt displayed changes the power-trans-
mitting capacity of the flexible connector.

Center Distance. Center distance is
shown on an automated drawing as the distance between the centers of driver and
driven shafts.

Normally, flexible connectors are selected from one of three separate groups for
belts, identifiable by the size and shape.
Figure 16.5 represents the design intent of
belt sizes for the following:(16.5)

(a) light duty
(b) standard industrial
(c) super
(d) cogged
(e) steel-cable
(f) open-end
(g) link-V
(h) double-V
(i) narrow
(j) wide range.

Within each group a computer display is poss-

Figure 16.3 Use of flat belts on power sheaves. (Courtesy Raybestos-Manhattan, Inc.)

ible, for example the standard group (b)
belt sizes are currently designated by the
letters A,B,C,D, and E as shown in Figure
16.6. Each of the different lettered belts
has specific size limitations indicated by
the output display of Figure 16.5. Some
belt manufacturers sizes vary slightly from
the computer model shown. Belts are one
type of flexible connector that is manufactured in specific lengths, although they
can be purchased as a single strand 16.5 (f).

Another group of belts that can be displayed is (c) high capacity. These are used
where standard belts may not perform well
because of high horsepower or loading conditions. Also, heat, moisture, or other
similar conditions may require a different
display. Sometimes, there is no room for a
standard drive, and the reduced section of
the high capacity belts allows it to fit into the smaller space. By comparing the outputs shown in Figures 16.6 the drafter can

Figure 16.4 Use of a timing belt and its
cross-section. (Courtesy Diamond Chain Co.)

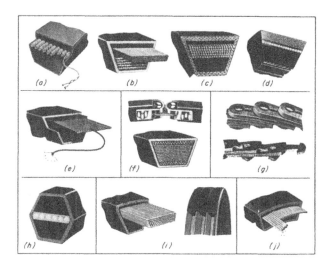

Figure 16.5 Belt types. (Courtesy Industrial Fasteners Institute)

of smaller belts for lighter duty (a) and with smaller drive pulleys. The two types commonly displayed are the L and M series. The 2L through 5L belts are similar in cross-section to standard belts, and are displayed as factored standard flexible connectors. The 3,5,7 and 11 M belts are made with a different configuration allowing them to flex more easily and are displayed as factored (reduced size) high capacity connector belts. The values of these types of flexible connectors are important when the draftsperson considers the application. Figure 16.7 is an output of the light duty series shown in Figure 16.5 (a). Note the comparison with the standard series.

note that the standard belt is considerably wider than it is high, while the high capacity belts are almost as high as they are wide. Because of the difference in belts and how they must be displayed in an automated drawing, separate display sheaves are used for the standard and high capacity.

In addition to the standard and high capacity belt displays, there are a number

16.3 PULLEY AND BELT COMBINATIONS

While there are two major and two minor types of fexible connectors commonly displayed, there are three types of sheaves displayed with them. Of course, the sheaves for standard and high capacity belts are two of the three display types. The third type is a combination sheave used with both type A and B of the standard belts. Combination sheaves are frequently used in com-

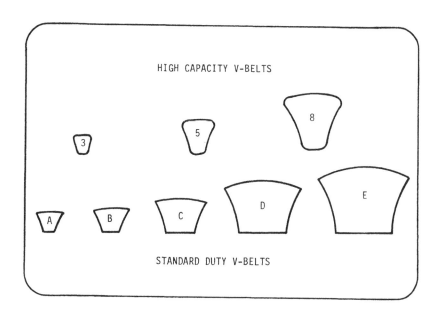

Figure 16.6 Standard and high capacity computer display output.

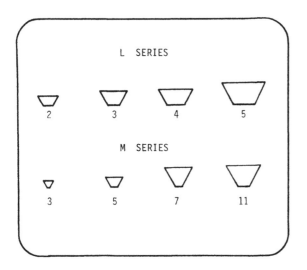

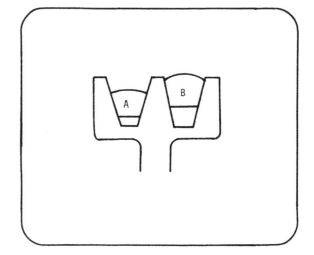

Figure 16.7   L and M series belts.

Figure 16.8   Combination A and B standard.

puter displays having both A and B-type
drives.   This allows interchangeability between drives as shown in Figure 16.8.

It should be noted in Figure 16.8 that
the outside diameter of the sheave is never
the same as the pitch diameter. This is
especially true when displaying combination
sheaves, because the pitch diameter is located at two different points, depending on
which belt is being displayed. In high-
capacity sheaves, the belt is made so that
it extends slightly beyond the outside diameter of the sheave. In instances where
grouped belts are displayed with standard
sheaves, the belt covers the sheave outside
diameter except on the two outside edges.

Variable-speed sheaves are displayed
in two general types; manual-adjust and
spring-loaded types. The manual-adjust
sheave has a smaller range of adjustment
than the spring-loaded type. Although they
may differ slightly when displayed, the
basic design and display principles can be
taken from an automated drafting manual.
Figure 16.9 represents a page from this
type manual to display two sets of pulleys
with connecting v-belts. The display program will determine the length of a v-belt
placed between two pulleys. The draftsperson may select the distance between the
pulleys and the size of each pulley.

First the operational constraints are
listed for the draftsperson, then the data
input for the graphical display is handled.
Several checks are made to see that impossible pulley combinations are not selected.
In Figure 16.10, the second page of the
drafting manual, shows the programming for
the plotting of the pulleys, determination
of v-belt lengths and the plotting of the
v-belt. In Figure 16.11, the third page of
the manual, the formats for all the display
options are listed and in Figure 16.12, the
last page of the manual, the display is
shown as it might appear on a DVST if the
combinations were:

    v-belt length = 43.1836
    diameter of smaller pulley = 4.
    diameter of larger pulley =8.
    distance between pulleys = 12.

16.4   VARIABLE SPEED TRANSMISSION

Computer-generated displays for flexible connectors have special tasks tp perform in addition to the previously stated
functions for the v-belt programming manual.
When flexible connectors are considered in
a larger sense - chains and sprockets, cables
and drums, or motors and couplings - a connector must accomplish several objectives

```
//MBUSH4 JOB (0923-1-012-MB-4F,:01,1),'BUSHELOW     BOX 37'         00000010
//S1 EXEC FTG1CL,PDS='MBUSH.LOAD',NAME=PULL,PLOTTER=SER281I         00000020
//*STEP1 EXEC FTG1CL,PDS='MBUSH.LOAD',NAME=PULLEY,PLOTTER=IBM3277   00000025
//*EXEC PREVIEW,PDS='MBUSH.LOAD',NAME=PULLEY                        00000030
//C.SYSIN DD *                                                      00000040
C*******************************************************************00000050
C* THIS PROGRAM WILL DETERMINE THE LENGTH OF A 'V' BELT PLACED BETWEEN *00000060
C* TWO PULLEYS.  THE USER WILL INPUT THE DISTANCE BETWEEN THE PULLEY  *00000070
C* SHAFTS AND THE DIAMETERS OF THE TWO PULLEYS.                       *00000080
C*******************************************************************00000090
                                                                   00000100
                                                                   00000110
C------------------------------------------------------------------00000120
C--                      CONSTRAINTS                            --00000130
C-- 1) THE DISTANCE BETWEEN THE PULLEY SHAFTS MUST BE BETWEEN 11 AND --00000140
C--    22 UNITS, INCLUSIVE.                                     --00000150
C-- 2) THE DIAMETER OF THE PULLEYS MUST BE BETWEEN 4 AND 11 UNITS, --00000160
C--    INCLUSIVE.                                               --00000170
C------------------------------------------------------------------00000180
C                                                                   00000190
C------------------------------------------------------------------00000200
C                         DATA INPUT                               00000210
C------------------------------------------------------------------00000220
C                                                                   00000230
C                                                                   00000240
      WRITE(3,212)                                                 00000250
   15 WRITE(3,10)                                                  00000260
      READ(1,11) DIST                                              00000270
      IF(DIST.GE.11..AND.DIST.LE.22.)GOTO 12                       00000280
      WRITE(3,13)                                                  00000290
      GOTO 15                                                      00000300
   12 WRITE(3,20)                                                  00000310
      READ(1,25) DIAA                                              00000320
      IF(DIAA.GE.4..AND.DIAA.LE.11.)GOTO 45                        00000330
      WRITE(3,30)                                                  00000340
      GOTO 12                                                      00000350
   45 WRITE(3,35)                                                  00000360
      READ(1,40) DIAB                                              00000370
      IF(DIAB.GE.4..AND.DIAB.LE.11.)GOTO 62                        00000380
      WRITE(3,31)                                                  00000390
C                                                                   00000400
C------------------------------------------------------------------00000410
C      CHECKS TO MAKE SURE THAT DIAMETER OF LARGER PULLEY IS GREATER 00000420
C      THAN DIAMETER OF SMALLER PULLEY.  IF IT IS NOT, THE PROGRAM   00000430
C      WILL AUTOMATICALLY EXCHANGE THE TWO VALUES.                   00000440
C------------------------------------------------------------------00000450
C                                                                   00000460
   62 IF(DIAB.LE.DIAA)GOTO 50                                      00000470
      WRITE(3,66)                                                  00000480
      A=DIAA                                                       00000490
      B=DIAB                                                       00000500
      DIAA=B                                                       00000510
      DIAB=A                                                       00000520
   50 RADA=DIAA/2.                                                 00000530
      RADB=DIAB/2.                                                 00000540
C                                                                   00000550
C------------------------------------------------------------------00000560
C      MAKES SURE THERE IS NO INTERFERENCE BETWEEN THE TWO PULLEYS. 00000570
```

Figure 16.9 Page one of automated drafting manual to display pulleys and v-belts.

```
C-------------------------------------------------------------------00000580
C                                                                    00000590
      TEST=RADA+RADB                                                 00000600
      IF(TEST.LT.DIST)GOTO 110                                       00000610
      WRITE(3,130)                                                   00000620
      GOTO 15                                                        00000630
  110 WRITE(3,112)                                                   00000640
                                                                     00000650
      CALL PLOTS                                                     00000660
      CALL FACTOR(.45)                                               00000670
      CALL PLOT(0.,15.,-3)                                           00000680
C                                                                    00000690
C-------------------------------------------------------------------00000700
C                     PLOTTING OF PULLEYS                            00000710
C-------------------------------------------------------------------00000720
C                                                                    00000730
      X=1.+DIAA                                                      00000740
      Y=1.+RADA                                                      00000750
      CALL CIRCL(X,Y,0.,360.,RADA,RADA,0.0)                          00000760
      X2=X+DIST+RADB-RADA                                            00000770
      CALL CIRCL(X2,Y,0.,360.,RADB,RADB,0.0)                         00000780
      X3=X-RADA                                                      00000790
      X4=X2-RADB                                                     00000800
      TLEN=.5                                                        00000810
      DASH=0.5                                                       00000820
      SPACE=0.00                                                     00000830
      ALINE=0.0                                                      00000840
      CALL CENTER(X3,Y,TLEN,0.,DASH,SPACE,ALINE)                     00000850
      CALL CENTER(X3,Y,TLEN,90.,DASH,SPACE,ALINE)                    00000860
      CALL CENTER(X4,Y,.5,270.,0.5,0.,0.)                            00000865
      CALL CENTER(X4,Y,TLEN,0.,DASH,SPACE,ALINE)                     00000870
C                                                                    00000890
C-------------------------------------------------------------------00000900
C                 DETERMINATION OF V BELT LENGTH                     00000910
C-------------------------------------------------------------------00000920
C                                                                    00000930
      PI=3.1415927                                                   00000940
      VTHET1=ASIN((DIAA-DIAB)/(2.*DIST))                             00000950
      TS=PI-(2.*VTHET1)                                              00000960
      TL=PI+(2.*VTHET1)                                              00000970
      G=4.*(DIST**2.)                                                00000980
      G1=(DIAA-DIAB)**2.                                             00000981
      G3=.5*((DIAA*TL)+(DIAB*TS))                                    00000982
      VL=SQRT(G-G1)+G3                                               00000983
C                                                                    00001000
C-------------------------------------------------------------------00001010
C                   PLOTTING OF THE V BELT                           00001020
C-------------------------------------------------------------------00001030
C                                                                    00001040
      XA=RADA*SIN(VTHET1)                                            00001050
      YA=RADA*COS(VTHET1)                                            00001060
      XB=RADB*SIN(VTHET1)                                            00001070
      YB=RADB*COS(VTHET1)                                            00001080
      VXA=1.+RADA+XA                                                 00001090
      VYA=RADA+YA+1.                                                 00001100
      VXB=1.+RADA+DIST+XB                                            00001110
      VYB=RADA+YB+1.                                                 00001120
      CALL PLOT(VXA,VYA,3)                                           00001130
```

Figure 16.10 Page two of programming manual for pulley and belt combinations.

```
        CALL PLOT(VXB,VYB,2)                                        00001140
        VYA1=RADA-YA+1.                                             00001150
        VYB1=RADA-YB+1.                                             00001160
        CALL PLOT(VXB,VYB1,3)                                       00001170
        CALL PLOT(VXA,VYA1,2)                                       00001180
        CALL PLOT(0.,-7.,-3)                                        00001190
        CALL NEWPEN(2)                                              00001195
        CALL SYMBOL(2.,0.,.4,'DISTANCE BETWEEN PULLEYS IS ',0.,28)  00001200
        CALL SYMBOL(2.,.6,.4,'DIAMETER OF LARGER PULLEY IS ',0.,29) 00001210
        CALL SYMBOL(2.,1.1,.4,'DIAMETER OF SMALLER PULLEY IS ',0.,30) 00001220
        CALL SYMBOL(2.,1.6,.4,'V BELT LENGTH IS ',0.,17)            00001230
                                                                    00001234
        CALL NEWPEN(3)                                              00001235
        CALL NUMBER(13.1,0.,.4,DIST,0.,4)                           00001240
        CALL NUMBER(13.5,.6,.4,DIAA,0.,4)                           00001250
        CALL NUMBER(14.1,1.1,.4,DIAB,0.,4)                          00001260
        CALL NUMBER(8.6,1.6,.4,VL,0.,4)                             00001270
                                                                    00001272
        CALL PLOT(0.,30.,999)                                       00001280
C                                                                   00001290
C------------------------------------------------------------------00001300
C                       FORMAT BLOCK                                00001310
C------------------------------------------------------------------00001320
C                                                                   00001330
  212 FORMAT(5X,'THIS PROGRAM WILL DETERMINE THE LENGTH OF A V BELT',/, 00001340
     +5X,'PLACED BETWEEN TWO PULLEYS.  THE DISTANCE BETWEEN THE PULLEYS'00001350
     +,/,5X,'IS BETWEEN 11 AND 22 UNITS, INCLUSIVE.  THE PULLEY',/,5X,'D00001360
     +IAMETERS MUST BE BETWEEN 4 AND 11 UNITS, INCLUSIVE.',///,5X,'THE 00001370
     +ANSWER WILL BE OUTPUTED AS A PICTORIAL REPRESENTATION OF THE',/,5X00001380
     +,'SYSTEM',///)                                                00001390
   10 FORMAT(5X,'INPUT DISTANCE BETWEEN PULLEY SHAFTS.',/)          00001400
   11 FORMAT(F4.2)                                                  00001410
   13 FORMAT(5X,'DISTANCE BETWEEN PULLEY SHAFTS MUST BE BETWEEN 11 AND 200001420
     *2 UNITS.',/)                                                  00001430
   20 FORMAT(5X,'INPUT DIAMETER OF LARGER PULLEY.',/)               00001440
   25 FORMAT(F4.2)                                                  00001450
   35 FORMAT(5X,'INPUT DIAMETER OF SMALLER PULLEY.',/)              00001460
   40 FORMAT(F4.2)                                                  00001470
   30 FORMAT(5X,'PULLEY DIAMETER MUST BE BETWEEN 4 AND 11 UNITS.',/)00001480
   31 FORMAT(5X,'PULLEY DIAMETER MUST BE BETWEEN 4 AND 11 UNITS.',/)00001490
   66 FORMAT(5X,'LARGE PULLEY DIAMETER IS SMALLER THAN SMALL PULLEY DIAM00001500
     *ETER.  STANDARD FIX UP TAKEN.',/)                            00001510
  130 FORMAT(5X,'THIS SET OF PARAMETERS WILL CAUSE INTERFERENCE BETWEEN 00001520
     *THE TWO PULLEYS.',/)                                          00001530
  112 FORMAT(5X,'ALL DATA HAS BEEN ACCEPTED.',/)                    00001540
  124 STOP                                                          00001550
        END                                                        00001560
                                                                    00001570
        SUBROUTINE CENTER(X,Y,TLEN,THETA,DASH,SPACE,ALINE)         00001580
        THETA=3.14159/180.*THETA                                   00001590
        TOTAL=DASH +SPACE+ALINE                                     00001600
        NUM=TLEN/2./TOTAL                                           00001610
        CALL PLOT(X,Y,3)                                            00001620
        X1=X                                                        00001630
        Y1=Y                                                        00001640
        DO 3 I=1,NUM                                                00001650
        X1=X1-DASH/2.*COS(THETA)                                    00001660
        Y1=Y1-DASH/2.*SIN(THETA)                                    00001670
```

Figure 16.11 Page three of drafting manual. (Courtesy DLR Associates)

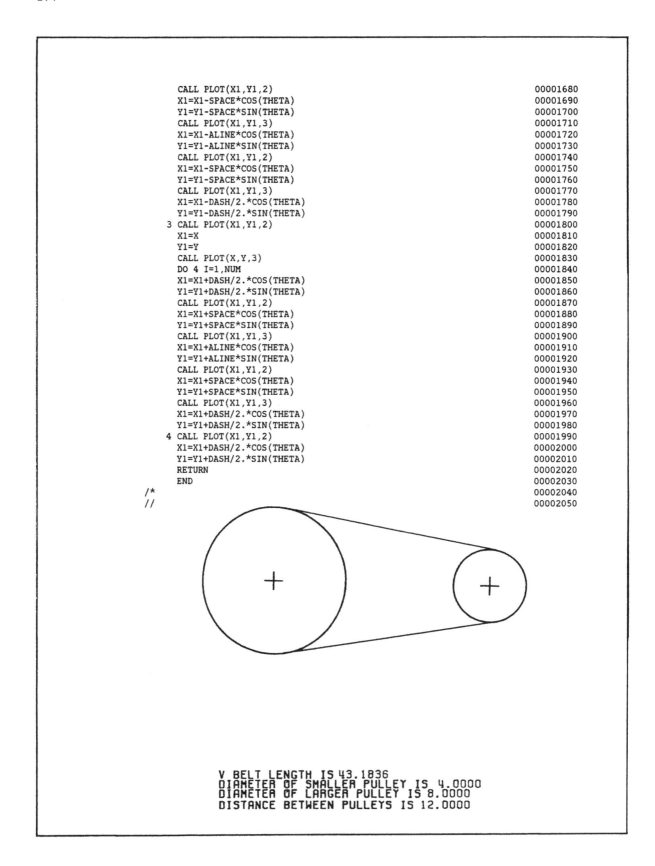

```
                CALL PLOT(X1,Y1,2)                              00001680
                X1=X1-SPACE*COS(THETA)                          00001690
                Y1=Y1-SPACE*SIN(THETA)                          00001700
                CALL PLOT(X1,Y1,3)                              00001710
                X1=X1-ALINE*COS(THETA)                          00001720
                Y1=Y1-ALINE*SIN(THETA)                          00001730
                CALL PLOT(X1,Y1,2)                              00001740
                X1=X1-SPACE*COS(THETA)                          00001750
                Y1=Y1-SPACE*SIN(THETA)                          00001760
                CALL PLOT(X1,Y1,3)                              00001770
                X1=X1-DASH/2.*COS(THETA)                        00001780
                Y1=Y1-DASH/2.*SIN(THETA)                        00001790
              3 CALL PLOT(X1,Y1,2)                              00001800
                X1=X                                            00001810
                Y1=Y                                            00001820
                CALL PLOT(X,Y,3)                                00001830
                DO 4 I=1,NUM                                    00001840
                X1=X1+DASH/2.*COS(THETA)                        00001850
                Y1=Y1+DASH/2.*SIN(THETA)                        00001860
                CALL PLOT(X1,Y1,2)                              00001870
                X1=X1+SPACE*COS(THETA)                          00001880
                Y1=Y1+SPACE*SIN(THETA)                          00001890
                CALL PLOT(X1,Y1,3)                              00001900
                X1=X1+ALINE*COS(THETA)                          00001910
                Y1=Y1+ALINE*SIN(THETA)                          00001920
                CALL PLOT(X1,Y1,2)                              00001930
                X1=X1+SPACE*COS(THETA)                          00001940
                Y1=Y1+SPACE*SIN(THETA)                          00001950
                CALL PLOT(X1,Y1,3)                              00001960
                X1=X1+DASH/2.*COS(THETA)                        00001970
                Y1=Y1+DASH/2.*SIN(THETA)                        00001980
              4 CALL PLOT(X1,Y1,2)                              00001990
                X1=X1+DASH/2.*COS(THETA)                        00002000
                Y1=Y1+DASH/2.*SIN(THETA)                        00002010
                RETURN                                          00002020
                END                                             00002030
        /*                                                      00002040
        //                                                      00002050
```

V BELT LENGTH IS 43.1836
DIAMETER OF SMALLER PULLEY IS 4.0000
DIAMETER OF LARGER PULLEY IS 8.0000
DISTANCE BETWEEN PULLEYS IS 12.0000

Figure 16.12 Page four of manual and graphical display. (Courtesy DLR Associates)

while performing its work.

1. dampening vibration
2. dampening or absorbing torque
3. insulating the connections

Most connectors combine several of these basic purposes in their design. For example, some flexible connectors will do all of the above, whereas others may do only the basic tasks (tranmitting power, compensating for misalignment, and end float). Still other connectors will not accept misalignment between the shafts.

Several devices accomplish a similar purpose in a different way. One of these is the variable-speed reducer shown in Figure 16.13. Variable-speed reducers perform many functions in the computer-aided display of fexible connectors. A few applications include the control of various speeds in machines such as turning, blending, and mixing. In addition, they are used to assure a smooth, equal speed for the power train. In a simple display case (Figure 16.14) the speed adjustment is accomplished by a variable-speed pulley on the output shaft of a motor.

The selection of a specific type of variable-speed drive to be displayed on a computer output device is governed by many factors. The drive may be displayed by the amount of variable speed required, the speed range, the drive horsepower, or the operating conditions present. Selection and display is often based on the size of the unit, weight, operating life, reliability of operation, speed response, or cost.

The display of the unit is relatively simple. The input shaft of the drive is directly connected to the motor rotor and is supported on either side of the drive by bearings. Usually, the display area of the input shaft that the pulley contacts will be displayed as splined or keyed. The spline assures that positive power is transmitted between the shaft and the adjustable sheave.

The input pulley flange width is displayed as adjustable and controls the output speed of the drive by causing the belt to change its position on the pulley flanges.

Figure 16.13 Variable-speed reducer. (Courtesy Lewellen Manufacturing, Columbus)

With the belt at the lower position on the flanges, as shown in Figure 16.14A, the reduction ratio would be greatest when the output shaft is rotating at its slowest speed. As the flanges of the pulley are brought closer together, the belt changes its position to the outer edges of the flanges. This is accomplished by the pressure against the belt. The change in position of the belt increases the output

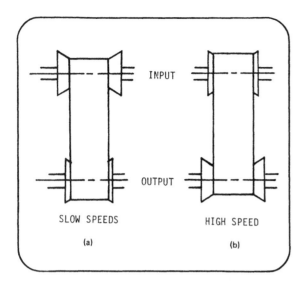

Figure 16.14 Simple display of drive.

Figure 16.16 Sprocket used with roller chain. (Courtesy Kodak Co.)

speed of the drive shaft. Like the adjustable v-belts discussed earlier, the driven pulley of the variable-speed drive is spring-loaded. This helps maintain constant pressure on the belt as it contacts the driven pulley. At the same time as the belt contacts the input pulley, it is forced to the other extremities of the flanges, as shown in Figure 16.14B.

16.5 CHAINS AND SPROCKETS

Chain drives, unlike v-belt drives, are not displayed with friction to aid in tranmitting motion. As shown in Figure 16.15, their means of transmitting motion is positive and similar.

Figure 16.15 Chain-driven variable-speed.

In the case of a chain and sprocket, the means of transmitting motion is like gear tooth contact. Because of the positive transfer of motion between the sprocket, shown in Figure 16.16 and a suitable chain, these types of flexible connectors are 96% efficient. The chain serves as a flexible connection between the driver and the driven sprockets, as shown in Figure 16.17. This allows them to be placed some distance apart. The chain weight forms its own takeup on the loose or slack side of the drive. This eliminates the adjustment that is required in a v-belt drive to maintain the proper friction contact. But chains do stretch, and occasionally the takeup has to be adjusted or a link or two removed from the chain. Another important feature with automated drafting display is that a chain drive can be positioned in any part of the drawing (detailed, production or assembly) with only minor programming or display pro-

Figure 16.17 Sprocket and chain drive system. (Courtesy Industrial Fasteners Institute)

blems. This is accomplished by the link design, which allows the draftsperson to place the strand of chain in position and ten display it together.

16.6 DISPLAY TERMS

Like v-belt drives, chain drives have specific terms describing their various display components. Some of these are quite similar to those for v-belt drives, whereas others are considerably different. A few of the more common terms are:

Driver sprocket. Usually, the smaller of the two sprockets and one having the highest rpm.

Driven sprocket. Usually, the larger of the two sprockets and the one having slower rpm.

Chain pitch. The distance (in display units) from the center of one connecting pin to the center of the next. In chains having a solid link, the chain pitch is on alternate spacing.

Center distance. The number of display units between the centers of driver and the driven shafts.

Chain length. The distance from the centerline of the connecting pin at one end of the strand to the empty connecting hole at the opposite end. Chains can be specified in pitches or display units.

Chain rating. The load in pounds that the chain will satisfactorily handle over extended periods of time; also called the recommended working load.

Ultimate strength. The strength of the chain before it will break. This is not a governing factor in the selection of the chain. However, it gives the shock capacity of the chain.

Pitch diameter. A theoretical circle described by the center link of the chain

as it passes over the sprocket. The pitch
diameter of a sprocket is usually below the
top of the tooth or the outside diameter
of the sprocket. On drives displaying a
shortened tooth, such as silent chains,
the pitch diameter may extend beyond the
top of the tooth form.

16.7 DISPLAY OF THE CHAIN

Because of their wide use throughout
industry, standard displays have been developed for roller type chains. The roller
chain is so named because the rollers that
contact the sprocket teeth revolve around
a bushing. This turning action allows the
roller to make a rolling contact rather
than a sliding contact with the sprocket
teeth. Roller chains are made up of simple display geometry as shown in Figure 16.
18. A plate construction is connected by
pins with rollers between the side plates.
The side plates are displayed as either the
pin link or the roller link. The pin links
are located outside the roller links and
connect the roller links together. Because
of this alternate pin/roller link combination, the chain is normally inventoried
in an even number of pitches. If even
pitches can not be used, as offset connecting link, sometimes called a half-link, can
be displayed to make up one pitch.

16.8 REVIEW PROBLEMS

1. Use the program shown in Figures
16.9 through 16.12 to display Figure 16.19.
2. Repeat problem one and display
Figure 16.20.
3. Use the image processor from Chapter 5 and display Figure 16.6.
4. Repeat problem 3 for Figure 16.7.
5. Repeat problem 3 for Figure 16.8.
6. Repeat problem 3 for Figure 16.14.
7. Repeat problem 3 for Figure 16.18.
8. Repeat problem 3 for Figure 16.1.
9. Repeat problem 3 for Figure 16.2.
10. Repeat problem 3 for Figure 16.15.

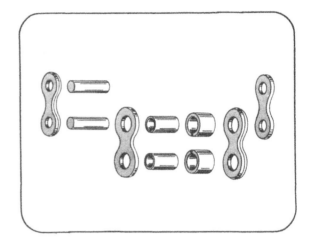

Figure 16.18   Roller Chain.

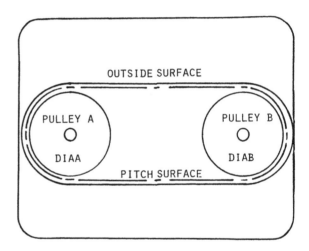

Figure 16.19   Pitch surface display.

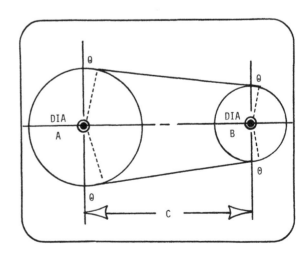

Figure 16.20   Belt length display.

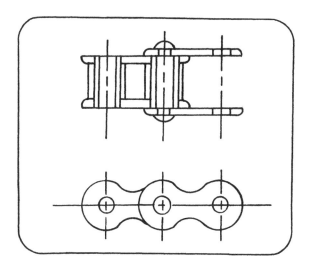

Figure 16.21 Chain assembly.

16.9 CHAPTER SUMMARY

    When the distance between the driving shaft in an automated drawing or display and the driven shaft is too great to be connected by gears and cams; a flexible connector is used. The most common replacement for a set of gears is a pulley and belt as demonstrated in Figure 16.2.

    For the purposes of automated drafting displays, the number of flexible connectors was reduced. Just as in the case of gears and cams, where only a few types were used; the connectors chosen were:

1. V-belt and pulleys
2. Flat belts and timing drives
3. Variable speed drives
4. Chains and sprockets

    By using just these few, a detailed display program could be introduced to the reader. Many types of display programs exist for flexible connectors, the reader is encouraged to locate others and practice the display options that exist.

17

Vector Analysis

Automated vector analysis is the study of graphic statics used by a designer or drafter. The application of graphic statics to the solution of structural problems has been in wide use by designers and drafters for many years. The addition of automated drafting to this method is fairly new and is called automated vector geometry (AVG). Like the automated auxiliary view construction shown in Chapter 9, it contains powerful tools for problem solution. AVG representation of the forces which act in various members of a structural framework possesses many advantages over manual solution; the key advantage, beyond presenting a graphical picture of the stresses, are that most problems can be solved with the speed and accuracy of the computer.

With the AVG approach, stresses may be obtained much more accurately than the various memebers can be sized, since in sizing we must select, from a handbook, members capable of withstanding loads equal to or greater than the design load. Using AVG, the designer computes a size and then applies a factor of safety when designing any and all structures.

Problems in statics are customarily solved by either graphical or algebraic methods. In this chapter the author has assumed that the reader is familiar with one or the other. With no previous background in statics, the reader will gain little from the study of how to automate it. Before proceeding further, review the following common terms of graphic statics shown in Table 17.1.

17.1 COMMON GRAPHIC STATICS TERMS

Many excellent references for studying the terminology listed in Table 17.1 exist. The reader should consult a reference source if any of the terms listed are unclear. We shall begin the study of automated vector analysis by building directly upon the skills learned in Chapter 9.

Line segments were labeled starting point and ending point in earlier chapters because they were not vectors. They were known as scalars. In this chapter we shall deal with line segments known as vectors.

Table 17.1
Common terms of graphic statics[*]

Statics	Force	Elements force
Vector	Tension	Compression
Shear	Equilibrium	Equilibrant
Magnitude	Structural	Coplanar
Noncoplanar	Concurrent	Nonconcurrent
Resultant	Moment	Couples
Funicular poly	Space diagram	Stress diagram
Load line	Reactions	Free-body

[*]Computer-aided Kinetics -- Marcel Dekker

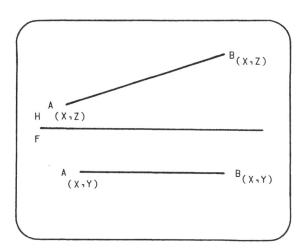

Figure 17.1 Scalar display.

This is the name applied to a line of scaled
length which represents the magnitude and
direction of a force. An arrowhead placed
on the line, usually at the end, shows the
sense (which way the force acts). The next
noticeable difference is that vectors use
BOW'Σ notation for labeling. For example,
scalar notation for a line segment looks:

$$A \overline{\qquad\qquad} B_{(X,Y)}$$
$$_{(X,Y)}$$

while AVG notation using Bow's notation is:

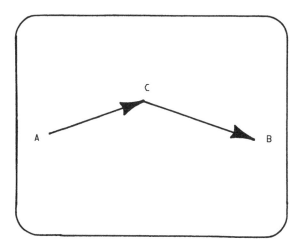

Figure 17.2 Vector display

You will note that the space around the
vector has been labeled. Both are referenced AB, the first representing a frontal
projection of a line segment. The use of
the scalar notation

$$A_{(X,Y)} \quad \text{and} \quad B_{(X,Y)}$$

indicated that locations X and Y were used
to define the location. A complete description must include another orthogonal view,
as shown in Figure 17.1. The horizontal or
top view contains the true length of the
scalar AB in Figure 17.1. In the case of
a vector it, too, will appear orthogonal
as a directed line segment which may or may
not be in true length in any of the princi-

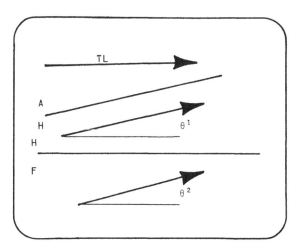

Figure 17.3 Simple two-vector diagram.

pal views. In Figure 17.2 this vector may
be described as;

$$AB = TL\ \theta^1, \theta^2$$

here vector notation is used to say: a vector AB is TL vector units long {magnitude}
and the angles from the polar references
are 1 and 2.

Vector notation should not be confused with Bow's noation. Bow's notation is
the labeling of a diagram to read around
two or more vectors called joints. The
reading is made in a clockwise direction
for convenience.

17.2 RESULTANTS

If a group of vectors describing a system are in equilibrium, the system is said
to be balanced. The resultant of this system is always zero. Systems are either
balanced or unbalanced. Vectors represent
abstract quantities of the physical system.
Two or more vectors acting together to describe a system are required. To solve a
problem in which vectors are used, a resultant is often found. It is found by the use
of two diagrams. One is called the space
diagram; it shows the relationship in the
physical system and indicates how the vectors are applied. The second is called the
stress diagram and is built from the space
diagram to determine characteristics of the
system, i.e., balanced or unbalanced. Suppose we display a diagram as shown in Figure 17.3? If the two vectors act simultaneously at point A, the result will be a path
shown as dashed from A to B in Figure 17.4.

The resultant then is the result of
two or more vectors acting at the same time
upon a system. If the vectors act independently, the path taken will be A to C to B
or along the vector lines. When two vectors
are contained in the system, a principle of
the parallelogram of forces is employed in
finding the resultant of two vectors acting
on a body. By using the same example the
body would be moved directly along the dia-
gonal connection of the parallelogram in

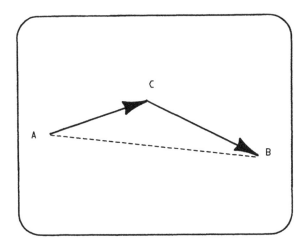

Figure 17.4 Resultant of Figure 17.3.

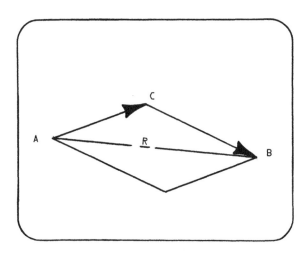

Figure 17.5 Resultant by parallelogram.

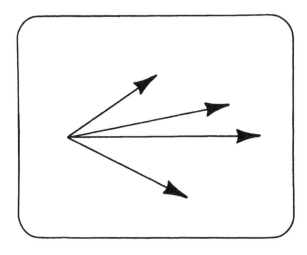

Figure 17.6 Space diagram for concurrent
coplanar system of vectors.

Figure 17.5. A computer graphics program can be used to display all coplanar diagrams like those shown in Figures 17.3 through 17.5. Figure 17.7 is a listing of this program type.

17.3 RESULTANT OF A SYSTEM OF FORCES

Figure 17. 6 illustrates a space diagram for a system of concurrent coplanar forces of given magnitude and directions. This system can be displayed on a DVST by a variation of the program shown.

After the system is displayed we may want to find the single force (resultant) which will have the same effect as the four vectors shown in Figure 17.6. One method of solution would be to consider the system as two sets of double vectors to be added. The method for this is shown in a series of display figures. The procedure would be to repeatedly apply the principle of the parallelogram of forces. Doing this, we can determine a single force which will produce the same result as the four vectors. We should input the forces in a clockwise order and start with vectors A and B. The coordinate locations of A, B and the point of concurrency are input to the program as shown in Figure 17.8.

Now R and C are combined without erasing the screen as shown in Figure 17.9. R_1 and D are added together to find the resultant of the total system as shown in Figure 17.10. Obviously the system is not in equilibrium because R_2 has direction and magnitude. The use of the parallelogram program for solving complex systems is not recommended. An advanced program should be used for the construction of a **stress** diagram. A stress diagram uses the coordinate locations of the space diagram and redraws the diagram by combining one vector to another (maintaining true direction and length). This is much quicker than constructing a series of parallelograms.

The **tip-to-tail** technique shown in Figure 17.11 clearly shows any resultant or opening. The system is balanced or in equilibrium if the stress diagram closes. If an

```
C  *******************************************
C  *                                         *
C  *   THIS IS AN AUTOMATED VECTOR DISPLAY *
C  *   PROGRAM FOR THE SOULTION OF THE PAR-*
C  *   ALLELOGRAM OF FORCES.  INPUT THE FO-*
C  *   LLOWING INFORMATION:                  *
C  *      X & Y LOCATION OF A,B, & C        *
C  *      A = POINT OF APPLICATION          *
C  *      B = END POINT OF SYSTEM           *
C  *      C = MIDPOINT OF SYSTEM            *
C  *******************************************
      CALL INITT(240)
      DIMENSION X(3), Y(3)
      READ(1,*)X(1),Y(1),X(2),Y(2), X(3), Y(3)
C  DISPLAY SYSTEM OF VECTORS
      CALL AROHD(X(1),Y(1),X(3), Y(3),.125,0.,16)
      CALL AROHD(X(3),Y(3),X(2), Y(2),.125,0.,16)
      CALL DASHP(X(1),Y(1),.1)
C  COMPUTE VECTOR LENGTHS
      X1 = X(3)-X(1)
      X2 = X(2)-X(3)
      X3 = X(2)-X(1)
      Y1 = Y(3)-Y(1)
      Y2 = Y(2)-Y(3)
      Y3 = Y(2)-Y(1)
      TL =(X1**2+Y1**2)**.5
      TLA=(X2**2+Y2**2)**.5
      TLR=(X3**2+Y3**2)**.5
C  DISPLAY VECTOR QUANTITIES
      PRINT(3,*)TL,TLA,TLR
      CALL FINITT(0,0)
      STOP
      END
```

Figure 17.7 Program for resultant display.

opening occurs, then a resultant exists or an equilibrant is needed to close and balance the system.

17.4 RESULTANT, POINT OF APPLICATION

The general term for the process of replacing a group of vectors by a single vector is composition. This process is the addition of two or more vectors. The opposite process, that of replacing a single vector by two or more vectors having the

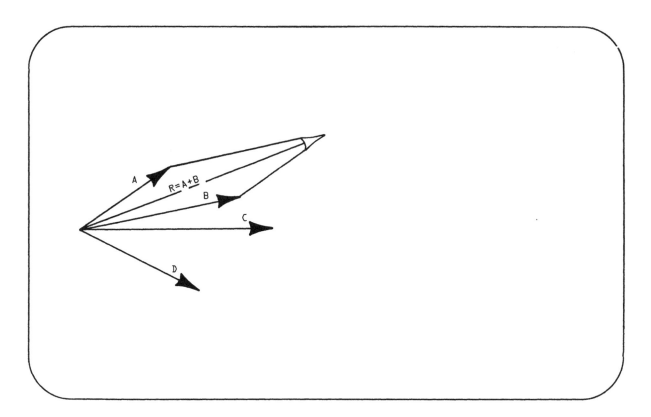

Figure 17.8   Sum of vectors A and B.

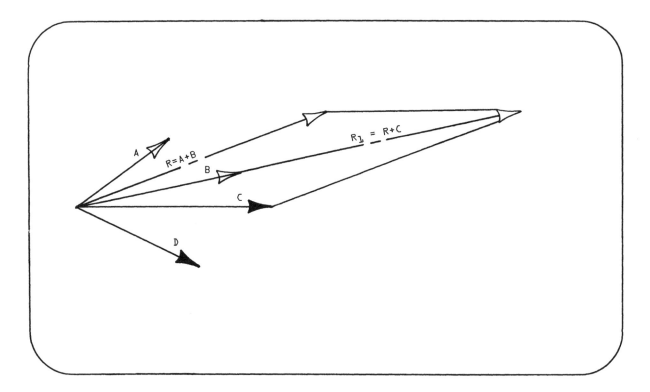

Figure 17.9   Sum of vectors R and C.

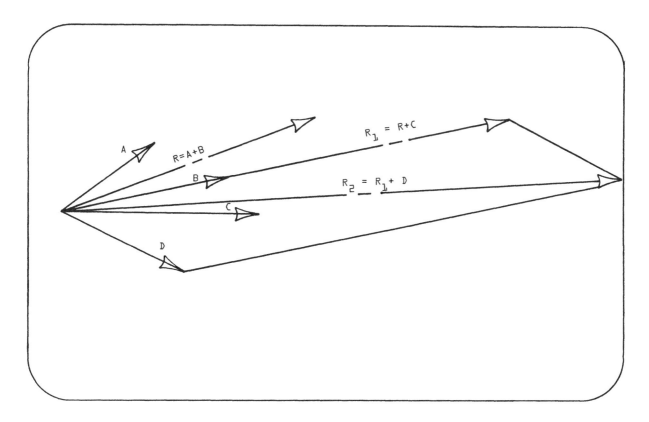

Figure 17.10   Sum of vectors R_1 and D.

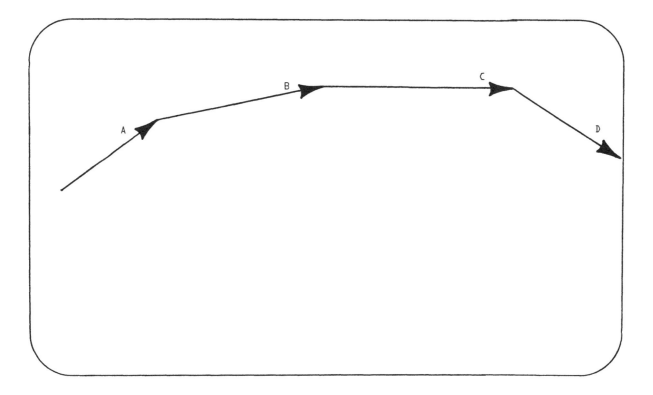

Figure 17.11   Stress diagram of vector system.

same effect, is called resolution. Each
vector in the new system is called a component of the given vector. Resolution and
composition are useful techniques when you
study nonconcurrent systems of vectors.

Nonconcurrent, coplanar forces may be
combined to show a point of application.
The combination or addition will illustrate
a common resultant. In Figure 17.12A, four
vectors are separated by Bow's notation: A,
B,C,D and E. They are not concurrent, and
they do not act in parallel. By using the
DVST and the advanced program for addition
of vectors, the tips are connected to each
tail as shown in Figure 17.12B. This clearly shows that a resultant is present. The
magnitude of the resultant can be computed
and displayed, but the point of application
has not been located   The graphics input
(GIN) mode of the DVST can be used to locate a convenient point beside the plotted
information as shown in Figure 17.12C. This
point is called a pole point. From the pole
point, CALL PLOTS are connected to each tip
and tail.

A call to alpha mode can now be used
for convenient labeling of points and diagrams used so far in the solution process
as shown in Figure 17.13. The last step in
the location of the point of application
would be the construction of the funicular
polygon in the GIN mode (step 2, Figure 17.
13). The resultant has already been computed for magnitude and displayed for direction; with the aid of the GIN mode, the
point of application was found. With the
alpha mode all labeling of the construction
steps was completed as shown in step 3 of
Figure 17.13.

17.5 EQUILIBRANTS

If a single vector is added to an unbalanced system to produce equilibrium, that
vector is known as an equilibrant. The
equilibrant in an unbalanced system will
always have the same location and magnitude as the resultant of that system but
will have the opposite sense.

Many statics problems can be worked

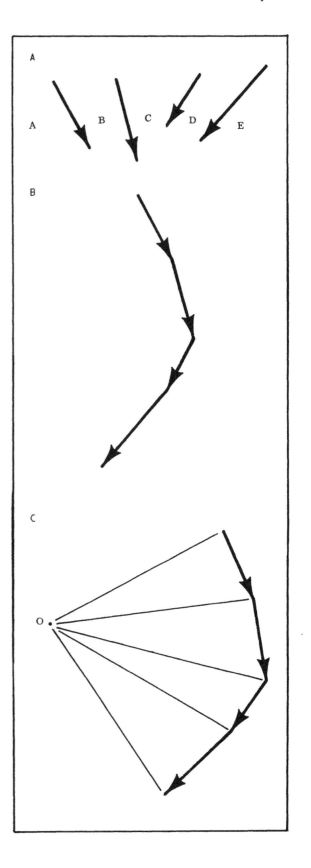

Figure 17.12 Nonconcurrent, coplanar forces.

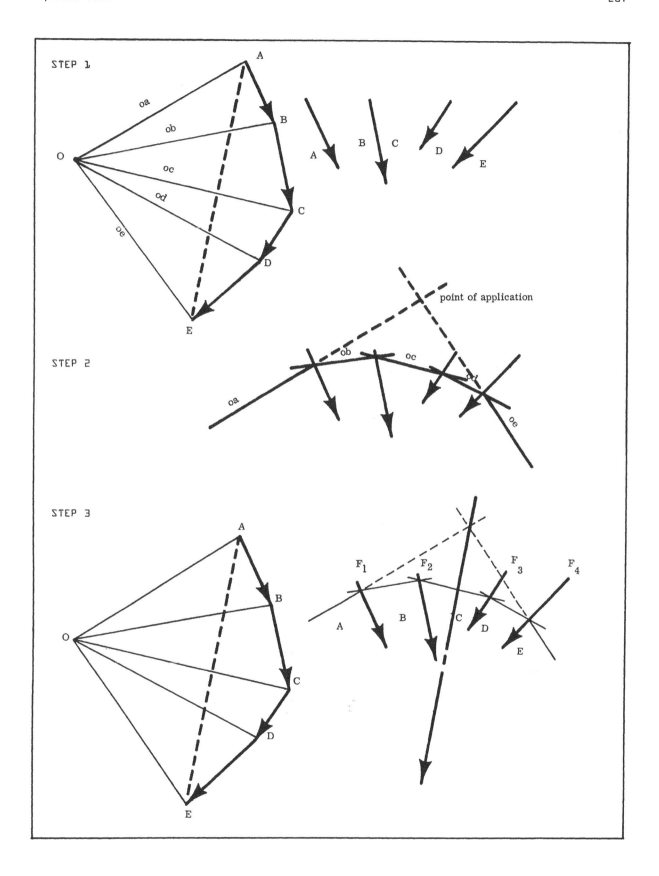

Figure 17.13 Point of application and resultant for a system of vectors.

17.7 VECTOR DIAGRAMS

The vector or stress diagram is displayed for determining certain characteristics of the system. The stress diagram
represents the magnitudes and directions
of the forces called for in the space diagram. The type of program listed in section 17.6 for displaying systems in equilibrium can be used to present stress diagrams of a tip to tail nature as shown in
Figure 17.15, if they are concurrent and
coplanar. Nonconcurrent, coplanar systems
as illustrated in Figure 17.16 may use all
three.

The subroutine vector is used to display the stress diagram for noncoplanar
systems of forces. The solution of a noncoplanar system of vectors to determine the
state of equilibrium is similar to coplanar methods. When concurrent vectors are

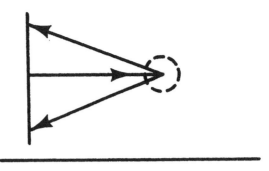

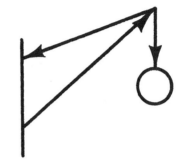

Figure 17.14 Support frame display.

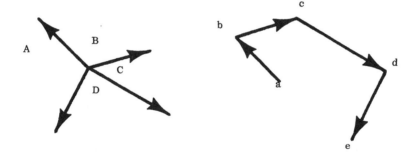

Figure 17.15 Stress diagram as a system and then as a tip-to-tail diagram.

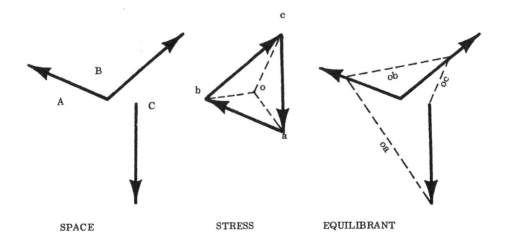

SPACE   STRESS   EQUILIBRANT

Figure 17.16 Types of diagrams displayed by interactive means.

coplanar vectors is listed in Figure 17.20.

The bulk of vector geometry problems fall into the noncoplanar category. Therefore, computer assistance and graphical displays play an important role in the visualization of a statics problem. Bow's notation also comes in handy in labeling more than one view of a single system of vectors. This is illustrated in Figure 17.21. By this method, letters or numbers are placed on both sides of each line of action shown in the space diagram, and each vector is described by the two letters between which it lies. The diagram shown in Figure 17.21 can be displayed directly by

```
CALL INITT(240)
CALL VECTOR(.5,3.,1.5,2.,1.,2.25,2.5,
+.5)
CALL VECTOR(.5,3.,3.5,1.5,2.,2.25,2.5,
+.5)
CALL VECTOR(.5,3.5,4.25,2.25,1.,2.25,
+2.5,.5)
CALL FINITT(0,0)
STOP
END
```

or in another example as shown in Figure 17.17, the case of nonconcurrent vectors, a similar use of CALL VECTOR can be used as

```
CALL VECTOR(.5,3.,2.,1.5,1.,1.5,2.5,
+2.5)
CALL VECTOR(.5,3.,3.,1.5,1.,2.75,2.25
+,2.5)
CALL VECTOR(.5,3.,3.6,1.,.5,4.1,2.8,2
+.25)
```

Conventional interactive display subroutines can be used to display a space diagram for a system of noncoplanar forces. The use of a space diagram in setting a problem is important; a series of coplanar and non-coplanar examples will illustrate this. Figure 17.18 shows a weight lodged between two inclined surfaces. Figure 17.19 illustrates a jib crane supporting a load of 2 Kips (2000 pounds). Figure 17.14 illustrates a type of support frame commonly used in overhanging loads.

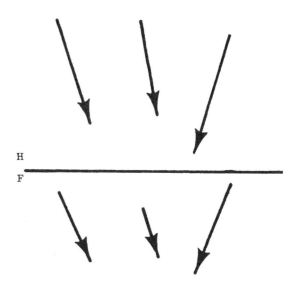

Figure 17.17   Display from program.

Figure 17.18   Display from interactive graphics display program.

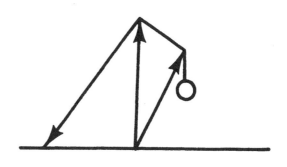

Figure 17.19   Jib crane diagram.

more general type of image processor that
is part of an interactive graphics program
written for coplanar-type problems. It is
used by typing a series of commands as outlined in Chapter 5 where:

```
    LOGON
    USERID/PASSWD
    CALL 'ACS.USERID.DEMO(VECTORS)'
```

will produce the diagrams shown in Figures
17.1 Through 17.13 and 17.22 through 17.24.

Before leaving the concept of computer
displays for balanced systems, it is important to mention that a complete treatment
of equilibrium has not been presented here.
Only a few ideas and only one method for
displaying has been presented, a typical
list of examples of equilibrium diagrams
might include

1. Forces required to resist wedging
2. Stresses in jib cranes
3. Stresses caused by deflected weights
4. Stresses in cable problems
5. Forces in struts and braces
6. Solving for reactions
7. Inverse proportion force problems
8. Simple truss stresses
9. Weights on ladders and supports
10. Forces required to move objects

17.6 SPACE DIAGRAMS

The use of simple diagrams in space
has been illustrated in the two previous
chapter sections (17.4 and 17.5). To explain the two concepts and solve static
problems, two separate diagrams were displayed. One diagram, called the space diagram, will now be expanded to include nonconcurrent, noncoplanar uses. The other
diagram, called a stress or vector diagram,
will be covered in the detail in the next
chapter section (17.7).

All computer displays have thus far
been 2-dimensional (coplanar), described
in X and Y locations. To display noncoplanar diagrams, X,Y, and Z locations must
be known. A subroutine for displaying non-

```
C  ******************************************
C  *                                        *
C  *  THIS SUBROUTINE WILL DISPLAY A 3-D     *
C  *  VECTOR IN TWO VIEWS ON THE FACE OF     *
C  *  A DVST SCREEN.   THE USER SUPPLIES:    *
C  *  X,Y...LOCATION OF HF REFERENCE LINE    *
C  *  XTIP,YTIP,ZTIP... AROHD LOCATION       *
C  *  XTAIL,YTAIL,ZTAIL.. END OF VECTOR      *
C  ******************************************
       SUBROUTINE VECTOR(X,Y,XTIP,YTIP,ZTIP
      +XTAIL,YTAIL,ZTAIL)
C  DISPLAY HF REFERENCE LINE
       CALL PLOT(X,Y,3)
       CALL PLOT(X+8.,Y,2)
       CALL SYMBOL(X,Y+.1,.125,'H',0.,1)
       CALL SYMBOL(X,Y-.25,.125,'F',0.,1)
C  DISPLAY FRONTAL PROJECTION OF VECTOR
       CALL AROHD(XTAIL,YTAIL,XTIP,YTIP,.125
      +,0.,16)
       ZTAIL=ZTAIL+Y
       ZTIP=ZTIP+Y
C  DISPLAY HORIZONTAL PROJECTION OF VECTOR
       CALL AROHD(XTAIL,ZTAIL,XTIP,ZTIP,.125
      +,0.,16)
       RETURN
       END
```

Figure 17.20   Subroutine to display vectors.

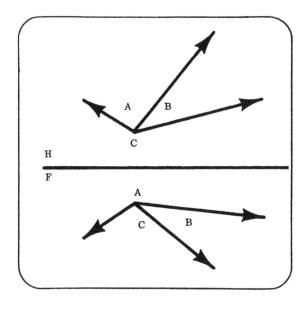

Figure 17.21 Display of vector subroutine.

because an equilibrant can be added to a
system or the system is already in a state
of equilibrium.

For example, let us assume that the
wheel in Figure 17.22 is to be pushed over
a 6-inch block. The horizontal force tend-
ing to push the wheel is applied level with
the centerline of the wheel. A stress diagram of the concurrent system is displayed
on the DVST screen in Figure 17.23,
because we know the system is in equilibrium. The diagram in Figure 17.23 indicates
the direction and magnitudes of the forces
in the balanced system. As shown in the
space diagram, the lines of action of the
three forces meet in a common point called
concurrency, and the stress diagram closes
(equilibrium). A slight increase in horizontal force will produce motion.

The next question that might be asked
is whether or not the centerline is the
most ideal place to apply a pushing force
if labor saving is important. Two other
points of application are selected and an-
alyzed as shown in Figure 17.24. A combination of space diagrams and stress diagrams
are displayed as described in earlier chapter material so will not be repeated here.

The equilibrium diagram shown in Figures 17.22 and 17.24 is sometimes called a
free-body diagram and is displayed from a

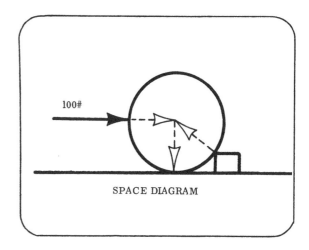

Figure 17.22   Example problem.

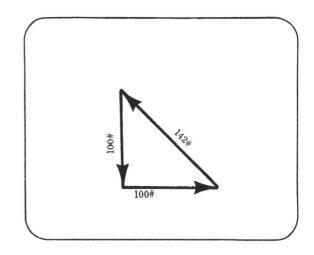

Figure 17.23   Stress diagram of 17.14.

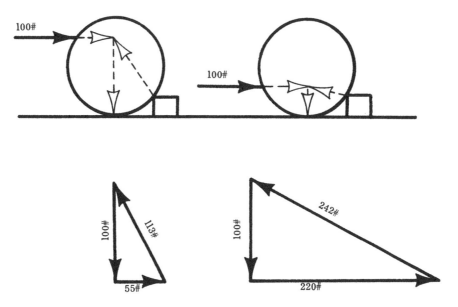

Figure 17.24   Combination free-body and stress diagrams.

present, the parallelogram method shown in
Figure 17.25 can be used.

Figure 17.26 was displayed with call
vectors and routed to a plotter. In this
example the resultant was located in both
the horizontal and frontal planes of projection. A resultant was then determined
by finding R(TL). Figure 17.27 illustrates
stresses in a system of concurrent, noncoplanar members.

17.8 SIMPLE STRUCTURE ANALYSIS

Figure 17.28 illustrates the combination of space and stress diagrams to determine the stresses in a simple structure.
The loads are laid out in a straight line
(load line). This is displayed first, at

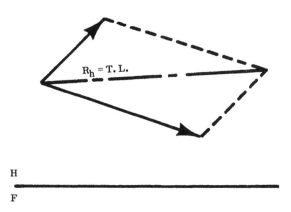

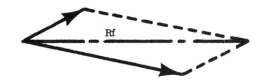

Figure 17.25 Parallelogram system displayed.

Figure 17.26 Plotter output of concurrent, noncoplanar vector system with resultant.

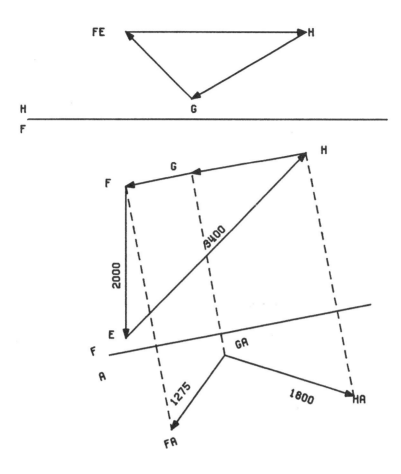

Figure 17.27 Plotter output of stresses in a system of concurrent, noncoplanar members.

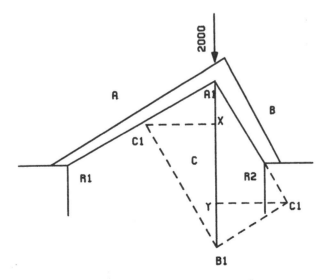

COMBINED SPACE AND STRESS DIAGRAM

Figure 17.28 Plotter output of combination diagrams.

a scale suitable for the DVST screen, and
then the stress diagram is displayed around
it. By providing a calculation section inside this type of display program, the forces in the members can be determined and
plotted on a digital plotter. The tendency
for the structure to slide at R1 and R2 is
shown by C1X and YC1. This stress must be
balanced by a tie member (equilibrant)
across the base.

Figure 17.29 shows a ladder leaning
against a wall. This stress diagram shows
the approximate pressure against the wall
and the floor. Calculated values will be
approximate since coefficients of friction
between the different materials will be
left out. The stress has been combined
with the space diagram to show the weight
of the ladder by the vector AB and the pressure against the wall by CB. The vector CB
also represents the horizontal thrust at
the foot of the ladder.

Figure 17.30 shows a simple form of
roof truss with supporting loads. To deter-

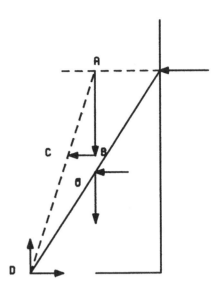

LADDER PROBLEM

Figure 17.29 Stress diagram.

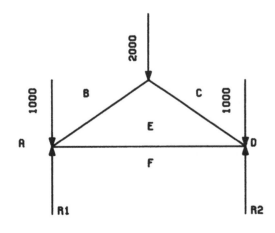

SPACE DIAGRAM

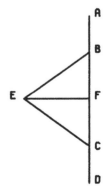

STRESS DIAGRAM

Figure 17.30 Roof truss analysis.

mine the stresses in the members, it is
necessary to designate the loads and members in some convenient method. This should
be done with Bow's notation. The labels
for members must be read in a clockwise direction. For example, the left rafter member at the joint R1 is BE. If read about
the top, this same member would be EB.

When determining the size of a member
required to carry a given load, it is important to know the kind of stress (tension
or compression) of the force and the length
of the required member. Stress diagrams for
simple structures contain both tension and
compression vectors. A tension vector tends
to cause stretching, separation, or pulling
apart. It is labeled with a minus sign and
is read away from a joint. A compression
vector causes shortening, the state of being
compressed. It is labeled with a plus sign
and is read into a joint.

17.9 TRUSS DESIGN AND ANALYSIS

When AVG is used to design a truss,
computer displays are used to determine

stresses in the various members of the truss.
Any structure must be designed to carry its
own weight in addition to the specified
weights or loads. For convenience in this
section, we shall assume that the weight of
the truss is included with the given loading.
The weights of roofing members can also be
determined by the following,

Steel: $W=1/2*A*L*(L+1/10*L)$

Wood: $W=3/4*A*L*(L+1/10*L)$

where

L = span in feet

A = distance in feet between adjacent
    trusses

Figure 17.31 shows a typical solution
when determining stresses in the members of
a truss. The truss is anchored at both ends
and is subjected to dead loads concentrated
at points where the truss members frame or
(connect) together. When analyzing stresses
by AVG, the space diagram is displayed first.
Bow's notation is applied, and the stress
diagram is written as follows:

1. Start by displaying the load line
and dividing it proportionally between the
reactions at both fixed ends.

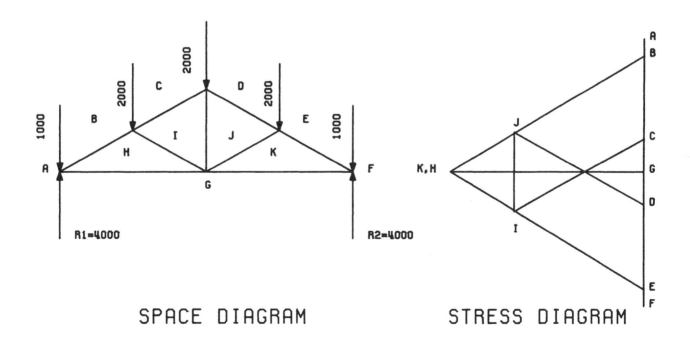

Figure 17.31 Truss design and analysis.

2. The load line is vertical, since
the loads act vertically. The length of
the load line represents the sum of the
loads when displayed to scale.

3. The loads in this example are also
symmetrical, the reactions R1 and R2 are,
therefore, equal.

4. The first line in the stress diagram should be displayed horizontally through
point G and toward the left edge of the DVST
so it cuts vector BH displayed through B on
the load line and is parallel to truss member BH in the space diagram.

5. All vectors in the stress diagram
should be labeled by a call aplha as they
are plotted (remember that the DVST is used
to preview information sent to the pen plotter). The load line is labeled, load by
load, before displaying the vectors representing the truss members.

6. Each joint in the space diagram,
from left to right, must be matched by a
closed polygon in the stress diagram. The
completed stress diagram is a series of
closed polygons.

7. Each vector in the stress diagram
is parallel to the corresponding member in
the space diagram. While each vector in
the stress diagram is automatically at the'
same scale as used to plot the load line,
individual member stresses can be calculated and displayed for our use.

17.10 REVIEW PROBLEMS

AVG software is available from some
turnkey vendors, but is not available to
general computer users as described throughout this textbook. Therefore, short subprograms will need to be supplied by your
instructor. In Chapters 5 (Image Processing) and 6 (Multiview Projection) we learned how to create free-body diagrams for
real-world objects. These types of diagrams
are definitely the most time-consuming part
of the AVG analysis. In each of the problems assigned for this chapter, you will
know how to construct simple diagrams. You
will need practice in displaying stress type
diagrams.

1. Begin this practice with a simple
beam (call rect) supported at both ends.
Display three loads on the beam (call vector). Combine these vectors tip-to-tail.
Does the diagram close? Is the system in
equilibrium?

2. The system displayed in problem 1
was a coplanar problem. Using this same
problem, we can solve for the reactions at
the right and left ends of the beam.

3. Display Figures 17.14 through 17.
16 and practice these types of problems.

4. Display Figure 17.20 and solve.

5. Display Figure 17.21 and solve.

6. Display Figure 17.22 and solve.

7. Practice the solution steps shown
in Figures 17.23 through 17.27 with information supplied by your instructor.

8. Display Figure 17.28 and solve for
a load of 4000 pounds.

9. Display Figure 17.30 and solve for
loads of 2000, 3000, and 2000 pounds.

10. Display Figure 17.31 and solve for
loads of .5,1,2,1,.5 kips.

17.11 CHAPTER SUMMARY

AVG (automated vector geometry) is the
study of graphic statics used by a designer
or draftsperson. The representation of the
forces which act in various members of a
structural framework possesses many advantages over the manual solution. With this
approach, stresses may be obtained much more
accurately than the various members can be
sized, since in sizing we must select, from
a handbook, members capable of withstanding
loads equal to or greater than the design
load.

Using AVG, the designer computes a
size and then applies a factor of safety
when designing any and all structures. The
chapter relates a complete procedure for
displaying all the common vector geometry
situations that a draftsperson may need to
know in order to assist an engineer. While
it is true that few drafters will be doing
creative design, drafting is the science
of design documentation, and as such AVG
is an important automated drafting tool.

18

CHARTS AND GRAPHS

The use of an automated drafting system to produce charts or graphs during the design process has been well documented by a number of computer graphics hardware companies. The most common form of documentation has been a collection of programs grouped together under an operating system (Tektronix Advanced Graphing Package AGII, for example). Either charts or graphs can be produced.

Webster's Dictionary defines a chart as a sheet showing facts graphically or in tabular form or as a graph showing changes in temperature, variation in population, or prices. A graph is defined as a diagram representing the relationship between two or more factors by means of a series of connected points or by bars, curves, or lines.

It is obvious that the editors of the dictionary regard the two terms as synonymous. Most recent engineering graphics texts do not distinguish between the two. For our discussion, a distinguishing difference will be made for computer applications and we will define the two as:

CHART: A more or less pictorial com-

puter graphics presentation of facts. It should be pointed out that facts presented by charts are easy to read and are quite meaningful to the lay person. Therefore, charts are seen in newspapers, magazines, and the like.

GRAPH: A presentation of data plotted in one of the many formats where each point is connected with the following adjacent point by a line. There may be more than one set of data per sheet. Graphs are used by engineers to illustrate trends, to predict, to develop an equation for a certain behavior, to present results of test data obtained in experiments, and to correlate the observations of natural phenomena.

18.1 ANALYSIS OF DESIGN DATA

Before a proposed design is accepted, it must be subjected to a careful analysis. During this process the computer data provided must be evaluated and interpreted by the engineer. Most frequently, data are submitted in numerical form, and interpretation is often a lengthy and difficult pro-

cedure.   Thus, to ensure that the engineer
and each member of the design team understands all aspects of the project, it is
convenient to convert numerical computer
data to a more customary form which will
permit ready understanding.

Before selecting the type of chart or
graph to illustrate the design data, consideration must be given to its use.   If
it is to be used to determine numerical
vlaues or reading numbers, it would be a
quantitative chart or graph.   If it is used
to present comparative relationships, it is
called qualitative.   Since there are many
types of programs available to present
data, the purpose must be established before the graph or chart is created.

18.2   CREATING A BASIC GRAPH

To create a graph it is necessary to
prepare the data which are to be plotted.
Two data lists will be necessary; one for
the horizontal (X) values and one for the
vertical (Y) values.   The lists will include in the first position, the number of
data values to be plotted.   The rest of the
positions in the list contain the data values.   Each list will be called an array.

```
DIMENSION XDATA(7), YDATA(7)
DATA XDATA/6.,1.,2.,3.,4.,5.,6./
DATA YDATA/6.,211.,114.,306.,453.,291.
+,325./
```

The first line is a standard FORTRAN dimension statement assigning the names XDATA
and YDATA to the two arrays and setting
the length of each to 7.   The second line
contains the number of data points (6)
followed by the actual data values for the
X axis.   The third line sets the Y.   Once
the data are in a usable form, five call
statements will produce the graph:

```
CALL INITT(240)
CALL BINITT
CALL CHECK(XDATA,YDATA)
CALL DSPLAY(XDATA,YDATA)
CALL FINITT(0,700)
```

The call to INITT initializes the Tektronix terminal.   The parameter 240 indicates that transmission is at 240 characters
per second.   BINITT begins the AGII package.
CHECK has two arguments, the names of the
data arrays to be plotted.   DSPLAY plots the
graph on the face of the DVST.   FINITT ter-
minates use of the AGII package.

It is often desirable to plot two sets
of data on the same plot.   With regard to
adding a second data curve to the plot we
just created, refer to Figure 18.1.   Since
the second curve will be plotted on the same
graph as the first, the widest range of data
should be plotted first.   The same horizontal
values will be used, so only one array for
another curve (YDATA2) has been added to the
program.   An additional call statement has
been added in Figure 18.1 immediately after
the call of DSPLAY.   It will plot the second
curve:

```
CALL CPLOT(XDATA,YDATA2)
```

many options are available for reproducing
the graph.   Figure 18.1 represents a screen
copy from a hard copy unit.   Both the source
program and graph can be copied by inserting
another program statement such as:

```
CALL HCOPY
```

and a thermo-type copy of the DVST is made.
This type of reproduction is quick, but its
appearance is not suitable for a technical
report or business report.   A call for plotter output will improve the quality of the
reproduction.   Figure 18.2 is an example of
a pen plotter graph.   In this example the
data points have been labeled by:

```
CALL SYMBL(N)
```

where N equals an integer 0 to 11.   Figure
18.3 introduces the use of the twelve symbols for labeling data points.

The eight examples in Figures 18.4 through 18.11 have been selected to illustrate
some of the display techniques that can be
used for automated charts and graphs.

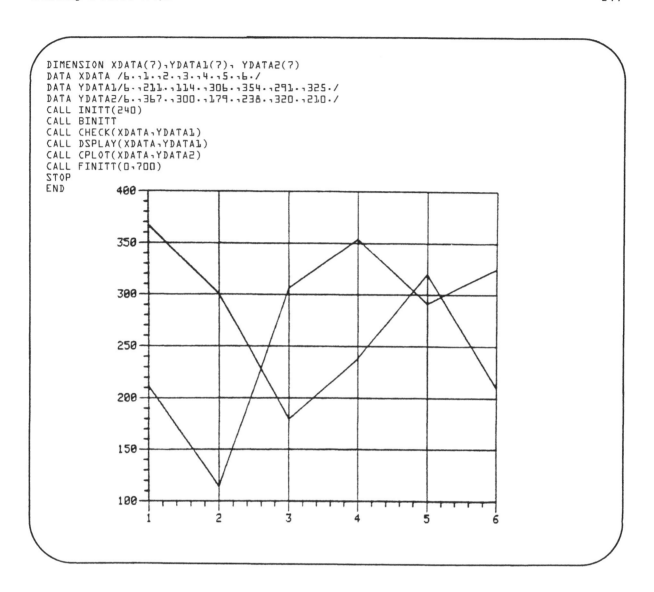

```
DIMENSION XDATA(7),YDATA1(7), YDATA2(7)
DATA XDATA /6.,1.,2.,3.,4.,5.,6./
DATA YDATA1/6.,211.,114.,306.,354.,291.,325./
DATA YDATA2/6.,367.,300.,179.,238.,320.,210./
CALL INITT(240)
CALL BINITT
CALL CHECK(XDATA,YDATA1)
CALL DSPLAY(XDATA,YDATA1)
CALL CPLOT(XDATA,YDATA2)
CALL FINITT(0,700)
STOP
END
```

Figure 18.1 Line graph from Tektronix terminal. (Courtesy Tektronix, Inc., Information Display Division.)

18.3 LINEAR GRAPHS

The majority of engineering graphs are plotted in rectangular coordinate format. This format is generally grided to form 1/20 inch squares (Figures 18.8 and 18.10). All others are plotted in centimeters or in 1/8 , 1/10. or 1/4 inch squares. The larger grid makes the graph easier to read. Variables selected for the abscissa (X) and the ordinate (Y) are chosen showing the independent variables as X and the dependent variables as Y. It is important to chose the correct scaling since this has an effect on the slope of the curve. The slope of the curve provides a visual impression of the degree of change in the dependent variable for a given increment of the independent variable. Always try to create the correct impression when programming graphs for the computer display.

The range of scales should ensure effective and efficient use of the area available. If a chart is quantitative, the intersection of the axis need not be at the origin (zero) of the coordinates. However,

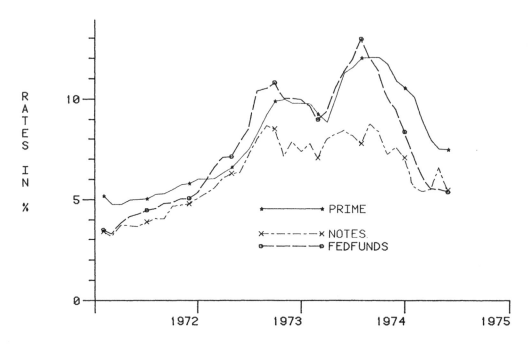

Figure 18.2 Graph from pen plotter. (Courtesy Tektronix Inc.)

```
       DIMENSION XARRAY(4), YARRAY(7)
       DATA XARRAY/-1.,6.,22.,2.4/
       DATA YARRAY/6.,211.,114.,306.,354.,291.,325./
       CALL INITT(240)
       CALL BINITT
C *****DRAW A SYMBOL AT ALL DATA POINTS*************
       CALL SYMBL(6)
       CALL CHECK(XARRAY,YARRAY)
       CALL DSPLAY(XARRAY,YARRAY)
       CALL TINPUT(I)
       CALL FINITT(0,700)
       STOP
       END
```

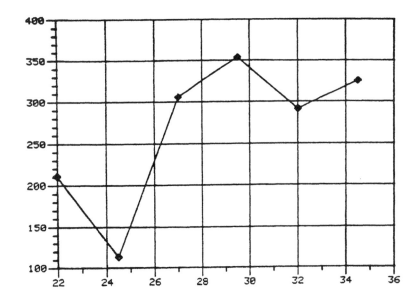
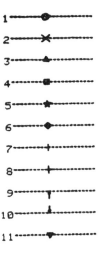

Figure 18.3 Program, symbols, and graph. (Courtesy Tektronix Information Display Division.)

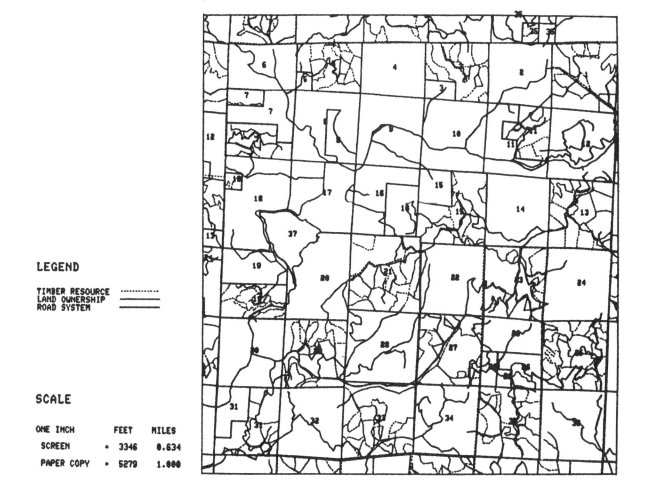

Figure 18.4 Hard copy of chart showing timber resources. (Courtesy Tektronix Inc.)

if the chart is qualitative, both the ordinate and the abscissa generally should have
the value of zero at the intersection. The
arithmetic scale numbers should be 1,2,3...;
2.4,6, ...; or 5,10.15, ... Of course
these units may be multiples of 0.1, 0.01,
and 10, 100, and 1000. The use of too many
digits in the scale numbers should be avoided. The scale caption can be designated as
"100s", "millions", or 10^5.

The plotted points on the graph should
be identified. If more than one curve is
plotted on a sheet, then a different SYMBL
may be used; refer to Figure 18.3. After
all points have been plotted the curve must
be displayed by CALL DSPLAY. If the data
are continuous, the curve will be smooth,
and if discontinuous, each adjacent point

should be connected with a straight line.
When displaying the curve for continuous
data the curve should be an average of the
plotted points. This is done by a short
linear from:

```
DATA YARRAY/-1.,6.,1.,2./
```

The first element is -1, indicating the method of expansion (linear). The second element is the number of data points to be plotted, while the third is the first value to
be displayed. The last element is the
amount by which the value is to be incre-
mented for each data point. Therefore, the
DATA statement would cause the values 1,3,
5,7,9. and 11 to be selected.

After the curve has been determined it

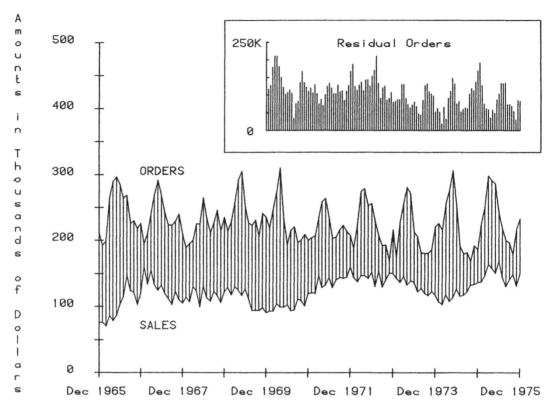

Figure 18.5 Display of data with shading routine. (Courtesy Tektronix Inc.)

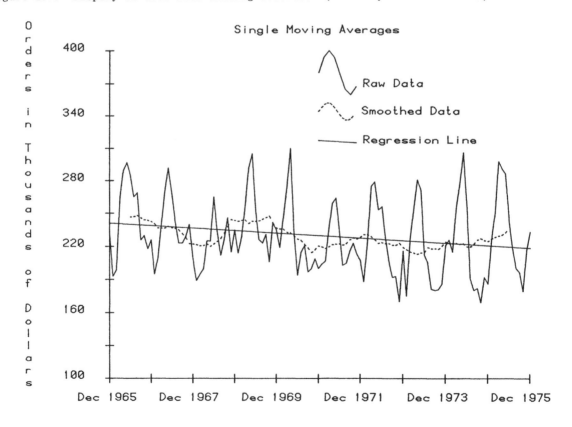

Figure 18.6 Three sets of data; raw, smoothed, and averaged. (Courtesy Tektronix Inc.)

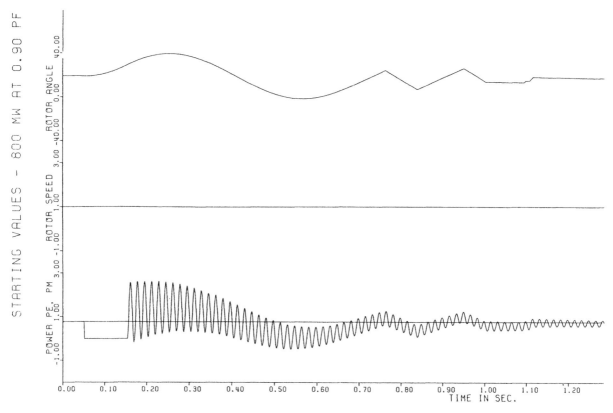

Figure 18.7 CALCOMP output of three independent sets of data.

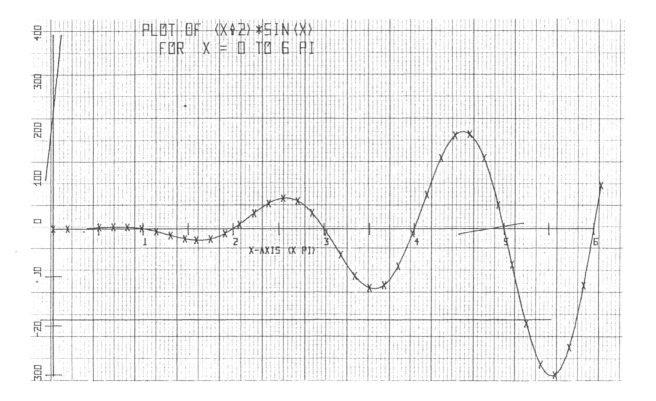

Figure 18.8 Display of equation information.

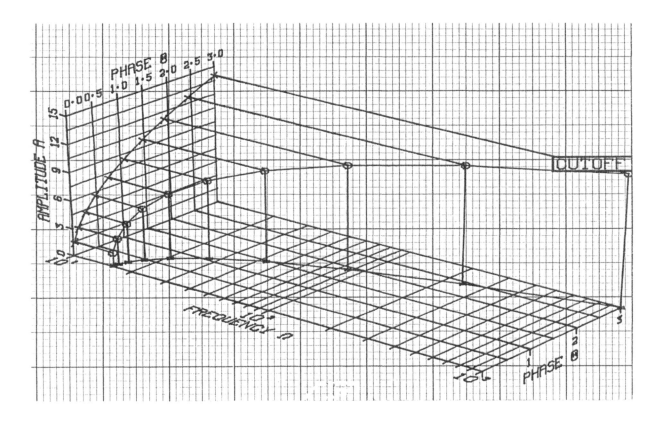

Figure 18.9   Pictorial chart of 3-dimensional Bode diagram.

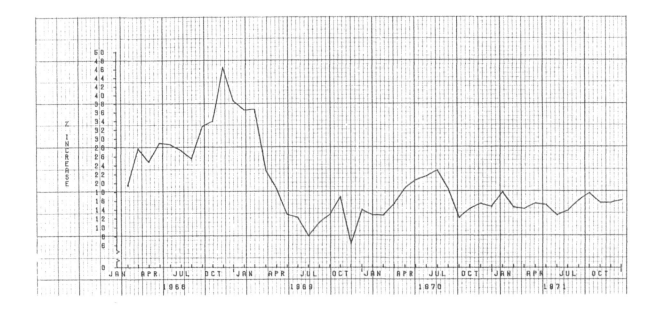

Figure 18.10 Use of multiline in a graph.

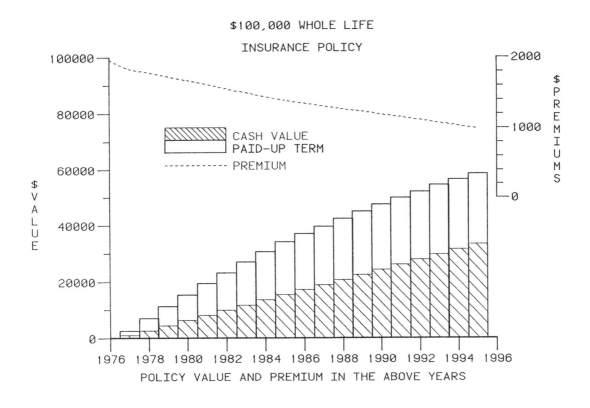

Figure 18.11 Bar chart and use of shading. (Courtesy Tektronix Inc.)

should then be displayed according to line
type. This may be done by

        CALL LINE(IVALUE)

where IVALUE = 0 unless changed by a call
to LINE. The following IVALUES may be used

    0   SOLID   ———————————————
    1   LONG DASH - - - - - - - - -
    2   SHORT DASH ------------------
    3   DASH DOT -.-.-.-.-.-.-.-.-
    4   DOT   ....
   -1   NO LINE
   -2   VERTICAL BAR |||||||||||||||
   -3   HORIZONTAL BAR —— —— ——
   -4   POINT PLOT oooooooooooooooooo

    Multiple curves on a linear graph make
excellent use of different line types. A
simple mrthod of displaying a second curve
was shown in Figure 18.1. Here the second
curve fit nicely on the scales selected.
However, if the second curve has data values
which extend outside the first data limits,
the curve will be clipped at the edge of
the screen. See Figure 18.12. To prevent
this from occuring, the user may make all
the calls to display the first curve followed by calls to DINITX or DINITY. An axis
location change is necessary to prevent the
new label values from being printed over the
old ones; these techniques will be discussed
in the remaining sections of this chapter.
For now we are concerned with

        CALL DINITX
        CALL DINITY

These subroutines reinitialize the label
values of the X and Y axis, allowing the
drawing of an additional curve with new
label values. All values related to labeling the axes are set to zero so that new
values can be computed for the display of
an additional curve. Figure 18.13 shows
the same sets of data plotted with a CALL
CALL DINITY to avoid the clip at the top.

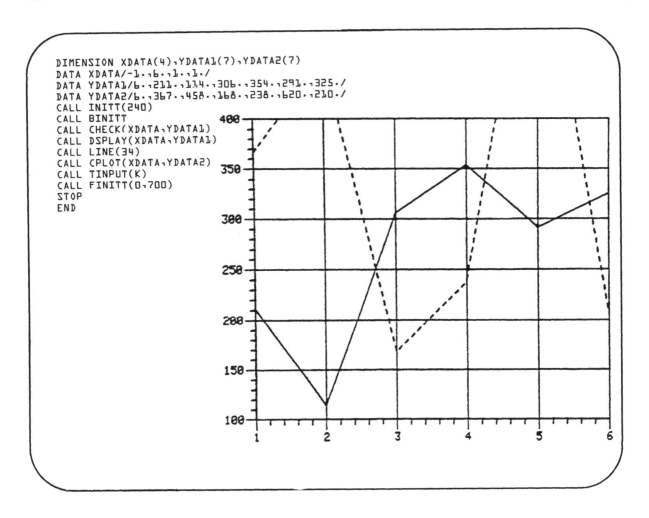

```
DIMENSION XDATA(4),YDATA1(7),YDATA2(7)
DATA XDATA/-1.,6.,1.,1./
DATA YDATA1/6.,211.,114.,306.,354.,291.,325./
DATA YDATA2/6.,367.,458.,168.,238.,620.,210./
CALL INITT(240)
CALL BINITT
CALL CHECK(XDATA,YDATA1)
CALL DSPLAY(XDATA,YDATA1)
CALL LINE(34)
CALL CPLOT(XDATA,YDATA2)
CALL TINPUT(K)
CALL FINITT(0,700)
STOP
END
```

Figure 18.12   Example of clipping.   (Courtesy Tektronix, Inc., Information Display Division)

By studying Figure 18.13. you will see that CALL SLIMX(200,800) has been used. This is the screen window location. If no reference is made to window size, then the window is the size of the screen. On a Tektronix terminal there are 1024 visible addressable units along the X axis and 781 on the Y axis with 128 units to the inch. The unit designations are used to show screen location. A full screen plotting surface can be described as 0 to 1023 in X and 0 to 780 in Y. The limits of the plotting area may be set directly in screen units by calling either SLIMX or SLIMY in the form

CALL SLMIX(NX,NY)

where Nx is the X-axis screen minimum in display units and NY is the X-axis maximum. This would limit the width of the plotting window. SLIMY would be called to set the height of the plotting window.

In most cases the 'window techniques' described above can be useful in the arrangement of linear graphs. In some cases the range of the data is so large that the plotting window becomes extremely small. Figure 18.14 is a photograph of three graphs displayed on a Tektronix screen; here setting and resetting windows has allowed the graphs to be displayed as one. In the case of the lower right-hand bar chart, the information has been compressed into a window size unsuitable for presentation. For proper display the same information is given in Figure 18.11.

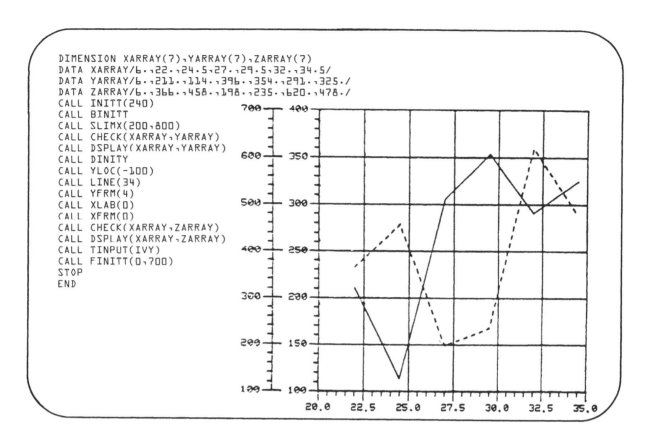

```
DIMENSION XARRAY(7),YARRAY(7),ZARRAY(7)
DATA XARRAY/6.,22.,24.5,27.,29.5,32.,34.5/
DATA YARRAY/6.,211.,114.,396.,354.,291.,325./
DATA ZARRAY/6.,366.,458.,198.,235.,620.,478./
CALL INITT(240)
CALL BINITT
CALL SLIMX(200,800)
CALL CHECK(XARRAY,YARRAY)
CALL DSPLAY(XARRAY,YARRAY)
CALL DINITY
CALL YLOC(-100)
CALL LINE(34)
CALL YFRM(4)
CALL XLAB(0)
CALL XFRM(0)
CALL CHECK(XARRAY,ZARRAY)
CALL DSPLAY(XARRAY,ZARRAY)
CALL TINPUT(IVY)
CALL FINITT(0,700)
STOP
END
```

Figure 18.13   Proper 'windowing' to avoid clipping.   (Courtesy Tektronix Inc.)

Figure 18.14   Tektronix terminal with AGII displays.   (Courtesy Tektronix Inc.)

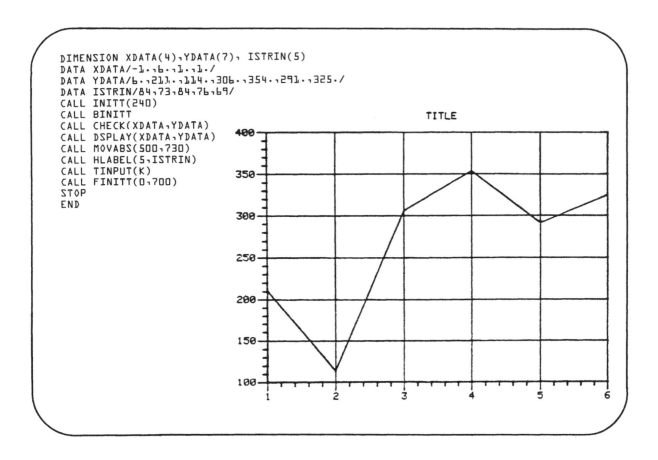

```
DIMENSION XDATA(4),YDATA(7), ISTRIN(5)
DATA XDATA/-1.,b.,1.,1./
DATA YDATA/b.,211.,114.,306.,354.,291.,325./
DATA ISTRIN/84,73,84,76,69/
CALL INITT(240)
CALL BINITT
CALL CHECK(XDATA,YDATA)
CALL DSPLAY(XDATA,YDATA)
CALL MOVABS(500,730)
CALL HLABEL(5,ISTRIN)
CALL TINPUT(K)
CALL FINITT(0,700)
STOP
END
```

Figure 18.15   Labels and titles.   (Courtesy Tektronix Inc.)

To avoid overcrowding a graph, data setting limits may be used:

```
CALL MNMX(ARRAY,AMIN,AMAX)
```

Where ARRAY is the data array for which the minimum and maximum are to be found, AMIN is the minimum array value and AMAX is the maximum array value. MNMX is used internally to determine data limits and may be useful to the user in setting the minimum and maximum values for multiple overlapping curves.

18.4   LABELING THE AXES

Axes may be labeled by establishing an array of the ASCII equivalents of the characters to be displayed. These equivalents may be selected from Table 18.1. The user then moves to the starting point and calls either HLABEL (horizontal label routine) or VLABEL to display the axis label. The instruction

```
CALL HLABEL(5,ISTRIN)
```

would display the five characters of TITLE contained in the string ISTRIN horizontally. VLABEL is used the same way for 90-degree orientation or vertical labels. Figure 18. 15 is a display of this. A more direct test of the HLABEL routine might be the alphabet displayed.

```
DIMENSION ISTRG(26)
DATA ISTRG/65,66,67,68,69,70,71,72,73,
+74,75,76,77,78,79,80,81,82,83,84,85,
+86,87,88,89,90/
CALL INITT(240)
CALL BINITT
CALL HLABEL(26,ISTRG)
CALL FINITT(0,700)
STOP
END
```

Table 18.1
ASCII Equivalent Chart

CONTROL				HIGH X & Y GRAPHIC INPUT				LOW X				LOW Y			
NUL	0	DLE	16	SP	32	0	48	@	64	P	80	\	96	p	112
SOH	1	DC1	17	!	33	1	49	A	65	Q	81	a	97	q	113
STX	2	DC2	18	"	34	2	50	B	66	R	82	b	98	r	114
ETX	3	DC3	19	#	35	3	51	C	67	S	83	c	99	s	115
EOT	4	DC4	20	$	36	4	52	D	68	T	84	d	100	t	116
ENQ	5	NAK	21	%	37	5	53	E	69	U	85	e	101	u	117
ACK	6	SYN	22	&	38	6	54	F	70	V	86	f	102	v	118
BEL BELL	7	ETB	23	'	39	7	55	G	71	W	87	g	103	w	119
BS BACK SPACE	8	CAN	24	(	40	8	56	H	72	X	88	h	104	x	120
HT	9	EM	25	)	41	9	57	I	73	Y	89	i	105	y	121
LF LINE FEED	10	SUB	26	*	42	:	58	J	74	Z	90	j	106	z	122
VT	11	ESC	27	+	43	;	59	K	75	[	91	k	107	{	123
FF	12	FS	28	,	44	<	60	L	76	\	92	l	108	|	124
CR RETURN	13	GS	29	-	45	=	61	M	77	]	93	m	109	}	125
SO	14	RS	30	.	46	>	62	N	78	^	94	n	110	~	126
SI	15	US	31	/	47	?	63	O	79	_	95	o	111	RUBOUT (DEL)	127

A direct test for the VLABEL routine using digit plotting is as follows:

```
DIMENSION KSTRG(22)
DATA KSTRG/48,32,49,32,50,32,51,32,52,
+32,53,32,54,32,55,32,56,32,57,32,58/
CALL INITT(240)
CALL BINITT
CALL VLABEL(22,KSTRG)
CALL FINITT(0,700)
STOP
END
```

and the output would look like:

```
0
1
2
3
4
5
6
7
8
9
```

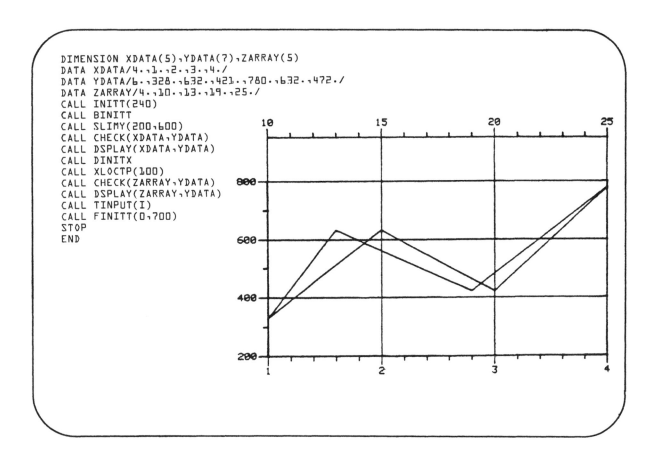

Figure 18.16   Axis location variation.   (Courtesy Tektronix Inc.)

Alphanumeric labels can be displayed with calls to HLABEL and VLABEL. For special character generation inside graph labels, CALL NOTATE should be used as

```
CALL NOTATE(IX,IY,LENCHR,ISTRIN)
```

where IX,IY replaces CALL MOVABS. It designates the beginningpoint of the label, while ISTRIN is the string of characters in integers from 0 to 127.

Axis labels may also be justified as right, left or center by the use of:

```
CALL JUSTER(LENGTH,ISTRIN,KEY,IFILL)
```

where LENGTH is the total length of the character string, including fill characters

```
DATA ISTRIN/84,73,84,76,69/
CALL JUSTER(9,ISTRIN,0,32)
```

(spaces). ISTRIN is the array of characters selected from Table 18.1 in ASCII equivalent. KEY is an integer that designates if the string is to be right, left, or center justified where:

+ = right justified
- = left justified
0 = centered

IFILL is the character selected from Table 18.1 used as filler (usually 32 which is a space). JUSTER provides the information necessary for a call to NOTATE, which displays the label.

18.5   PLOTTING THE DATA

To plot the data and complete the graph, the location of the axes must be known. So this is done by calling XLOC and YLOC; these

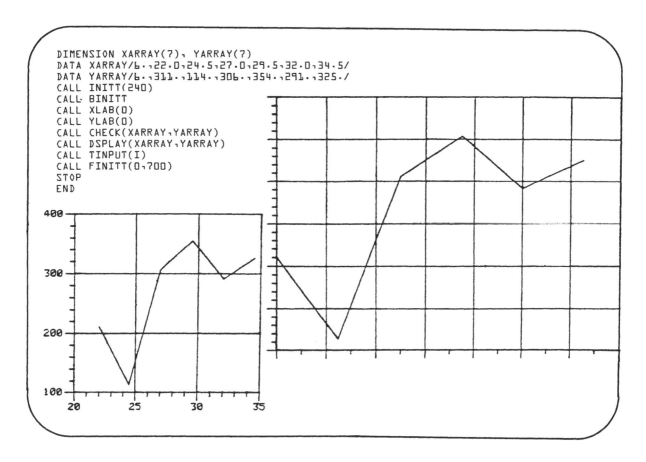

```
DIMENSION XARRAY(7), YARRAY(7)
DATA XARRAY/6.,22.0,24.5,27.0,29.5,32.0,34.5/
DATA YARRAY/6.,311.,114.,306.,354.,291.,325./
CALL INITT(240)
CALL BINITT
CALL XLAB(0)
CALL YLAB(0)
CALL CHECK(XARRAY,YARRAY)
CALL DSPLAY(XARRAY,YARRAY)
CALL TINPUT(I)
CALL FINITT(0,700)
STOP
END
```

Figure 18.17   Changing axes.   (Courtesy Tektronix Inc.)

subroutines specify the location of the axes with tick marks and labels.   The ualues placed in these subroutines are in display units, with a positive value indicating the distance above the X-axis screen minimum or to the right of the Y-axis screen minimum and a negative value indicating the distance in the opposite direction.   The instruction

          CALL XLOC(50)

will change the location of the X-axis by 50 display units to the right of its present location.

     subroutines to establish the original location of the axes in relation to the right and upper edges of the screen are

          CALL XLOCTP(IVALUE)
          CALL YLOCTP(IVALUE)

Figure 18.16 demonstrates the use of XLOCTP.

     Graphical data are plotted in reference to the graduation of the axes.   These graduations are called tick marks.   X and Y axes may be graduated differently by the use of

          CALL XLAB(IVALUE)
          CALL YLAB(IVALUE)

where IVALUE is an arbitrary integer which refers to a lable type shown in Table 18.2.

     Figure 18.17 illustrates two of the types shown in the table.   Tick marks may be changed in the following ways:

1.   Density   XDEN,YDEN
2.   Intervals   XTICS,YTICS
3.   Length   XLEN,YLEN
4.   FORM   XFRM,YFRM
5.   Interval and form   XMTCS,YMTCS

Table 18.2

IVALUE Integer and type

Integer	Type
0	None
1	Default value
2	Logarithmic
3	Days
4	Weeks
5	Periods
6	Months
7	Quarters
8	Years
Any negative	User written

After the axes have been graduated the graph can be placed anywhere on the DVST screen. This is done by

    CALL PLACE(LIT)

where LIT is a literal string of three characters specifying the window location desired; standard (STD), upper half (UPH), lower half (LOH), upper left quarter (UL4), upper right quarter (UR4), lower left quarter (LL4), lower right quarter (LR4), upper left sixth (UL6), upper center sixth (UC6), upper right sixth (UR6), lower left sixth (LL6), lower center sixth (LC6), and lower right sixth (LR6). Screen locations for each of these is shown in Figure 18.18.

18.6 LOGARITHMIC GRAPHS

Logarithmic plots may be created by entering data in the manner described in the earlier sections of this chapter and specifying YTYPE or XTYPE of 2. XTYPE and YTYPE are described as

    .. CALL XTYPE(IVALUE)
    CALL YTYPE(IVALUE)

where IVALUE is 2 if a logarithmic graph is desired. Figure 18.19 shows the output that is needed to create a comparison of a logarithmic and linear graph. Figure 18.20 indicates log output on the Y axis only.

Several routines have been developed

for use with log graphs or semilog graph outputs:

1. Least significant digit- XLSIG,YLSIG
2. Transformation - XTYPE,YTYPE
3. TICK marks - XWIDTH,YWIDTH
4. Remote exponent - XEPON,YEPON
5. Label frequency - XSTEP,YSTEP
6. Staggered labels - XSTAG,YSTAG
7. Exponent type - XETYP,YETYP

18.7 BAR GRAPHS

Bar graphs may be displayed by using one of two routines

    CALL HBARS(ISHADE,IWBAR,IDBAR)
    CALL VBARS(ISHADE,IWBAR,IDBAR)

where ISHADE is the integer value of the type of shading to fill the bar. Figure 18.21 illustrates the type of cross-hatching available. IWBAR is the width of the bar in display units. The width of the bar must be greater than 1. while 0 will result in a default value of 40 display units. IDBAR is the distance between the shading lines in display units. Figure 18.11 is an example of this type of output. Figure 18.22 is a display using the VBARST routine.

18.8 REVIEW PROBLEMS

1. Using Figure 18.1 as a guide, input the program shown in the upper left hand corner and add a third line to the graph with YDATA3/6.,300.,250.,225.,300.,250.,200./

2. Write the display program for Figure 18.2 and make a hard copy of the DVST screen contents.

3. Enter the program from Figure 18.3. Change the SYMBL from 6 to 3.

4. Write the display program for Figure 18.4.

5. Write the display program for Figure 18.5.

6. Write the display program for Figure 18.6.

7. Write the display program for Figure 18.7.

Figure 18.18   Six locations for graphical output.   (Courtesy Tektronix Inc.)

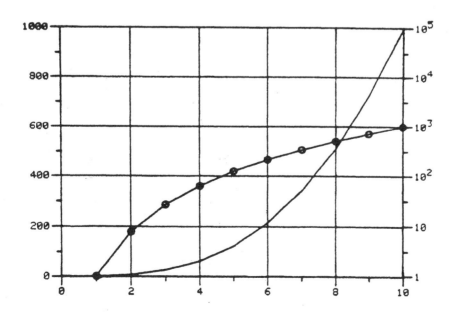

Figure 18.19   Logarithmic verus linear graph.   (Courtesy Tektronix Inc.)

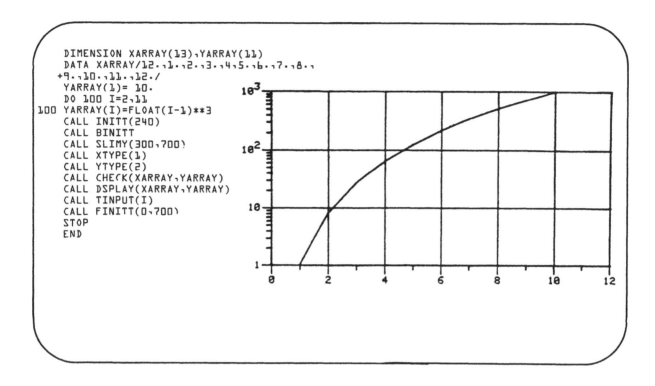

Figure 18.20 Logarithmic plot.   (Courtesy Tektronix Inc.)

```
      DIMENSION IARRAY(15)
      CALL INITT(240)
      CALL BINITT
      ISYMB=0
      DO 99 J=1,2
      MINY=100+300*(J-1)
      MAXY=MINY+250
      DO 99 I=1,8
      MINX=50+100*(I-1)
      MAXX=MINX+80
      CALL FILBOX(MINX,MINY,MAXX,MAXY,ISYMB,20)
      SYMB=FLOAT(ISYMB)
      CALL FFORM(SYMB,3,0,IARRAY,32)
      CALL MOVABS(MINX+30,MINY-30)
      CALL HLABEL(15,IARRAY)
   99 ISYMB=ISYMB+1
      CALL TINPUT(KEY)
      STOP
      END
```

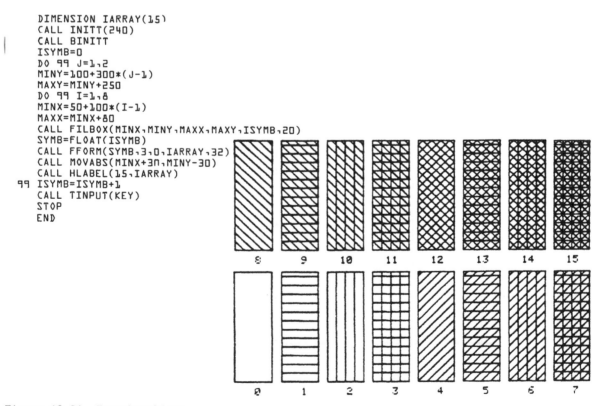

Figure 18.21   Crosshatchings.

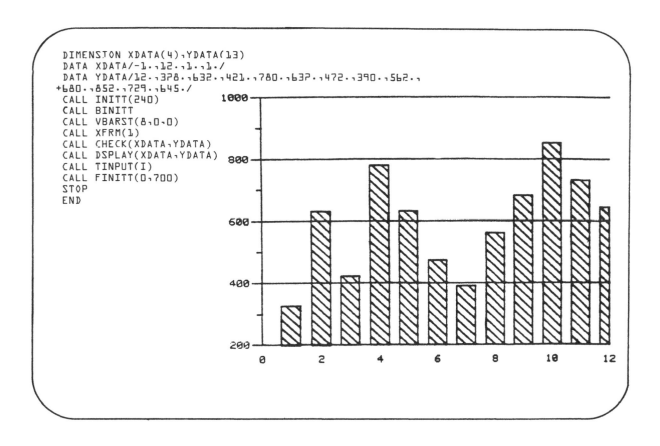

Figure 18.22   Bar chart program and output.   (Courtesy Tektronix Inc.)

8.   Write the display program for Figure 18.8.

9.   Write the display program for Figure 18.9.

10.   Write the display program for Figure 18.10.

11.   Write the display program for Figure 18.11.

12.   Write the display program for Figure 18.12 so that three sets of data are represented.

13.   Modify the program shown in Figure 18.13 so that three sets of data are not clipped.

14.   Write a display program to output three different graphs as shown in Figure 18.14.   Use any of the earlier graphs if you wish.

15.   Modify the program shown in Figure 18.15 so that the "title" is replaced by "I T E M S   S O L D   V S P R I C E".

16.   Use the program shown in Figure 18.16 to output the graph shown.

17.   Modify the program shown in Figure 18.17 to output a third size graph.

18.   Write the display program for Figure 18.19.

19.   Use the display method shown in Figure 18.21 to write an interactive program for selecting the crosshatching.

20.   Modify the program shown in Figure 18.22 with another crosshatching shown in Figure 18.21.

18.9   CHAPTER SUMMARY

A review of all the subroutines necessary to display graphs is now in order.

DATA PLOTTING		AXIS PLOTTING	
LINE	SIZES	NEAT	X-YTICS
SYMBL	SIZEL	ZERO	X-YLEN
STEPS	PLACE	X-YLOC	X-YFRM
INFIN		X-YLAB	XDMIN
NUMBER		X-YDEN	XDMAX

INDEX